学习量子比特（影印版）
Dancing with Qubits

Robert S. Sutor 著

南京　东南大学出版社

图书在版编目(CIP)数据

学习量子比特:影印版:英文/(美)罗伯特·S.
萨托(Robert S. Sutor)著. —南京:东南大学出版社,2020.8
书名原文:Dancing with Qubits
ISBN 978-7-5641-8974-7

Ⅰ.①学… Ⅱ.①罗… Ⅲ.①量子计算机-英文
Ⅳ.①TP385

中国版本图书馆 CIP 数据核字(2020)第 115994 号
图字:10-2020-160 号

学习量子比特(影印版)

出版发行:东南大学出版社
地　　址:南京四牌楼 2 号　　邮编:210096
出　版　人:江建中
网　　址:http://www.seupress.com
电子邮件:press@seupress.com
印　　刷:常州市武进第三印刷有限公司
开　　本:787 毫米×980 毫米　16 开本
印　　张:32
字　　数:627 千字
版　　次:2020 年 8 月第 1 版
印　　次:2020 年 8 月第 1 次印刷
书　　号:ISBN 978-7-5641-8974-7
定　　价:119.00 元

本社图书若有印装质量问题,请直接与营销部联系。电话(传真):025-83791830

Packt>

www.packt.com

Subscribe to our online digital library for full access to over 7,000 books and videos, as well as industry leading tools to help you plan your personal development and advance your career. For more information, please visit our website.

Why subscribe?

- Spend less time learning and more time coding with practical eBooks and Videos from over 4,000 industry professionals
- Learn better with Skill Plans built especially for you
- Get a free eBook or video every month
- Fully searchable for easy access to vital information
- Copy and paste, print, and bookmark content

Did you know that Packt offers eBook versions of every book published, with PDF and ePub files available? You can upgrade to the eBook version at www.Packt.com and as a print book customer, you are entitled to a discount on the eBook copy. Get in touch with us at customercare@packtpub.com for more details.

At www.Packt.com, you can also read a collection of free technical articles, sign up for a range of free newsletters, and receive exclusive discounts and offers on Packt books and eBooks.

To Judith, Katie, and William,
to whom my debt is beyond computation.

Contributors

About the author

Robert S. Sutor has been a technical leader and executive in the IT industry for over 30 years. More than two decades of that have been spent in IBM Research in New York. During his time there, he worked on or led efforts in symbolic mathematical computation, optimization, AI, blockchain, and quantum computing. He is the co-author of several research papers and the book *Axiom: The Scientific Computation System* with the late Richard D. Jenks.

He also was an executive on the software side of the business in areas including emerging industry standards, software on Linux, mobile, and open source. He's a theoretical mathematician by training, has a Ph.D. from Princeton University, and an undergraduate degree from Harvard College. He started coding when he was 15 and has used most of the programming languages that have come along.

I want to thank:

- My wife, Judith Hunter, and children, Katie and William, for their love and humor while this book was being written,
- John Kelly, Arvind Krishna, Dario Gil, Jay Gambetta, Jamie Thomas, Tom Rosamilia, and Ken Keverian for their leadership of the IBM Q program and their personal support,
- the following for their conversations, insight, and inspiration regarding the breadth of quantum computing science, technology, business, and ecosystem:

Abe Asfaw, Alexis Harrison, Ali Javadi, Amanda Carl, Andrew Cross, Anthony Annunziata, Antonio Corcoles-Gonzalez, Antonio Mezzacapo, Aparna Prabhakar, Bill Minor, Brian Eccles, Carmen Recio Valcarce, Chris Lirakis, Chris Nay, Christine Ouyang, Christine Vu, Christopher Schnabel, Denise Ruffner, Doug McClure, Edwin Pednault, Elena Yndurain, Eric Winston, Frederik Flöther, Hanhee Paik, Heather Higgins, Heike Riel, Ingolf Wittmann, Ismael Faro, James Wootten, Jeanette Garcia, Jenn Glick, Jerry Chow, Joanna Brewer, John Gunnels, Jules Murphy, Katie Pizzolato, Lev Bishop, Liz Durst,

Luuk Ament, Maika Takita, Marco Pistoia, Mark Ritter, Markus Brink, Matthias Steffen, Melissa Turesky, Michael Gordon, Michael Osborne, Mike Houston, Pat Gumann, Paul Kassebaum, Paul Nation, Rajeev Malik, Robert Loredo, Robert Wisnieff, Sarah Sheldon, Scott Crowder, Stefan Woerner, Steven Tomasco, Suzie Kirschner, Talia Gershon, Vanessa Johnson, Vineeta Durani, Wendy Allan, Wendy Cornell, and Zaira Nazario

- the many authors whose works I reference throughout the book, and
- the Packt production and editorial team including Andrew Waldron, Tom Jacob, and Ian Hough.

Any errors or misunderstandings that appear in this book are mine alone.

About the reviewer

Jhonathan Romero is a quantum computing scientist and entrepreneur. Born in Barranquilla, Colombia, he received a Ph.D. in Chemical Physics from Harvard University, after earning B.S. and M.S. degrees in Chemistry from the National University of Colombia. His research has focused on the development of algorithms for quantum simulation and artificial intelligence on near-term quantum devices. Jhonathan is one of the co-founders and research scientists at Zapata Computing, a company pioneering the development of quantum algorithms and software for commercial applications. He has authored several publications in computational chemistry and quantum computing.

Contents

Contents

List of Figures

Preface

*Everything we call real is made of things
that cannot be regarded as real.*

Niels Bohr [1]

When most people think about computers, they think about laptops or maybe even the bigger machines like the servers that power the web, the Internet, and the cloud. If you look around, though, you may start seeing computers in other places. Modern cars, for example, have anywhere from around 20 computers to more than 100 to control all the systems that allow you to move, brake, monitor the air conditioning, and control the entertainment system.

The smartphone is the computer many people use more than anything else in a typical day. A modern phone has a 64-bit processor in it, whatever a "64-bit processor" is. The amount of memory used for running all those apps might be 3Gb, which means 3 gigabytes. What's a "giga" and what is a byte?

All these computers are called *classical computers* and the original ideas for them go back to the 1940s. Sounding more scientific, we say these computers have a *von Neumann architecture*, named after the mathematician and physicist John von Neumann.

It's not the 1940s anymore, obviously, but more than seventy years later we still have the modern versions of these machines in so many parts of our lives. Through the years, the "thinking" components, the processors, have gotten faster and faster. The amount of memory has also gotten larger so we can run more—and bigger—apps that do some pretty sophisticated things. The improvements in graphics processors have given us better and better games. The amount of storage has skyrocketed in the last couple of decades, so we can have more and more apps and games and photos and videos on devices we carry around with us. When it comes to these classical computers and the way they have developed, "more is better."

We can say similar things about the computer servers that run businesses and the Internet around the world. Do you store your photos in the cloud? Where is that exactly? How many photos can you keep there and how much does it cost? How quickly can your photos and all the other data you need move back and forth to that nebulous place?

It's remarkable, all this computer power. It seems like every generation of computers will continue to get faster and faster and be able to do more and more for us. There's no end in sight for how powerful these small and large machines will get to entertain us, connect us to our friends and family, and solve the important problems in the world.

Except . . . that's false.

While there will continue to be some improvements, we will not see anything like the doubling in processor power every two years that happened starting in the mid-1960s. This doubling went by the name of *Moore's Law* and went something like "every two years processors will get twice as fast, half as large, and use half as much energy."

These proportions like "double" and "half" are approximate, but physicists and engineers really did make extraordinary progress for many years. That's why you can have a computer in a watch on your wrist that is more powerful than a system that took up an entire room forty years ago.

A key problem is the part where I said processors will get half as large. We can't keep making transistors and circuits smaller and smaller indefinitely. We'll start to get so small that we approach the atomic level. The electronics will get so crowded that when we try to tell part of a processor to do something a nearby component will also get affected.

There's another deeper and more fundamental question. Just because we created an architecture over seventy years ago and have vastly improved it, does that mean all kinds of problems can eventually be successfully tackled by computers using that design? Put another way, why do we think the kinds of computers we have now might eventually be suitable for solving every possible problem? Will "more is better" run out of steam if we keep to the same kind of computer technology? Is there something wrong or limiting about our way of computing that will prevent our making the progress we need or desire?

Depending on the kind of problem you are considering, it's reasonable to think the answer to the last question if somewhere between "probably" and "yes."

That's depressing. Well, it's only depressing if we can't come up with one or more new types of computers that have a chance of breaking through the limitations.

That's what this book is about. Quantum computing as an idea goes back to at least the early 1980s. It uses the principles of quantum mechanics to provide an entirely new kind of computer architecture. Quantum mechanics in turn goes back to around 1900 but especially to the 1920s when physicists started noticing that experimental results were not matching what theories predicted.

However, this is not a book about quantum mechanics. Since 2016, tens of thousands of users have been able to use quantum computing hardware via the cloud, what we call quantum cloud services. People have started programming these new computers even though the way you do it is unlike anything done on a classical computer.

Why have so many people been drawn to quantum computing? I'm sure part of it is curiosity. There's also the science fiction angle: the word "quantum" gets tossed around enough in sci-fi movies that viewers wonder if there is any substance to the idea.

Once we get past the idea that quantum computing is new and intriguing, it's good to ask "ok, but what is it really good for?" and "when and how will it make a difference in my life?" I discuss the use cases experts think are most tractable over the next few years and decades.

It's time to learn about quantum computing. It's time to stop thinking classically and to start thinking *quantumly,* though I'm pretty sure that's not really a word!

For whom did I write this book?

This book is for anyone who has a very healthy interest in mathematics and wants to start learning about the physics, computer science, and engineering of quantum computing. I review the basic math, but things move quickly so we can dive deeply into an exposition of how to work with qubits and quantum algorithms.

While this book contains a lot of math, it is not of the definition-theorem-proof variety. I'm more interested in presenting the topics to give you insight on the relationships between the ideas than I am in giving you a strictly formal development of all results.

Another goal of mine is to prepare you to read much more advanced texts and articles on the subject, perhaps returning here to understand some core topic. You do not need to be a physicist to read this book, nor do you need to understand quantum mechanics beforehand.

At several places in the book I give some code examples using Python 3. Consider these to be extra and not required, but if you do know Python they may help in your understanding.

Many of the examples in this book come from the IBM Q quantum computing system. I was an IBM Q executive team member during the time I developed this content.

What does this book cover?

Before we jump into understanding how quantum computing works from the ground up, we need to take a little time to see how things are done classically. In fact, this is not only for the sake of comparison. The future, I believe, will be a hybrid of classical and quantum computers.

The best way to learn about something is start with basic principles and then work your way up. That way you know how to reason about it and don't rely on rote memorization or faulty analogies.

1 – Why Quantum Computing?

In the first chapter we ask the most basic question that applies to this book: why quantum computing? Why do we care? In what ways will our lives change? What are the use cases to which we hope to apply quantum computing and see a significant improvement? What do we even mean by "significant improvement"?

I – Foundations

The first full part covers the mathematics you need to understand the concepts of quantum computing. While we will ultimately be operating in very large dimensions and using complex numbers, there's a lot of insight you can gain from what happens in traditional 2D and 3D.

2 – They're Not Old, They're Classics

Classical computers are pervasive but relatively few people know what's inside them and how they work. To contrast them later with quantum computers, we look at the basics along with the reasons why they have problems doing some kinds of calculations. I introduce the simple notion of a bit, a single 0 or 1, but show that working with many bits can eventually give you all the software you use today.

3 – More Numbers than You Can Imagine

The numbers people use every day are called real numbers. Included in these are integers, rational numbers, and irrational numbers. There are other kinds of numbers, though, and structures that have many of the same algebraic properties. We look at these to lay the groundwork to understand the "compute" part of what a quantum computer does.

4 – Planes and Circles and Spheres, Oh My

From algebra we move to geometry and relate the two. What is a circle, really, and what does it have in common with a sphere when we move from two to three dimensions? Trigonometry becomes more obvious, though that is not a legally binding statement. What you thought of as a plane becomes the basis for understanding complex numbers, which are key to the definition of quantum bits, usually known as *qubits*.

5 – Dimensions

After laying the algebraic and geometric groundwork, we move beyond the familiar two- and three-dimensional world. Vector spaces generalize to many dimensions and are essential for understanding the exponential power that quantum computers can harness. What can you do when you are working in many dimensions and how should you think about such operations? This extra elbow room comes into play when we consider how quantum computing might augment AI.

6 – What Do You Mean "Probably"?

"God does not play dice with the universe," said Albert Einstein.

This was not a religious statement but rather an expression of his lack of comfort with the idea that randomness and probability play a role in how nature operates. Well, he didn't get that quite right. Quantum mechanics, the deep and often mysterious part of physics on which quantum computing is based, very much has probability at its core. Therefore, we cover the fundamentals of probability to aid your understanding of quantum processes and behavior.

II – Quantum Computing

The next part is the core of how quantum computing really works. We look at quantum bits— qubits—singly and together, and then create circuits that implement algorithms. Much of this is the ideal case when we have perfect fault-tolerant qubits. When we really create quantum computers, we must deal with the physical realities of noise and the need to reduce errors.

7 – One Qubit

At this point we are finally able to talk about qubits in a nontrivial manner. We look at both the vector and Bloch sphere representations of the quantum states of qubits. We define superposition, which explains the common cliché about a qubit being "zero and one at the same time."

8 – Two Qubits, Three

With two qubits we need more math, and so we introduce the notion of the tensor product, which allows us to explain entanglement. Entanglement, which Einstein called "spooky action at a distance," tightly correlates two qubits so that they no longer act independently. With superposition, entanglement gives rise to the very large spaces in which quantum computations can operate.

9 – Wiring Up the Circuits

Given a set of qubits, how do you manipulate them to solve problems or perform calculations? The answer is you build circuits for them out of gates that correspond to reversible operations. For now, think about the classical term "circuit board." I use the quantum analog of circuits to implement algorithms, the recipes computers use for accomplishing tasks.

10 – From Circuits to Algorithms

With several simple algorithms discussed and understood, we next turn to more complicated ones that fit together to give us Peter Shor's 1995 fast integer factoring algorithm. The math is more extensive in this chapter, but we have everything we need from previous discussions.

11 – Getting Physical

When you build a physical qubit, it doesn't behave exactly like the math and textbooks say it should. There are errors, and they may come from noise in the environment of the quantum system. I don't mean someone yelling or playing loud music, I mean fluctuating temperatures, radiation, vibration, and so on. We look at several factors you must consider when you build a quantum computer, introduce Quantum Volume as a whole-system metric of the performance of your system, and conclude with a discussion of the most famous quantum feline.

This book concludes with a chapter that moves beyond today.

12 – Questions about the Future

If I were to say, "in ten years I think quantum computing will be able to do . . . ," I would also need to describe the three or four major scientific breakthroughs that need to happen before then. I break down the different areas in which we're trying to innovate in the science and engineering of quantum computing and explain why. I also give you some guiding principles to distinguish hype from reality. All this is expressed in terms of motivating questions.

References

[1]   Karen Barad. *Meeting the Universe Halfway. Quantum Physics and the Entanglement of Matter and Meaning.* 2nd ed. Duke University Press Books, 2007.

What conventions are used in this book?

When I want to highlight something important that you should especially remember, I use this kind of box:

> This is very important.

This book does not have exercises but it does have questions. Some are answered in the text and others are left for you as thought experiments. Try to work them out as you go along. They are numbered within chapters.

> **Question 0.0.1**
>
> Why do you ask so many questions?

Code samples and output are presented to give you an idea about how to use a modern programming language, Python 3, to experiment with basic ideas in quantum computing.

```
def obligatoryFunction():
    print("Hello quantum world!")

obligatoryFunction()

Hello quantum world!
```

Numbers in brackets (for example, [1]) are references to additional reading materials. They are listed at the end of each chapter in which the bracketed number appears.

> **To learn more**
>
> Here is a place where you might see a reference to learn more about some topic. [1]

Get in touch

Feedback from our readers is always welcome.

General feedback: If you have questions about any aspect of this book, mention the book title in the subject of your message and email us at customercare@packtpub.com.

Errata: Although we have taken every care to ensure the accuracy of our content, mistakes do happen. If you have found a mistake in this book we would be grateful if you would report this to us. Please visit, http://www.packt.com/submit-errata, selecting your book, clicking on the Errata Submission Form link, and entering the details.

Piracy: If you come across any illegal copies of our works in any form on the Internet, we would be grateful if you would provide us with the location address or website name. Please contact us at copyright@packt.com with a link to the material.

If you are interested in becoming an author: If there is a topic that you have expertise in and you are interested in either writing or contributing to a book, please visit http://authors.packtpub.com.

. . .

Now let's get started by seeing why we should look at quantum computing systems to try to solve problems that are intractable with classical systems.

1

Why Quantum Computing?

Nature isn't classical, dammit,
and if you want to make a simulation of nature,
you'd better make it quantum mechanical.

Richard Feynman [5]

In his 1982 paper "Simulating Physics with Computers," Richard Feynman, 1965 Nobel Laureate in Physics, said he wanted to "talk about the possibility that there is to be an *exact* simulation, that the computer will do *exactly* the same as nature." He then went on to make the statement above, asserting that nature doesn't especially make itself amenable for computation via classical binary computers.

In this chapter we begin to explore how quantum computing is different from classical computing. Classical computing is what drives smartphones, laptops, Internet servers, mainframes, high performance computers, and even the processors in automobiles.

We examine several use cases where quantum computing may someday help us solve problems that are today intractable using classical methods on classical computers. This is to motivate you to learn about the underpinnings and details of quantum computers I discuss throughout the book.

No single book on this topic can be complete. The technology and potential use cases are moving targets as we innovate and create better hardware and software. My goal here is to prepare you to delve more deeply into the science, coding, and applications of quantum computing.

Topics covered in this chapter

1.1 The mysterious quantum bit

Suppose I am standing in a room with a single overhead light and a switch that turns the light on or off. This is just a normal switch, and so I can't dim the light. It is either fully on or fully off. I can change it at will, but this is the only thing I can do to it. There is a single door to the room and no windows. When the door is closed I cannot see any light.

I can stay in the room or I may leave it. The light is always on or off based on the position of the switch.

Now I'm going to do some rewiring. I'm replacing the switch with one that is in another part of the building. I can't see the light at all but, once again, its being on or off is determined solely by the two positions of the switch.

If I walk to the room with the light and open the door, I can see whether it is lit or dark. I can walk in and out of the room as many times as I want and the status of the light is still determined by that remote switch being on or off. This is a "classical" light.

Now let's imagine a *quantum* light and switch, which I'll call a "qu-light" and "qu-switch," respectively.

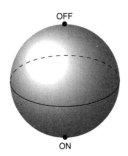

When I walk into the room with the qu-light it is always on or off, just like before. The qu-switch is unusual in that it is shaped like a sphere with the topmost point (the "north pole") being OFF and the bottommost (the "south pole") being ON. There is a line etched around the middle.

The interesting part happens when I cannot see the qu-light, when I am in the other part of the building with the qu-switch.

I control the qu-switch by placing my index finger on the qu-switch sphere. If I place my finger on the north pole, the qu-light is definitely off. If I put it on the south, the qu-light is definitely on. You can go into the room and check. You will always get these results.

If I move my finger anywhere else on the qu-switch sphere, the qu-light may be on or off when you check. If you do not check, the qu-light is in an indeterminate state. It is not dimmed, it is not on or off, it just exists with some probability of being on or off when seen. This is unusual!

The moment you open the door and see the qu-light, the indeterminacy is removed. It will be on or off. Moreover, if I had my finger on the qu-switch, the finger would be forced to one or other of the poles corresponding to the state of the qu-light when it was seen.

The act of observing the qu-light forced it into either the on or off state. I don't have to see the qu-light fixture itself. If I open the door a tiny bit, enough to see if any light is shining or not, that is enough.

If I place a video camera in the room with the qu-light and watch it when I try to place my finger on the qu-switch, it behaves just like a normal switch. I will be prevented from touching the qu-switch at anywhere other than the top or bottom. Since I'm making up this example, assume some sort of force field keeps me away from anywhere but the poles!

If you or I are not observing the qu-light in any way, does it make a difference where I touch the qu-switch? Will touching it in the northern or southern hemisphere influence whether it will be on or off when I observe the qu-light?

Yes. Touching it closer to the north pole or the south pole will make the probability of the qu-light being off or on, respectively, be higher. If I put my finger on the circle between the poles, the equator, the probability of the light being on or off will be exactly 50 – 50.

What I just described is called a *two-state quantum system*. When it is not being observed, the qu-light is in a *superposition* of being on and off. We explore superposition in section 7.1.

While this may seem bizarre, evidently nature really works this way. Electrons have a property called "spin" and with this they are two-state quantum systems. The photons that make up light itself are two-state quantum systems. We return to this in section 11.3 when we look at polarization (as in Polaroid® sunglasses).

More to the point of this book, however, a *quantum bit*, more commonly known as a *qubit*, is a two-state quantum system. It extends and complements the classical computing notion of bit, which can only be 0 or 1. The qubit is the basic information unit in quantum computing.

This book is about how we manipulate qubits to solve problems that currently appear to be intractable using just classical computing. It seems that just sticking to 0 or 1 will not be sufficient to solve some problems that would otherwise need impractical amounts of time or memory.

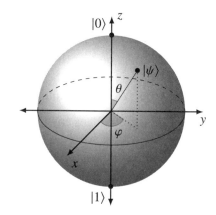

With a qubit, we replace the terminology of on or off, 1 or 0, with $|1\rangle$ and $|0\rangle$, respectively. Instead of qu-lights, it's qubits from now on.

In the diagram on the right, the position of your finger on the qu-switch is now indicated by two angles, θ and φ. The picture itself is called a Bloch sphere and is a standard representation of a qubit, as we shall see in section 7.5.

1.2 I'm awake!

What if we could do chemistry inside a computer instead of in a test tube or beaker in the laboratory? What if running a new experiment was as simple as running an app and having it complete in a few seconds?

For this to really work, we would want it to happen with full *fidelity*. The atoms and molecules as modeled in the computer should behave **exactly** like they do in the test tube. The chemical reactions that happen in the physical world would have precise computational analogs. We would need a fully faithful simulation.

If we could do this at scale, we might be able to compute the molecules we want and need. These might be for new materials for shampoos or even alloys for cars and airplanes. Perhaps we could more efficiently discover medicines that are customized to your exact physiology. Maybe we could get better insight into how proteins fold, thereby understanding their function, and possibly creating custom enzymes to positively change our body chemistry.

Is this plausible? We have massive supercomputers that can run all kinds of simulations. Can we model molecules in the above ways today?

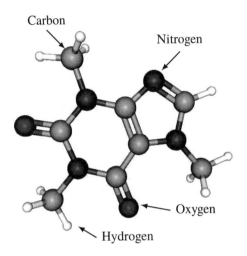

Let's start with $C_8H_{10}N_4O_2$ – 1,3,7-Trimethylxanthine. This is a very fancy name for a molecule which millions of people around the world enjoy every day: **caffeine**. An 8 ounce cup of coffee contains approximately 95 mg of caffeine, and this translates to roughly 2.95×10^{20} molecules. Written out, this is

$$295,000,000,000,000,000,000 \text{ molecules.}$$

A 12 ounce can of a popular cola drink has 32 mg of caffeine, the diet version has 42 mg, and energy drinks often have about 77 mg. [11]

Question 1.2.1

How many molecules of caffeine do you consume a day?

These numbers are large because we are counting physical objects in our universe, which we know is very big. Scientists estimate, for example, that there are between 10^{49} and 10^{50} atoms in our planet alone. [4]

To put these values in context, one thousand $= 10^3$, one million $= 10^6$, one billion $= 10^9$, and so on. A gigabyte of storage is one billion bytes, and a terabyte is 10^{12} bytes.

Getting back to the question I posed at the beginning of this section, can we model caffeine exactly in a computer? We don't have to model the huge number of caffeine molecules in a cup of coffee, but can we fully represent a single molecule at a single instant?

Caffeine is a small molecule and contains protons, neutrons, and electrons. In particular, if we just look at the energy configuration that determines the structure of the molecule and the bonds that hold it all together, the amount of information to describe this is staggering. In particular, the number of bits, the 0s and 1s, needed is approximately 10^{48}:

$$10,000,000,000,000,000,000,000,000,000,000,000,000,000,000,000,000.$$

From what I said above, this is comparable to 1% to 10% of the number of atoms on the Earth.

This is just one molecule! Yet somehow nature manages to deal quite effectively with all this information. It handles the single caffeine molecule, to all those in your coffee, tea, or soft drink, to every other molecule that makes up you and the world around you.

How does it do this? We don't know! Of course, there are theories and these live at the intersection of physics and philosophy. We do not need to understand it fully to try to harness its capabilities.

We have no hope of providing enough traditional storage to hold this much information. Our dream of exact representation appears to be dashed. This is what Richard Feynman meant in his quote at the beginning of this chapter: "Nature isn't classical."

However, 160 qubits (quantum bits) could hold $2^{160} \approx 1.46 \times 10^{48}$ bits while the qubits were involved in computation. To be clear, I'm not saying how we would get all the data into those qubits and I'm also not saying how many more we would need to do something interesting with the information. It does give us hope, however.

Richard Feynman at the California Institute of Technology in 1959. Photo is in the public domain. ℗

In the classical case, we will never fully represent the caffeine molecule. In the future, with enough very high quality qubits in a powerful enough quantum computing system, we may be able to perform chemistry in a computer.

1.3 Why quantum computing is different

I can write a little app on a classical computer that can simulate a coin flip. This might be for my phone or laptop.

Instead of heads or tails, let's use 1 and 0. The routine, which I call **R**, starts with one of those values and randomly returns one or the other. That is, 50% of the time it returns 1 and 50% of the time it returns 0. We have no knowledge whatsoever of how **R** does what it does. When you see "**R**," think "random."

This is called a "fair flip." It is not weighted to slightly prefer one result or the other. Whether we can produce a truly random result on a classical computer is another question. Let's assume our app is fair.

If I apply **R** to 1, half the time I expect that same value and the other half 0. The same is true if I apply **R** to 0. I'll call these applications **R**(1) and **R**(0), respectively.

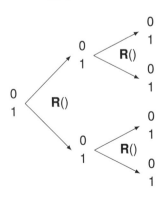

If I look at the result of **R**(1) or **R**(0), there is no way to tell if I started with 1 or 0. This is just as in a secret coin flip where I can't tell whether I began with heads or tails just by looking at how the coin has landed. By "secret coin flip," I mean that someone else does it and I can see the result, but I have no knowledge of the mechanics of the flip itself or the starting state of the coin.

If **R**(1) and **R**(0) are randomly 1 and 0, what happens when I apply **R** twice?

I write this as **R**(**R**(1)) and **R**(**R**(0)). It's the same answer: random result with an equal split. The same thing happens no matter how many times we apply **R**. The result is random and we can't reverse things to learn the initial value. In the language of section 4.1, **R** is not *invertible*.

Now for the quantum version. Instead of **R**, I use **H**, which we learn about in section 7.6. It too returns 0 or 1 with equal chance but it has two interesting properties:

1. It is reversible. Though it produces a random 1 or 0 starting from either of them, we can always go back and see the value with which we began.
2. It is its own reverse (or *inverse*) operation. Applying it two times in a row is the same as having done nothing at all.

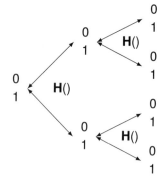

There is a catch, though. You are not allowed to look at the result of what **H** does if you want to reverse its effect.

If you apply **H** to 0 or 1, peek at the result, and apply **H** again to that, it is the same as if you had used **R**. If you observe what is going on in the quantum case at the wrong time, you are right back at strictly classical behavior.

To summarize using the coin language: if you flip a quantum coin and then **don't look at it**, flipping it again will yield the heads or tails with which you started. If you do look, you get classical randomness.

Question 1.3.1

Compare this behavior with that of the qu-switch and qu-light in section 1.1.

A second area where quantum is different is in how we can work with simultaneous values. Your phone or laptop uses bytes as the individual units of memory or storage. That's where we get phrases like "megabyte," which means one million bytes of information.

A byte is further broken down into eight bits, which we've see before. Each bit can be 0 or 1. Doing the math, each byte can represent $2^8 = 256$ different numbers composed of eight 0s or 1s, but it can only hold *one value at a time*.

Eight qubits can represent all 256 values *at the same time*.

This is through superposition, but also through *entanglement*, the way we can tightly tie together the behavior of two or more qubits. This is what gives us the (literally) exponential growth in the amount of working memory that we saw with a quantum representation of caffeine in section 1.2. We explore entanglement in section 8.2.

1.4 Applications to artificial intelligence

Artificial intelligence and one of its subsets, machine learning, are extremely broad collections of data-driven techniques and models. They are used to help find patterns in information, learn from the information, and automatically perform more "intelligently." They also give humans help and insight that might have been difficult to get otherwise.

Here is a way to start thinking about how quantum computing might be applicable to large, complicated, computation-intensive systems of processes such as those found in AI and elsewhere. These three cases are in some sense the "small, medium, and large" ways quantum computing might complement classical techniques:

1. There is a single mathematical computation somewhere in the middle of a software component that might be sped up via a quantum algorithm.
2. There is a well described component of a classical process that could be replaced with a quantum version.
3. There is a way to avoid the use of some classical components entirely in the traditional method because of quantum, or the entire classical algorithm can be replaced by a much faster or more effective quantum alternative.

As I write this, quantum computers are not "big data" machines. This means you cannot take millions of records of information and provide them as input to a quantum calculation. Instead, quantum may be able to help where the number of inputs are modest but the computations "blow up" as you start examining relationships or dependencies in the data. Quantum, with its exponentially growing working memory, as we saw in the caffeine example in section 1.2, may be able to control and work with the blow up. (See section 2.7 for a discussion of exponential growth.)

In the future, however, quantum computers may be able to input, output, and process much more data. Even if it is just theoretical now, it makes sense to ask if there are quantum algorithms that can be useful in AI someday.

Let's look at some data. I'm a big baseball fan, and baseball has a lot of statistics associated with it. The analysis of this even has its own name: "sabermetrics."

Year	GP	AB	R	H	2B	3B	HR	RBI	BB	SO
2019	136	470	105	136	27	2	41	101	110	111
2018	162	587	94	156	25	1	46	114	74	173
2017	152	542	73	132	24	0	29	92	41	145
2016	140	490	84	123	26	5	27	70	50	109
2015	162	634	66	172	32	4	25	83	26	108
2014	148	545	110	153	35	1	29	79	74	144

where

GP = Games Played

AB = At Bats

R = Runs scored

H = Hits

2B = 2 Base hits (doubles)

3B = 3 Base hits (triples)

HR = Home Runs

RBI = Runs Batted In

BB = Bases on Balls (walks)

SO = Strike Outs

Figure 1.1: Baseball player statistics by year

Suppose I have a table of statistics for a baseball player given by year as shown in Figure 1.1. We can make this look more mathematical by creating a matrix of the same data.

$$
\begin{bmatrix}
2019 & 136 & 470 & 105 & 136 & 27 & 2 & 41 & 101 & 110 & 111 \\
2018 & 162 & 587 & 94 & 156 & 25 & 1 & 46 & 114 & 74 & 173 \\
2017 & 152 & 542 & 73 & 132 & 24 & 0 & 29 & 92 & 41 & 145 \\
2016 & 140 & 490 & 84 & 123 & 26 & 5 & 27 & 70 & 50 & 109 \\
2015 & 162 & 634 & 66 & 172 & 32 & 4 & 25 & 83 & 26 & 108 \\
2014 & 148 & 545 & 110 & 153 & 35 & 1 & 29 & 79 & 74 & 144
\end{bmatrix}
$$

Given such information, we can manipulate it using machine learning techniques to make predictions about the player's future performance or even how other similar players may do. These techniques make use of the matrix operations we discuss in chapter 5.

There are 30 teams in Major League Baseball in the United States. With their training and feeder "minor league" teams, each major league team may each have more than 400 players throughout their systems. That would give us over 12,000 players, each with their complete player histories. There are more statistics than I have listed, so we can easily get greater than 100,000 values in our matrix.

In the area of entertainment, it's hard to make an estimate of how many movies exist, but it is well above 100,000. For each movie, we can list features such as whether it is a comedy or a drama or a romance or an action film, who each of the actors are, who each of the directorial and production staff are, geographic locations shown in the fim, languages used, and so on. There are hundreds of such features and *millions of people who have watched the films!*

For each person, we can also add features such as whether they like or dislike a kind of movie, actor, scene location, or director. Using all this information, which film should I recommend to you on a Saturday night in December based on what you and people similar to you like?

Think of each feature or each baseball player or film as a dimension. While you may think of two and three dimensions in nature, in AI we might have thousands or millions of dimensions.

Matrices as above for AI can grow to millions of rows and entries. How can we make sense of them to get insights and see patterns? Aside from manipulating that much information, can we even eventually do the math on classical computers quickly and accurately enough?

While it was originally thought that quantum algorithms might offer exponential improvements of such classical recommender systems, a 2019 "quantum-inspired algorithm" by Ewin Tang showed a classical method to gain such a huge improvement. [17] An example of being exponentially faster is doing something in 6 days instead of $10^6 = 1$ million days. That's approximately 2,740 years.

Tang's work is a fascinating example of the interplay of progress in both classical and quantum algorithms. People who develop algorithms for classical computing look to quantum computing, and vice versa. Also, any particular solution to a problem many include classical and quantum components.

Nevertheless, many believe that quantum computing will show very large improvements for some matrix computations. One such example is the **HHL** algorithm, whose abbreviation comes from the first letters of the last names of its authors, Aram W. Harrow, Avinatan Hassidim, and Seth Lloyd. This is also an example of case number 1 above.

Algorithms such as these may find use in fields as diverse as economics and computational fluid dynamics. They also place requirements on the structure and density of the data and may use properties such as the condition number we discuss in subsection 5.7.6.

To learn more

When you complete this book you will be equipped to read the original paper describing the **HHL** algorithm and more recent surveys about how to apply quantum computing to linear algebraic problems. [7]

An important problem in machine learning is classification. In its simplest form, a *binary classifier* separates items into one of two categories, or buckets. Depending on the definitions of the categories, it may be more or less easy to do the classification.

Examples of binary categories include:

- book you like **or** book you don't like
- comedy movie **or** dramatic movie
- gluten-free **or** not gluten-free
- fish dish **or** chicken dish
- UK football team **or** Spanish football team

- hot sauce **or** very hot sauce
- cotton shirt **or** permanent press shirt
- open source **or** proprietary
- spam email **or** valid email
- American League baseball team **or** National League team

The second example of distinguishing between comedies and dramas may not be well designed since there are movies that are both.

Mathematically, we can imagine taking some data as input and classifying it as either +1 or −1. We take a reasonably large set of data and label it by hand as either being a +1 or −1. We then *learn* from this *training set* how to classify future data.

Machine learning binary classification algorithms include random forest, k-nearest neighbor, decision tree, neural networks, naive Bayes classifiers, and support vector machines (SVMs).

In the training phase, we are given a list of pre-classified objects (books, movies, proteins, operating systems, baseball teams, etc.). We then use the above algorithms to learn how to put a new object in one bucket or another.

The SVM is a straightforward approach with a clear mathematical description. In the two-dimensional case, we try to draw a line that separates the objects (represented by points in the plot to the right) into one category or the other.

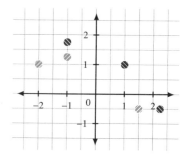

The line should maximize the gap between the sets of objects.

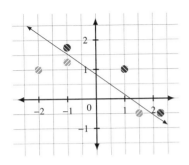

On the left is an example of a line that separates the light gray points below from the dark gray points above the line.

Given a new point, we plot it and determine whether it is above or below the line. That will classify it as dark gray or light gray, respectively.

Suppose we know that the point is correctly classified with those above the line. We accept that and move on.

If the point is misclassified, we add the point to the training set and try to compute a new and better line. This may not be possible.

In the plot to the right, I added a new light gray point above the line close to 2 on the vertical axis. With this extra point, there is no line we can compute to separate the points.

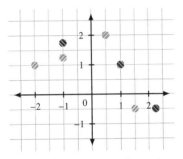

Had we represented the objects in three dimensions, we would try to find a plane that separated the points with maximum gap. We would need to compute some new amount that the points are above or below the plane. In geometric terms, if we are given x and y only, we somehow need to compute a z to work in that third dimension.

For a representation using n dimensions, we try to compute an $n - 1$ separating *hyperplane*. We look at two and three dimensions in chapter 4 and the general case in chapter 5.

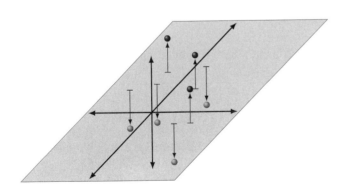

In this three-dimensional plot, I take the same values from the last two-dimensional version and lay the coordinate plane flat. I then add a vertical dimension. I push the light gray points below the plane and the dark gray ones above. With this construction, the coordinate plane itself separates the values.

While we can't separate the points in two dimensions, we can in three dimensions. This kind of mapping into a higher dimension is called the *kernel trick*. While the coordinate plane in this case might not be the ideal separating hyperplane, it gives you an idea of what we are trying to accomplish. The benefit of *kernel functions* (as part of the similarly named "trick") is that we can do far fewer explicit geometric computations than you might expect in these higher dimensional spaces.

It's worth mentioning now that we don't need to try quantum methods on small problems that are handled quite well using traditional means. We won't see any kind of quantum advantage until the problems are big enough to overcome the quantum circuit overhead versus classical circuits. Also, if we come up with a quantum approach that can be simulated easily on a classical computer, we don't really need a quantum computer.

A quantum computer with 1 qubit provides us with a two-dimensional working space. Every time we add a qubit, we double the number of dimensions. This is due to the properties of superposition and entanglement that I introduce in chapter 7. For 10 qubits, we get $2^{10} = 1024$ dimensions. Similarly, for 50 qubits we get $2^{50} = 1,125,899,906,842,624$ dimensions.

Remember all those dimensions for the features and baseball players and films? We want to use a sufficiently large quantum computer to do the AI calculations in a *quantum feature space*. This is the main point: handle the extremely large number of dimensions coming out of the data in a large quantum feature space.

There is a quantum approach that can generate the separating hyperplane in the quantum feature space. There is another that skips the hyperplane step and produces a highly accurate classifying kernel function. As the ability to entangle more qubits increases, the successful classification rate improves as well. [8] This is an active area of research: how can we use entanglement, which does not exist classically, to find new or better patterns than we can do with strictly traditional methods?

To learn more

There are an increasing number of research papers being written connecting quantum computing with machine learning and other AI techniques. The results are somewhat fragmented. The best book pulling together the state of the art is Wittek. [19].

I do warn you again that quantum computers cannot process much data now!

For an advanced application of machine learning for quantum computing and chemistry, see Torial et al. [18]

1.5 Applications to financial services

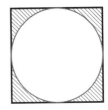

Suppose we have a circle of radius 1 inscribed in a square, which therefore has sides of length 2 and area $4 = 2 \times 2$. What is the area A of the circle?

Before you try to remember your geometric area formulas, let's compute it a different way using ratios and some experiments.

Suppose we drop some number N of coins onto the square and count how many of them have their centers on or inside the circle. If C is this number, then

$$\frac{\text{area of the circle}}{\text{area of the enclosing square}} = \frac{A}{4} \approx \frac{C}{N} = \frac{\text{number of coins that land in the circle}}{\text{total number of coins}}.$$

So $A \approx \frac{4C}{N}$.

There is randomness involved here: it's possible they all land inside the circle or, less likely, outside the circle. For $N = 1$, we do not get an accurate estimate of A at all because $\frac{C}{N}$ can only be 0 or 1.

Question 1.5.1

If $N = 2$, what are the possible estimates of A? What about if $N = 3$?

Clearly we will get a better estimate for A if we choose N large.

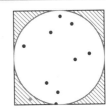

Using Python and its random number generator, I created 10 points whose centers lie inside the square. The plot on the right shows where they landed. In this case, $C = 9$ and so $A \approx 3.6$.

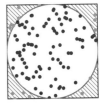

For $N = 100$ we get a more interesting plot with $C = 84$ and $A \approx 3.36$. Remember, if different random numbers had been generated then this number would be different.

The final plot is for $N = 500$. Now $C = 387$ and $A \approx 3.096$.

The real value of A is $\pi \approx 3.1415926$. This technique is called *Monte Carlo sampling* and goes back to the 1940s.

Using the same technique, here are approximations of A for increasingly large N. Remember, we are using random numbers and so these numbers will vary based on the sequence of values used.

N	10	100	1,000	10,000	100,000	1,000,000	10,000,000
A	3.6	3.36	3.148	3.1596	3.14336	3.141884	3.1414132

That's a lot of runs, the value of N, to get close to the real value of π. Nevertheless, this example demonstrates how we can use Monte Carlo sampling techniques to approximate the value of something when we may not have a formula. In this case we estimated A. For the example we ignored our knowledge that the formula for the area of a circle is πr^2, where r is the circle's radius.

In section 6.7 we work through the math and show that if we want to estimate π within 0.00001 with probability at least 99.9999%, we need $N \geq 82,863,028$. That is, we need to use more than 82 million points! So it is possible to use a Monte Carlo method here but it is not efficient.

In this example, we know the answer ahead of time by other means. If we do not know the answer and do not have a nice neat formula to compute, Monte Carlo methods can be a useful tool. However, the very large number of samples needed to get decent accuracy makes the process computationally intensive. If we can reduce the sample count significantly, we can compute a more accurate result much faster.

Given that the title of this section mentions "finance," I now note, perhaps unsurprisingly, that Monte Carlo methods are used in computational finance. The randomness we use to calculate π translates over into ideas like uncertainties. Uncertainties can then be related to probabilities, which can be used to calculate the risk and rate of return of investments.

Instead of looking at whether a point is inside or outside a circle, for the rate of return we might consider several factors that go into calculating the risk. For example,

- market size,
- share of market,
- selling price,
- fixed costs,
- operating costs,
- obsolescence,
- inflation or deflation,
- national monetary policy,
- weather, and
- political factors and election results.

For each of these or any other factor relevant to the particular investment, we quantify them and assign probabilities to the possible results. In a weighted way, we combine all possible combinations to compute risk. This is a function that we cannot calculate all at once, but we can use Monte Carlo methods to estimate. Methods similar to, but more complicated than, the circle analysis in section 6.7, give us how many samples we need to use to get a result within a desired accuracy.

In the circle example, even getting reasonable accuracy can require tens of millions of samples. For an investment risk analysis we might need many orders of magnitude greater. So what do we do?

We can and do use High Performance Computers (HPC). We can consider fewer possibilities for each factor. For example, we might vary the possible selling prices by larger amounts. We can consult better experts and get more accurate probabilities. This could increase the accuracy but not necessarily the computation time. We can take fewer samples and accept less precise results.

Or we might consider quantum variations and replacements for Monte Carlo methods. In 2015, Ashley Montanaro described a quadratic speedup using quantum computing. [12] How much improvement does this give us? Instead of the 82 million samples required for the circle calculation with the above accuracy, we could do it in something closer to 9,000 samples. $(9055 \approx \sqrt{82000000}.)$

In 2019, Stamatopoulos *et al* showed methods and considerations for pricing financial options using quantum computing systems. [16] I want to stress that to do this, we will need much larger, more accurate, and more powerful quantum computers than we have as of this writing. However, like much of the algorithmic work being done on industry use cases, we believe we are getting on the right path to solve significant problems significantly faster using quantum computation.

By using Monte Carlo methods we can vary our assumptions and do scenario analysis. If we can eventually use quantum computers to greatly reduce the number of samples, we can look at far more scenarios much faster.

To learn more

David Hertz' original 1964 paper in the Harvard Business Review is a very readable introduction to Monte Carlo methods for risk analysis without ever using the phrase "Monte Carlo." [9] A more recent paper gives more of the history of these methods and applies them to marketing analytics. [6]

My goal with this book is to give you enough of an introduction to quantum computing so that you can read industry-specific quantum use cases and research papers. For example, to see modern quantum algorithmic approaches to risk analysis, see the articles by Woerner, Egger, et al. [20] [3]. Some early results on heuristics using quantum computation for transaction settlements are covered in Braine et al. [1]

1.6   What about cryptography?

You may have seen media headlines like

Quantum Security Apocalypse!!!

Y2K??? Get ready for Q2K!!!

Quantum Computing Will Break All Internet Security!!!

These breathless announcements are meant to grab your attention and frequently contain egregious errors about quantum computing and security. Let's look at the root of the concerns and insert some reality into the discussion.

RSA is a commonly used security protocol and it works something like this:

- You want to allow others to send you secure communications. This means you give them what they need to encrypt their messages before sending. You and only you can decrypt what they then give you.
- You publish a *public key* used to encrypt these messages intended for you. Anyone who has access to the key can use it.
- There is an additional key, your *private key*. You and only you have it. With it you can decrypt and read the encrypted messages. [15]

Though I phrased this in terms of messages sent to you, the scheme is adaptable for sending transaction and purchase data across the Internet, and storing information securely in a database.

Certainly if anyone steals your private key, there is a cybersecurity emergency. Quantum computing has nothing to do with physically taking your private key or convincing you to give it to a bad person.

What if I could compute your private key from the public key?

The public key for RSA looks like a pair of numbers (e, n) where n is a very larger integer that is the product of two primes. We'll call these primes numbers p and q. For example, if $p = 982451653$ and $q = 899809343$, then $n = pq = 884019176415193979$.

Your private key looks like a pair of integers (d, n) using the very same n as in the public key. It is the d part you must really keep secret.

Here's the potential problem: if someone can quickly factor n into p and q, then they can compute d. That is, fast integer factorization leads to breaking RSA encryption.

Though multiplication is very easy and can be done using the method you learned early in your education, factoring can be very, very hard. For products of certain pairs of primes, factorization using known classical methods could take hundreds or thousands of **years**.

Given this, unless d is stolen or given away, you might feel pretty comfortable about security. Unless, that is, there is another way of factoring involving non-classical computers.

In 1995, Peter Shor published a quantum algorithm for integer factorization that is almost exponentially faster than known classical methods. We analyze Shor's algorithm in section 10.6.

This sounds like a major problem! Here is where many of the articles about quantum computing and security start to go crazy. The key question is: **how powerful, and of what quality, must a quantum computing system be in order to perform this factorization?**

As I write this, scientists and engineers are building quantum computers with double digit numbers of *physical* qubits, hoping to get to triple digits in the next few years. For example, researchers have discussed qubit counts of 20, 53, 72, and 128. (Do note there is a difference between what people say they will have versus what they really have.) A physical qubit is the hardware implementation of the *logical* qubits we start discussing in chapter 7.

Physical qubits have noise that cause errors in computation. Shor's algorithm requires fully fault-tolerant, error corrected logical qubits. This means we can detect and correct any errors that occur in the qubits. This happens today in the memory and data storage in your laptop and smartphone. We explore quantum error correction in section 11.5.

As a rule of thumb, assume it will take 1,000 very good physical qubits to make one logical qubit. This estimate varies by researcher, degree of marketing hype, and wishful thinking, but I believe 1,000 is reasonable. We discuss the relationship between the two kinds of qubits in chapter 11. In the meanwhile, we are in the Noisy Intermediate-Scale Quantum, or NISQ, era. The term NISQ was coined by physicist John Preskill in 2018. [14]

It will take many physical qubits to make one logical qubit

A further estimate is that it will take $10^8 = 100$ million physical qubits to use Shor's algorithm to factor the values of n used in RSA today. That's approximately one hundred thousand logical qubits. On one hand, we have quantum computers with two or three digits worth of physical qubits. For Shor's algorithm to break RSA, we'll need eight digits worth. That's a huge difference.

These numbers may be too conservative, but I don't think by much. If anyone quotes you much smaller numbers, try to understand their motivation and what data they are using.

There's a good chance we won't get quantum computers this powerful until 2035 or much later. We may never get such powerful machines. Assuming we will, what should you do now?

First, you should start moving to so-called "post-quantum" or "quantum-proof" encryption protocols. These are being standardized at NIST, the National Institute of Standards and Technology, in the United States by an international team of researchers. These protocols can't be broken by quantum computing systems as RSA and some of the other classical protocols might be eventually.

You may think you have plenty of time to change over your transactional systems. How long will it take to do that? For financial institutions, it can take ten years or more to implement new security technology.

Of greater immediate importance is your data. Will it be a problem if someone can crack your database security in 15, 30, or 50 years? For most organizations the answer is a loud YES. Start looking at hardware and software encryption support for your data using the new post-quantum security standards now.

Finally, quantum or no quantum, if you do not have good cybersecurity and encryption strategies and implementations in place now, you are exposed. Fix them. Listen to the people who make quantum computing systems to get a good idea of if and when they might be used to break encryption schemes. All others are dealing with second- and third-hand knowledge.

> **To learn more**
>
> Estimates for when and if quantum computing may pose a cybersecurity threat vary significantly. Any study on the topic will necessarily need to be updated as the technology evolves. The most complete analysis as of the time this book was first published appears to be Mosca and Piani. [13]

1.7 Summary

In this first chapter we looked at what is motivating the recent interest in quantum computers. The lone 1s and 0s of classical computing bits are extended and complemented by the infinite states of qubits, also known as quantum bits. The properties of superposition and entanglement give us access to many dimensions of working memory that are unavailable to classical computers.

Industry use cases for quantum computing are nascent but the areas where experts believe it will be applicable sooner are chemistry, materials science, and financial services. AI is another area where quantum may boost performance for some kinds of calculations.

There has been confusion in traditional and social media about the interplay of security, information encryption, and quantum computing. The major areas of misunderstanding are the necessary performance requirements and the timeline.

In the next chapter, we look at classical bit-based computing to more precisely and technically explore how quantum computing may help us attack problems that are otherwise impossible today. In chapter 3 through chapter 6 we work through the mathematics necessary for you to see how quantum computing works. There is a lot to cover, but it is worth it to be able to go deeper than a merely superficial understanding of the "whats," "hows," and "whys" of quantum computing.

References

[1]   Lee Braine et al. *Quantum Algorithms for Mixed Binary Optimization applied to Transaction Settlement*. 2019. URL: https://arxiv.org/abs/1910.05788.

[2]   Yudong Cao et al. "Quantum Chemistry in the Age of Quantum Computing". In: *Chemical Reviews* (2019).

[3]   Daniel J. Egger et al. *Credit Risk Analysis using Quantum Computers*. 2019. URL: https://arxiv.org/abs/1907.03044.

[4]   FermiLab. *What is the number of atoms in the world?* 2014. URL: https://www.fnal.gov/pub/science/inquiring/questions/atoms.html.

[5]   Richard P. Feynman. "Simulating Physics with Computers". In: *International Journal of Theoretical Physics* 21.6 (June 1, 1982), pp. 467–488.

[6]   Peter Furness. "Applications of Monte Carlo Simulation in marketing analytics". In: *Journal of Direct, Data and Digital Marketing Practice* 13 (2 Oct. 27, 2011).

[7]   Aram W. Harrow, Avinatan Hassidim, and Seth Lloyd. "Quantum Algorithm for Linear Systems of Equations". In: *Physical Review Letters* 103 (15 Oct. 2009), p. 150502.

[8]   Vojtěch Havlíček et al. "Supervised learning with quantum-enhanced feature spaces". In: *Nature* 567.7747 (Mar. 1, 2019), pp. 209–212.

[9]   David B Hertz. "Risk Analysis in Capital Investment". In: *Harvard Business Review* (Sept. 1979).

[10]   Abhinav Kandala et al. "Hardware-efficient variational quantum eigensolver for small molecules and quantum magnets". In: *Nature* 549 (Sept. 13, 2017), pp. 242–247.

[11]   Rachel Link. *How Much Caffeine Do Coke and Diet Coke Contain?* 2018. URL: https://www.healthline.com/nutrition/caffeine-in-coke.

[12]   Ashley Montanaro. "Quantum speedup of Monte Carlo methods". In: *Proceedings of the Royal Society A: Mathematical, Physical and Engineering Sciences* 471.2181 (2015), p. 20150301.

[13]   Michele Mosca and Marco Piani. *Quantum Threat Timeline.* 2019. URL: https://globalriskinstitute.org/publications/quantum-threat-timeline/.

[14]   John Preskill. *Quantum Computing in the NISQ era and beyond.* URL: https://arxiv.org/abs/1801.00862.

[15]   R. L. Rivest, A. Shamir, and L. Adleman. "A Method for Obtaining Digital Signatures and Public-key Cryptosystems". In: *Commun. ACM* 21.2 (Feb. 1978), pp. 120–126.

[16]   Nikitas Stamatopoulos et al. *Option Pricing using Quantum Computers.* 2019. URL: https://arxiv.org/abs/1905.02666.

[17]   Ewin Tang. *A quantum-inspired classical algorithm for recommendation systems.* 2019. URL: https://arxiv.org/pdf/1807.04271.pdf.

[18]   Giacomo Torlai et al. *Precise measurement of quantum observables with neural-network estimators.* URL: https://arxiv.org/abs/1910.07596.

[19]   P. Wittek. *Quantum Machine Learning. What quantum computing means to data mining.* Elsevier Science, 2016.

[20]   Stefan Woerner and Daniel J. Egger. "Quantum risk analysis". In: *npj Quantum Information* 5 (1 Feb. 8, 2019), pp. 198–201.

I
Foundations

2

They're Not Old, They're Classics

No simplicity of mind, no obscurity of station, can escape the universal duty of questioning all that we believe.

William Kingdon Clifford

When introducing quantum computing, it's easy to say "It's completely different from classical computing in every way!" Well that's fine, but to what exactly are you comparing it?

We start things off by looking at what a classical computer is and how it works to solve problems. This sets us up to later show how quantum computing replaces even the most basic classical operations with ones involving qubits, superposition, and entanglement.

Topics covered in this chapter

2.1   What's inside a computer?

If I were to buy a laptop today, I would need to think about the following kinds of hardware options:

- size and weight of the machine
- quality of the display

- processor and its speed
- memory and storage capacity

Three years ago I built a desktop gaming PC. I had to purchase and assemble and connect:

- the case
- power supply
- motherboard
- processor
- internal memory
- video card with a graphics processing unit (GPU) and memory

- internal hard drive and solid state storage
- internal Blu-ray drive
- wireless network USB device
- display
- speakers
- mouse and keyboard

As you can see, I had to make many choices. In the case of the laptop, you think about why you want the machine and what you want to do, and much less about the particular hardware. You don't have to make a choice about the manufacturers of the parts nor the standards that allow those parts to work together.

The same is true for smartphones. You decide on the mobile operating system, which then may decide the manufacturer, you pick a phone, and finally choose how much storage you want for apps, music, photos, and videos.

Of all the components above, I'm going to focus mainly on four of them: the processor, the "brain" for general computation; the GPU for specialized calculations; the memory, for holding information during a computation; and the storage, for long-term preservation of data used and produced by applications.

All these live on or are controlled by the motherboard and there is a lot of electronic circuitry that supports and connects them. My discussion about every possible independent or integrated component in the computer is therefore not complete.

Let's start with the storage. In applications like AI, it's common to process many megabytes or gigabytes of data to look for patterns and to perform tasks like classification. This information is held on either hard disk drives, introduced by IBM in 1956, or modern solid-state drives.

> The smallest unit of information is a bit, and that can represent either the value 0 or the value 1.

The capacity of today's storage is usually measured in terabytes, which is $1,000 = 10^3$ gigabytes. A gigabyte is $1,000$ megabytes, which is $1,000$ kilobytes, which is $1,000$ bytes. Thus a terabyte is $10^{12} = 1,000,000,000$ bytes. A *byte* is 8 bits.

These are the base 10 versions of these quantities. It's not unusual to see a kilobyte being given as $1,024 = 2^{10}$ bytes, and so on. The values are slightly different, but close enough for most purposes.

The data can be anything you can think of – from music, to videos, to customer records, to presentations, to financial data, to weather history and forecasts, to the source for this book, for example. This information must be held reliably for as long as it is needed. For this reason this is sometimes called *persistent storage*. That is, once I put it on the drive I expect it to be there whenever I want to use it.

The drive may be within the processing computer or somewhere across the network. Data "on the cloud" is held on drives and accessed by processors there or pulled down to your machine for use.

What do I mean when I say that the information is reliably stored? At a high level I mean that it has backups somewhere else for redundancy. I can also insist that the data is encrypted so that only authorized people and processes have access. At a low level I want the data I store, the 0s and 1s, to always have exactly the values that were placed there.

Let's think about how plain characters are stored. Today we often use Unicode to represent over 100,000 international characters and symbols, including relatively new ones like emojis. [10] However, many applications still use the ASCII (also known as US-ASCII) character set, which represents a few dozen characters common in the United States. It uses 7 bits (0s or 1s) of a byte to correspond to these characters. Here are some examples:

Bits	Character	Bits	Character
0100001	!	0111111	?
0110000	0	1000001	A
0110001	1	1000010	B
0110010	2	1100001	a
0111100	<	1100010	b

We number the bits from right to left, starting with 0:

$$\text{character `a': } \begin{array}{ccccccc} 1 & 1 & 0 & 0 & 0 & 0 & 1 \\ \uparrow & \uparrow & \uparrow & \uparrow & \uparrow & \uparrow & \uparrow \\ 6 & 5 & 4 & 3 & 2 & 1 & 0 \end{array}$$

position: 6 5 4 3 2 1 0

If something accidentally changes bit 5 in 'a' from 1 to 0, I end up with an 'A'. If this happened in text you might say that it wasn't so bad because it was still readable. Nevertheless, it is not the data with which we started. If I change bit 6 in 'a' to a 0, I get the character '1'. Data errors like this can change the spelling of text and the values of numeric quantities like temperature, money, and driver license numbers.

Errors may happen because of original or acquired defects in the hardware, extreme "noise" in the operating environment, or even the highly unlikely stray bit of cosmic radiation. Modern storage hardware is manufactured to very tight tolerances with extensive testing. Nevertheless, software within the storage devices can often detect errors and correct them. The goal of such software is to both detect and correct errors quickly using as little extra data as possible. This is called *fault tolerance*.

The idea is that we start with our initial data, encode it in some way with extra information that allows us to tell if errors occurred and perhaps how to fix them, do something with the data, then decode and correct the information if necessary.

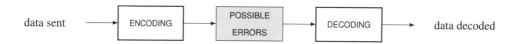

One strategy for trying to avoid errors is to store the data many times. This is called a *repetition code*. Suppose I want to store an 'a'. I could save it five times:

1100001   **0**100001   **1**100001   **1**100001   **1**100001

The first and second copies are different in bit 6 and so an error occurred. Since four of the five copies agree, we might "correct" the error by deciding the actual value is 1100001. However, who's to say the other copies also don't have errors? There are more efficient ways of detecting and correcting errors than repetition, but it is a central concept that underlies several other schemes.

Another way to possibly detect an error is to use an *even parity bit*. We append one more bit to the data: if there is an odd number of 1 bits, then we precede the data with a 1. For an even number of 1s, we place a 0 at the beginning.

$$1100001 \mapsto \mathbf{1}1100001$$
$$1100101 \mapsto \mathbf{0}1100101$$

If we get a piece of data and there are an odd number of 1s, we know at least one bit is wrong.

If errors continue to occur within a particular region of storage, then the controlling software can direct the hardware to avoid using it. There are three processes within in our systems that keep our data correct:

- *Error detection:* Discovering that an error has occurred.
- *Error correction:* Fixing an error with an understood degree of statistical confidence.
- *Error mitigation:* Preventing errors from happening in the first place through manufacturing or control.

To learn more

Many of the techniques and nomenclature of quantum error correction have their roots and analogs in classical use cases dating back to the 1940s. [2] [3] [8]

One of the roles of an operating system is to make persistent storage available to software through the handling of file systems. There are many schemes for file systems but you may only be aware that individual containers for data, the files, are grouped together into folders or directories. Most of the work regarding file systems is very low-level code that handles moving information stored in some device onto another device or into or out of an application.

From persistent storage, let's move on to the memory in your computer. I think a good phrase to use for this is "working memory" because it holds much of the information your system needs to use while processing. It stores part of the operating system, the running apps, and the working data the apps are using. Data or parts of the applications that are not needed immediately might be put on disk by a method called "paging." Information in memory can be gotten from disk, computed by a processor, or placed in memory directly in some other way.

The common rule is "more memory is good." That sounds trite, but low cost laptops often skimp on the amount or speed of the memory and so your apps run more slowly. Such laptops may have only 20% of the memory used by high-end desktop machines for video editing or games.

A memory location is accessed via an address:

0024								
0016		D						
0008					Q			
0000								

The byte of data representing the character "Q" is at address 0012 while we have a "D" at address 0017. If you have a lot of memory, you need to use very large numbers as addresses.

Next up to consider is the central processing unit, the CPU, within a classical computer. It's a cliché, but it really is like the brain of the computer. It controls executing a sequence of instructions that can do arithmetic, move data in and out of memory, use designated extra-fast memory called *registers* within the processor, and conditionally jump somewhere else in the sequence. The latest processors can help in memory management for interpreted programming languages and can even generate random numbers.

Physically, CPUs are today made from transistors, capacitors, diodes, resistors, and the pathways that connect them all into an *integrated circuit*. We differentiate between these electronic circuits and the logic circuits that are implemented using them.

In 1965, Gordon Moore extrapolated from several years of classical processor development and hypothesized that we would be able to double the speed of and number of transistors on a chip roughly every two years. [6] When this started to lose steam, engineers worked out how to put multiple processing units within a computer. These are called *cores*.

Within a processor, there can be special units that perform particular functions. A *floating-point unit* (FPU) handles fast mathematical routines with numbers that contain decimals. An *arithmetic logic unit* (ALU) provides accelerated hardware support for integer arithmetic. A

CPU is not limited to having only one FPU or ALU. More may be included within the architecture to optimize the chip for applications like High Performance Computing (HPC).

Caches within a CPU improve performance by storing data and instructions that might be used soon. For example, instead of retrieving only a byte of information from storage, it's often better to pull in several hundred or thousand bytes near it into fast memory. This is based on the assumption that if the processor is now using some data it will soon use other data nearby. Very sophisticated schemes have been developed to keep the cores busy with all the data and instructions they need with minimal waiting time.

You may have heard about 32- or 64-bit computers and chosen an operating system based on one or the other. These numbers represent the *word size* of the processor. This is the "natural" size of the piece of data that the computer usually handles. It determines how large an integer the processor can handle for arithmetic or how big an address it can use to access memory.

In the first case, for a 32-bit word we use only the first 31 bits to hold the number and the last to hold the sign. Though there are variations on how the number is stored, a common scheme says that if that sign bit is 1 the number is negative, and zero or positive if the bit is 0.

For addressing memory, suppose as above that the first byte in memory has address 0. Given a 32-bit address size, the largest address is $2^{32} - 1$. This is a total of 4,294,967,296 addresses, which says that 4 gigabytes is the largest amount of memory with which the processor can work. With 64 bits you can "talk to" much more data.

Question 2.1.1

What is the largest memory address a 64-bit processor can access?

In advanced processors today, data in memory is not really retrieved or placed via a simple integer address pointing to a physical location. Rather, a *memory management unit* (MMU) translates the address you give it to a memory location somewhere within your computer via a scheme that maps to your particular hardware. This is called *virtual* memory.

In addition to the CPU, a computer may have a separate graphics processing unit (GPU) for high-speed calculations involving video, particularly games and applications like augmented and virtual reality. The GPU may be part of the motherboard or on a separate plug-in card with 2, 4, or more gigabytes of dedicated memory. These separate cards can use a lot of power and generate much heat but they can produce extraordinary graphics.

Because a GPU has a limited number of highly optimized functions compared to the more general purpose CPU, it can be much faster, sometimes hundreds of times quicker at certain operations. Those kinds of operations and the data they involve are not limited to graphics. Linear algebra and geometric-like procedures make GPUs good candidates for some AI algorithms. [11] Even cryptocurrency miners use GPUs to try to find wealth through computation.

A quantum computer today does not have its own storage, memory, FPU, ALU, GPU, or CPU. It is most similar to a GPU in that it has its own set of operations through which it may be able to execute some special algorithms significantly faster. How much faster? I don't mean twice as fast, I mean thousands of times faster.

A GPU is a variation on the classical architecture while a quantum computer is something entirely different.

You cannot take a piece of classical software or a classical algorithm and directly run it on a quantum system. Rather, quantum computers work together with classical ones to create new hybrid systems. The trick is understanding how to meld them together to do things that have been intractable to date.

2.2 The power of two

For a system based on 0s and 1s, the number 2 shows up a lot in classical computing. This is not surprising because we use binary arithmetic, which is a set of operations on base 2 numbers.

Most people use base 10 for their numbers. These are also called decimal numbers. We construct such numbers from the symbols 0, 1, 2, 3, 4, 5, 6, 7, 8, and 9, which we often call digits. Note that the largest digit, 9, is one less than 10, the base.

A number such as 247 is really shorthand for the longer $2 \times 10^2 + 4 \times 10^1 + 7 \times 10^0$. For $1,003$ we expand to $1 \times 10^3 + 0 \times 10^2 + 0 \times 10^1 + 3 \times 10^0$. In these expansions we write a sum of digits between 0 and 9 multiplied by powers of 10 in decreasing order with no intermediate powers omitted.

We do something similar for binary. A binary number is written as a sum of bits (0 or 1) multiplied by powers of 2 in decreasing order with no intermediate powers omitted. Here are some examples:

$$0 = 0 \times 2^0$$
$$1 = 1 \times 2^0$$
$$10 = 1 \times 2^1 + 0 \times 2^0$$
$$1101 = 1 \times 2^3 + 1 \times 2^2 + 0 \times 2^1 + 1 \times 2^0$$

The "10" is the binary number 10 and not the decimal number 10. Confirm for yourself from the above that the binary number 10 is another representation for the decimal number 2. If the context doesn't make it clear whether binary or decimal is being used, I use subscripts like 10_2 and 2_{10} where the first is base 2 and the second is base 10.

If I allow myself two bits, the only numbers I can write are 00, 01, 10, and 11. 11_2 is 3_{10} which is $2^2 - 1$. If you allow me 8 bits then the numbers go from 00000000 to 11111111. The latter is $2^8 - 1$.

For 64 bits the largest number I can write is a string of sixty four 1s which is

$$2^{64} - 1 = 18,446,744,073,709,551,615.$$

This is the largest positive integer a 64-bit processor can use.

We do binary addition by adding bits and carrying.

$$0 + 0 = 0$$
$$1 + 0 = 1$$
$$0 + 1 = 1$$
$$1 + 1 = 0 \text{ carry } 1$$

Thus while $1 + 0 = 1$, $1 + 1 = 10$. Because of the carry we had to add another bit to the left. If we were doing this on hardware and the processor did not have the space to allow us to use that extra bit, we would be in an overflow situation. Hardware and software that do math need to check for such a condition.

2.3 True or false?

From arithmetic let's turn to basic logic. Here there are only two values: true and false. We want to know what kinds of things we can do with one or two of these values.

The most interesting thing you can do to a single logical value is to replace it with the other. Thus, the **not** operation turns true into false, and false into true:

not true = false

not false = true

For two inputs, which I call p and q, there are three primary operations **and**, **or**, and **xor**. Consider the statement "We will get ice cream only if you **and** your sister clean your rooms." The result is the truth or falsity of the statement "we will get ice cream."

If neither you nor your sister clean your rooms, or if only one of you clean your room, then the result is false. If both of you are tidy, the result is true, and you can start thinking about ice cream flavors and whether you want a cup or a cone.

Let's represent by (you, sister) the different combinations of you and your sister, respectively, having done what was asked. You will not get ice cream for (false, false), (true, false), or (false, true). The only combination that is acceptable is (true, true). We express this as

true **and** true = true

true **and** false = false

false **and** true = false

false **and** false = false

More succinctly, we can put this all in a table

p = you	q = your sister	p **and** q
true	true	true
true	false	false
false	true	false
false	false	false

where the first column has the values to the left of the **and**, and the second column has the values to the right of it. The rows are the values and the results. This is called a "truth table."

Another situation is where we are satisfied if at least one of the inputs is true. Consider "We will go to the movie if you **or** your sister feed the dog." The result is the truth or falsity of the statement "we will go to the movie."

p	q	p **or** q
true	true	true
true	false	true
false	true	true
false	false	false

Finally think about a situation where we care about one and only one of the inputs being true. This is similar to **or** except in the case (true, true) where the result is false. It is called an "exclusive or" and written **xor**. If I were to say, "I am now going to the restaurant or going to the library," then one of these can be true but not both, assuming I mean my very next destination.

Question 2.3.1

How would you state the inputs and result statement for this **xor** example?

---

p	q	p **xor** q
true	true	false
true	false	true
false	true	true
false	false	false

There are also versions of these that start with **n** to mean we apply **not** to the result. This means we apply an operation like **and**, **or**, or **xor** and then flip true to false or false to true. This "do something and then negate it" is not common in spoken or written languages but they are quite useful in computer languages.

nand is defined this way:

$$\text{true } \textbf{nand}\text{ false} = \textbf{not }(\text{true } \textbf{and}\text{ false}) = \text{true}$$

It has this table of values:

p	q	p **nand** q
true	true	false
true	false	true
false	true	true
false	false	true

I leave it to you to work out what happens for **nor** and **nxor**. These **n**-versions seem baroque and excessive now but several are used in the next section.

Question 2.3.2

Fill in the following tables based on the examples and discussion above.

p	q	p **nor** q
true	true	
true	false	
false	true	
false	false	

p	q	p **nxor** q
true	true	
true	false	
false	true	
false	false	

Instead of true and false we could have used 1 and 0, respectively, which hints at some connection between logic and arithmetic.

2.4   Logic circuits

Now that we have a sense of how the logic works, we can look at logic circuits. The most basic logic circuits look like binary relationships but more advanced ones implement operations for addition, multiplication, and many other mathematical operations. They also manipulate basic data. Logic circuits implement algorithms and ultimately the apps on your computer or device.

We begin with examples of the core operations, also called *gates*.

To me, the standard gate shapes used in the United States look like variations on spaceship designs.

Rather than use true and false, we use 1 and 0 as the values of the bits coming into and out of gates.

This gate has two inputs and one output. It is not reversible because it produces the same output with different inputs. Given the 0 output, we cannot know which example produced it. Here are the other gates we use, with example inputs:

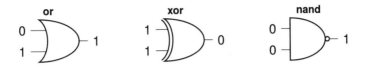

The symbol "⊕" is frequently used for the **xor** operation.

The **not** gate has one input and one output. It is reversible: if you apply it twice you get back what you started with.

$$0 \longrightarrow \!\!\!\!\triangleright\!\circ 1$$

People who study electrical engineering see these gates and the logic circuits you can build from them early in their training. This is not a book about electrical engineering. Instead, I want you to think of the above as literal building blocks. We plug them together, pull them apart, and make new logic circuits. That is, we'll have fun with them while getting comfortable with the notion of creating logic circuits that do what we want.

I connect the output of one **not** gate to the input of another to show you get the same value you put in. x can be 0 or 1.

If we didn't already have a special **nand** gate we could build it.

not (x **and** y) = x **nand** y

Note how we can compose gates to get other gates. We can build **and** from **nand** and **not**.

not (x **nand** y) = x **and** y

From this you can see that we could shrink the collection of gates we use if we desired. From this last example, it's technically redundant to have **and**, **nand**, and **not**, but it is convenient to have them all. Let's keep going to see what else we can construct.

Even though a gate like **nand** has two inputs, we can push the same value into each input. We show it this way:

What do we get when we start with 0 or 1 and run it through this logic circuit? 0 **nand** 0 = 1 and 1 **nand** 1 = 0. This behaves exactly like **not**! This means if we really wanted to, we could get rid of **not**. With this, having only **nand** we could drop **not** and **and**. It's starting to seem that **nand** has some essential building block property.

By stringing together four **nand** gates, we can create an **xor** gate. It takes three to make an **or** gate. It takes one more of each to make **nxor** and **nor** gates, respectively.

> Every logical gate can be constructed from a logic circuit of only **nand** gates. The same is true for **nor** gates. **nand** and **nor** gates are called *universal* gates because they have this property. [7]

It would be tedious and very inefficient to have all the basic gates replaced by multiple **nand** gates, but you can do it. This is how we would build **or**:

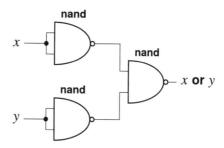

For the binary logic gates that have two inputs and one output, there are only eight possibilities, the four possible inputs 0/0, 0/1, 1/0, and 1/1 together with the outputs 0 and 1. Like **or** above, you can string together combinations of **nand**s to create any of these eight gates.

Question 2.4.1

Show how to create the **nor** gate only from **nand** gates.

We looked at these logic circuits to see how classical processing is done at a very low level. We return to circuits and the idea of universal gates again for quantum computing in section 9.2.

So far this has really been a study of classical gates for their own sakes. The behavior, compositions, and universality are interesting, but don't really do much fascinating yet. Let's do some math!

2.5   Addition, logically

Using binary arithmetic as we discussed in section 2.2,

$$0 + 0 = 0$$
$$1 + 0 = 1$$
$$0 + 1 = 1$$
$$1 + 1 = 0 \text{ carry } 1$$

Focusing on the value after the equal signs and temporarily forgetting the carrying in the last case, this is the same as what **xor** does with two inputs.

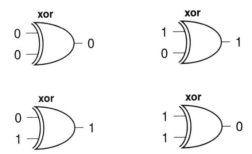

We did lose the carry bit but we limited ourself to having only one output bit. What gate operation would give us that 1 carry bit only if both inputs were also 1, and otherwise return 0? Correct, it's **and**! So if we can combine the **xor** and the **and** and give ourselves two bits of output, we can do simple addition of two bits.

Question 2.5.1

Try drawing a circuit that would do this before peeking at what follows. You are allowed to clone the value of a bit and send it to two different gates.

Question 2.5.2

Did you peek?

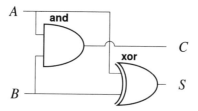

where A, B, S, and C are bits. The circuit takes two single input bits, A and B, and produces a 2-bit answer CD.

$$A + B = CS$$
$$0 + 0 = 00$$
$$1 + 0 = 01$$
$$0 + 1 = 01$$
$$1 + 1 = 10$$

We call S the *sum* bit and C the *carry-out* bit. This circuit is called a half-adder since, as written, it cannot be used in the middle of a larger circuit. It's missing something. Can you guess what it is?

A full-adder has an additional input that is called the *carry-in*. This is the carry bit from an addition that might precede it in the overall circuit. If there is no previous addition, the carry-in bit is set to 0.

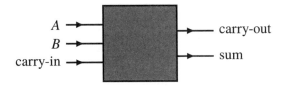

The square contains a circuit to handle the inputs and produce the two outputs.

Question 2.5.3

What does the circuit look like?

By extending this to more bits with additional gates, we can create classical processors that implement full addition. Other arithmetic operations like subtraction, multiplication, and division are implemented and often grouped within the *arithmetic logic unit* (ALU) of a classical processing unit.

For addition, the ALU takes multibit integer inputs and produces a multibit integer output sum. Other information can also be available from the ALU, such if the final bit addition caused an *overflow,* a carry-out that had no place to go.

ALUs contain circuits including hundreds or thousands of gates. A modern processor in a laptop or smartphone uses integers with 64 bits. Given the above simple circuit, try to estimate how many gates you would need to implement full addition.

Modern-day programmers and software engineers rarely deal directly with classical circuits themselves. Several layers are built on top of them so that coders can do what they need to do quickly.

If I am writing an app for a smartphone that involves drawing a circle, I don't need to know anything today about the low-level processes and circuits that do the arithmetic and cause the graphic to appear on the screen. I use a high-level routine that takes as input the location of the center of the circle, the radius, the color of the circle, and the fill color inside the circle.

At some point a person created that library implementing the high-level routine. Someone wrote the lower-level graphical operations. And someone wrote the very basic circuits that implement the primitive operations under the graphical ones.

Software is layered in increasing levels of abstraction. Programming languages like C, C++, Python, Java, and Swift hide the low-level details. Libraries for these languages provide reusable code that many people can use to piece together new apps.

There is always a bottom layer, though, and circuits live near there.

2.6 Algorithmically speaking

The word "algorithm" is often used generically to mean "something a computer does." Algorithms are employed in the financial markets to try to calculate the exact right moment and price at which to sell a stock or bond. They are used in artificial intelligence to find patterns in data to understand natural language, construct responses in human conversation, find manufacturing anomalies, detect financial fraud, and even to create new spice mixtures for cooking.

Informally, an algorithm is a recipe. Like a recipe for food, an algorithm states what inputs you need (water, flour, butter, eggs, etc.), the expected outcome (for example, bread), the sequence of steps you take, the subprocesses you should use (stir, knead, bake, cool), and what to do when a choice presents itself ("if the dough is too wet, add more flour").

We call each step an *operation* and give it a name as above: "stir," "bake," "cool," and so on. Not only do we want to overall process to be successful and efficient, but we construct the best possible operations for each action in the algorithm.

The recipe is not the real baking of the bread, you are doing the cooking. In the same way, an algorithm abstractly states what you should do with a computer. It is up to the circuits and higher-level routines built on circuits to implement and execute what the algorithm describes. There can be more than one algorithm the produces the same result.

Operations in computer algorithms might, for example, add two numbers, compare whether one is larger than another, switch two numbers, or store or retrieve a value from memory

Quantum computers do not use classical logical gates, operations, or simple bits. While quantum circuits and algorithms look and behave very differently from their classical counterparts, there is still the idea that we have data that is manipulated over a series of steps to get what we hope is a useful answer.

2.7 Growth, exponential and otherwise

Many people who use the phrase "exponential growth" use it incorrectly, somehow thinking it only means "very fast." Exponential growth involves, well, exponents. Here's a plot showing four kinds of growth: exponential, quadratic, linear, and logarithmic.

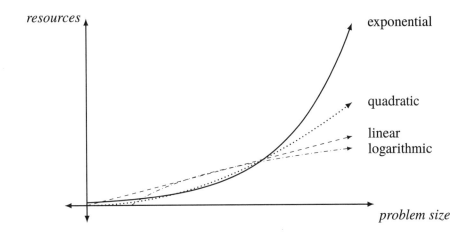

I've drawn them so they all intersect at a point but afterwards diverge. After the convergence, the logarithmic plot (dot dashed) grows slowly, the linear plot (dashed) continues as it did, the quadratic plot (dotted) continues upward as a parabola, and the exponential one shoots up rapidly.

Take a look at the change in the vertical axis, the one I've labeled *resources* with respect to the horizontal axis, labeled *problem size*. As the size of the problem increases, how fast does the amount of resources needed increase? Here a resource might be the time required for the algorithm, the amount of memory used during computation, or the megabytes of data storage necessary.

When we move a certain distance to the right horizontally for *problem size,* the logarithmic plot increases vertically at a rate proportional to the inverse of the size ($\frac{1}{problem\ size}$), the linear plot increases at a constant rate for *resources* that does not depend on *problem size*. The quadratic plot increases at a vertical rate that is proportional to *problem size*. The exponential plot increases at a rate that is proportional to its current *resources*.

The logarithm is only defined for positive numbers. The function $\log_{10}(x)$ answers the question "to what power must I raise 10 to get x?". When x is 10, the answer is 1. For x equals one million, the answer is 6. Another common logarithm function is $\log_2$ which substitutes 2 for 10 in these examples. Logarithmic functions grow *very* slowly.

Examples of growth are

$$resources = 2 \times \log_{10}(problem\ size)$$
$$resources = 4 \times (problem\ size)$$
$$resources = 0.3 \times (problem\ size)^2$$
$$resources = 7.2 \times 3^{\,problem\ size}$$

for logarithmic, linear, quadratic, and exponential, respectively. Note the variable *problem size* in the exponent in the fourth case.

This means that for large problem size, the exponential plot goes up rapidly and then more rapidly and so on. Things can quickly get out of hand when you have this kind of positive exponential growth.

If you start with 100 units of currency and get 6% interest compounded once a year, you will have $100(1+0.06)$ after one year. After two you will have $100(1+0.06)(1+0.06) = 100(1.06)^2$. In general, after t years you will have $100(1.06)^t$. This is exponential growth. Your money will double in approximately 12 years.

Quantum computers will not be used over the next several decades to fully replace classical computers. Rather, quantum computing may help make certain solutions doable in a short time instead of being intractable. The power of a quantum computer potentially grows exponentially with the number of quantum bits, or qubits, in it. Can this be used to control similar growth in problems we are trying to solve?

2.8   How hard can that be?

Once you decide to do something, how long does it take you? How much money or other resources does it involve? How do you compare the worst way of doing it with the best?

When you try to accomplish tasks on a computer, all these questions come to bear. The point about money may not be obvious, but when you are running an application you need to pay for the processing, storage, and memory you use. This is true whether you paid to get a more powerful laptop or have ongoing cloud costs.

To end this chapter we look at classical complexity. To start, we consider sorting and searching and some algorithms for doing them.

Whenever I hear "sorting and searching" I get a musical ear worm for Bobby Lewis' 1960 classic rock song "Tossin' and Turnin'." Let me know if it is contagious.

2.8.1   Sorting

Sorting involves taking multiple items and putting them in some kind of order. Consider your book collection. You can rearrange them so that the books are on the shelves in ascending alphabetic order by title. Or you can move them around so that they are in descending order by the year of publication. If more than one book was published in the same year, order them alphabetically by the first author's last name.

When we ordered by title, that title was the *key* we looked at to decide where to place the book among the others. When we considered it by year and then author, the year was the *primary key* and the author name was the *secondary key.*

Before we sort, we need to decide how we compare the items. In words, we might say "is the first number less than the second?" or "is the first title alphabetically before the second?". The response is either true or false.

Something that often snags new programmers is comparing things that look like numbers either numerically or *lexicographically.* In the first case we think of the complete item as a number while in the second we compare character by character. Consider 54 and 8. Numerically, the second is less than the first. Lexicographically, the first is less because the character 5 comes before 8.

Therefore when coding you need to convert them to the same format. If you were to convert -34,809 into a number, what would it be? In much of Europe the comma is used as the decimal point, while in the United States, the United Kingdom, and several other countries it is used to separate groups of three digits.

Now we look two ways of numerically sorting a list of numbers into ascending order. The first is called a "bubble sort" and it has several nice features, including simplicity. It is horribly inefficient, though, for large collections of objects that are not close to being in the right order.

We have two operations: *compare*, which takes two numbers and returns true if the first is less than the second, and false otherwise; and *swap*, which interchanges the two numbers in the list.

The idea of the bubble sort is to make repeated passes through the list, comparing adjacent numbers. If they are out of order, we swap them and continue through the list. We keep doing this until we make a full pass where no swaps were done. At that point the list is sorted. Pretty elegant and simple to describe!

Let's begin with a case where a list of four numbers is already sorted:

$$[-2, 0, 3, 7].$$

We compare -2 and 0. They are already in the correct order so we move on. 0 and 3 are also correct, so nothing to do. 3 and 7 are fine. We did three comparisons and no swaps. Because we did not have to interchange any numbers, we are done.

Now let's look at

$$[7, -2, 0, 3].$$

Comparing 7 and −2 we see the second is less than the first, so we swap them to get the new list

$$[-2, 7, 0, 3].$$

Next we look at 7 and 0. Again, we need to swap and get

$$[-2, 0, 7, 3].$$

Comparing 7 and 3 we have two more numbers that are out of order. Swapping them we get

$$[-2, 0, 3, 7].$$

So far we have done 3 comparisons and 3 swaps. Since the number of swaps is not zero, we pass through the list again. This time we do 3 more comparisons but no swaps and we are done. In total we did 6 comparisons and 3 swaps.

Now for the worst case with the list in reverse sorted order

$$[7, 3, 0, -2].$$

The first pass looks like

$[7, 3, 0, -2]$	swap first and second numbers
$[3, 7, 0, -2]$	swap second and third numbers
$[3, 0, 7, -2]$	swap third and fourth numbers
$[3, 0, -2, 7]$	

We did 3 comparisons and 3 swaps.

Second pass:

$[3, 0, -2, 7]$	swap first and second numbers
$[0, 3, -2, 7]$	swap second and third numbers
$[0, -2, 3, 7]$	compare third and fourth numbers but do nothing

We did 3 comparisons and 2 swaps.

Third pass:

$[0, -2, 3, 7]$	swap first and second numbers
$[-2, 0, 3, 7]$	compare second and third numbers but do nothing
$[-2, 0, 3, 7]$	compare third and fourth numbers but do nothing

We did 3 comparisons and 1 swap.

Fourth pass:

$[-2, 0, 3, 7]$ compare first and second numbers but do nothing

$[-2, 0, 3, 7]$ compare second and third numbers but do nothing

$[-2, 0, 3, 7]$ compare third and fourth numbers but do nothing

No swap, so we are done, and the usual 3 comparisons.

For this list of 4 numbers we did 12 comparisons and 6 swaps in 4 passes. Can we put some formulas behind these numbers for the worst case?

We had 4 numbers in completely the wrong order and so it took 4 full passes to sort them. For a list of length n we must do $n - 1$ comparisons, which is 3 in this example.

The number of swaps is the interesting number. On the first pass we did 3, on the second we did 2, the third we did 1, and the fourth we did none. So the number of swaps is

$$3 + 2 + 1 = (n - 1) + (n - 2) + \cdots + 1$$

where n is the length of the list. Obviously there is a pattern here.

There is a formula that can help us with this last sum. If we want to add up 1, 2, and so on through a positive integer m, we compute $\frac{m(m+1)}{2}$.

Question 2.8.1

Try this out for $m = 1$, $m = 2$, and $m = 3$. Now see if the formula still holds if we add in $m + 1$. That is, can you rewrite

$$\frac{m(m + 1)}{2} + m + 1$$

as

$$\frac{(m + 1)(m + 2)}{2} \ ?$$

If so, you have proved the formula by *induction*.

In our case, we have $m = n - 1$ and so we do a total of $\frac{(n-1)n}{2}$ swaps. For $n = 4$ numbers in the list, this is none other than the 6 swaps we counted by hand.

It may not seem so bad that we had to do 6 swaps for 4 numbers in the worst case, but what if we had 1,000 numbers in completely reversed order? Then the number of swaps would be

$$\frac{999 \times 1000}{2} = 499500$$

That's almost half a million swaps for a list of 1,000 numbers.

For 1 million numbers in the worst case it would be

$$\frac{999999 \times 1000000}{2} = 499999500000$$

which is 499 billion 999 million 500 thousand swaps. This is terrible. Rewriting

$$\frac{(n-1)n}{2} = \frac{n^2 - n}{2} = \frac{1}{2}n^2 - \frac{1}{2}n$$

we can see that the number of swaps grows with the *square* of the number of entries. In fact,

$$\text{number of swaps } \leq \frac{1}{2}n^2$$

for all $n \geq 1$. When we have this kind of situation, we say that our algorithm is $O(n^2)$, pronounced "big Oh of n squared."

More formally, we say that the number of interesting operations used to solve a problem on n objects is $O(f(n))$ if there is a positive real number c and an integer m such that

$$\text{number of operations } \leq cf(n)$$

once $n \geq m$, for some function f.

In our case, $c = \frac{1}{2}$, $f(n) = n^2$, and $m = 1$.

To learn more

In the area of computer science called *complexity theory*, researchers try to determine the best possible f, c, and m to match the growth behavior as closely as possible. [1, Chapter 3] [9, Section 1.4]

If an algorithm is $O(n^t)$ for some positive fixed number t then the algorithm is *of polynomial time*. Easy examples are algorithms that run in $O(n)$, $O(n^2)$, and $O(n^3)$ time.

If the exponent t is very large then we might have a very inefficient and impractical algorithm. Being of polynomial time really means that it is bounded above by something that is $O(n^t)$. We therefore also say that an $O(\log(n))$ algorithm is of polynomial time since it runs faster that an $O(n)$ algorithm.

Getting back to sorting, we are looking at the worst case for this algorithm. In the best case we do no swaps. Therefore when looking at a process we should examine the best case, the worst case, as well as the average case. For the bubble sort, the best case is $O(n)$ while the average and worst cases are both $O(n^2)$.

If you are coding up and testing an algorithm and it seems to take a very long time to run, you likely either have a bug in your software or you have stumbled onto something close to a worst case scenario.

Can we sort more efficiently than $O(n^2)$ time? How much better can we do?

By examination we could have optimized the bubble sort algorithm slightly. Look at the last entry in the list after each pass and think about how we could have reduced the number of comparisons. However, the execution time is dominated by the number of swaps, not the number of comparisons. Rather than tinkering with this kind of sort, let's examine another algorithm that takes a very different approach.

There are many sorting algorithms and you might have fun searching the web to learn more about the main kinds of approaches. You'll also find interesting visualizations about how the objects move around during the sort. Computer science students often examine the different versions when they learn about algorithms and data structures.

John von Neumann in the 1940s.
Photo subject to use via the Los
Alamos National Laboratory notice.

The second sort algorithm we look at is called a *merge sort* and it dates back to 1945. It was discovered by John von Neumann. [4]

To better see how the merge sort works, let's use a larger data set with 8 objects and we use names instead of numbers. Here is the initial list:

```
Katie   Bobby   William Atticus Judith  Gideon  Beatnik  Ruth
```

We sort this in ascending alphabetic order. This list is now neither sorted nor in reverse order, so this is an average case.

We begin by breaking the list into 8 groups (because we have 8 names), each containing only one item.

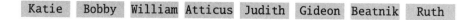

Trivially, each group is sorted within itself because there is only one name in it. Next, working from left to right pairwise, we create groups of two names where we put the names in the correct order as we merge.

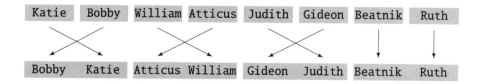

Now we combine the groups of two into groups of four, again working from left to right. We know the names are sorted within the group. Begin with the first name in the first pair. If it is less than the first name in the second pair, put it at the beginning of the group of four. If not, put the first name in the pair there.

Continue in this way. If a pair becomes empty, put all the names in the other pair at the end of the group of four, in order.

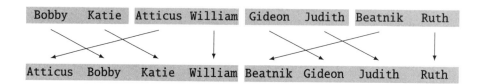

We finally create one group, merging in the names as we encounter them from left to right.

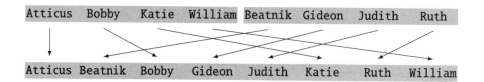

Among the variations of merge sort, this is called a *bottom-up* implementation because we completely break the data into chunks of size 1 and then combine.

For this algorithm we are not interested in swaps because we are forming new collections rather than manipulating an existing one. That is, we need to place a name in a new group no matter what. Instead the metric we care about is the number of comparisons. The analysis is non-trivial, and is available in algorithms books and on the web. The complexity of a merge sort is $O(n \log(n))$. [1]

This is a big improvement over $O(n^2)$. Forgetting about the constant in the definition of $O(\ )$ and using logarithms to base 10, for $n = 1000000 = 1$ million, we have $n^2 = 1000000000000 = 10^{12} = 1$ trillion, while $\log_{10}(1000000) \times 1000000 = 6000000 = 6 \times 10^6 = 6$ million. Would you rather do a million times more comparisons than there are names or 6 times the number of names?

In both bubble and merge sorts we end up with the same answer, but the algorithms used and their performance differ dramatically. Choice of algorithm is an important decision.

For the bubble sort we had to use only enough memory to hold the original list and then we moved around numbers within this list. For the merge sort in this implementation, we need the memory for the initial list of names and then we needed that much memory again when we moved to groups of 2. However, after that point we could reuse the memory for the initial list to hold the groups of 4 in the next pass.

By reusing memory repeatedly we can get away with using twice the initial memory to complete the sort. By being extra clever, which I leave to you and your research, this memory requirement can be further reduced.

Question 2.8.2

I just referred to storing the data in memory. What would you do if there were so many things to be sorted that the information could not all fit in memory? You would need to use persistent storage like a hard drive, but how?

Though I have focused on how many operations such as swaps and comparisons are needed for an algorithm, you can also look at how much memory is necessary and do a $O(\ )$ analysis of that. Memory-wise, bubble sort is $O(1)$ and merge sort is $O(n)$.

2.8.2 Searching

Here is our motivating problem: I have a collection S of n objects and I want to find out if a particular object *target* is in S. Here are some examples:

- Somewhere in my closet I think I have a navy blue sweater. Is it there and where? Here *target* = "my navy blue sweater."
- If I keep my socks sorted by the names of their predominant colors in my dresser drawer, are my blue argyle socks clean and available?
- I have a database of 650 volunteers for my charity. How many live in my town?
- I brought my kids to a magic show and I got volunteered to go up on stage. Where is the Queen of Hearts in the deck of cards the magician is holding?

With only a little thought, we can see that searching is at worst $O(n)$ unless you are doing something strange and questionable. Look at the first object. Is it *target*? If so, look at the second and compare. Continue if necessary until we get to the nth object. Either it is *target* or *target* is not in *S*. This is a *linear search* because we go in a straight line from the beginning to the end of the collection.

If *target* is in *S*, in the best case we find it the first time and in the worst it is the nth time. On average it takes $\frac{n}{2}$ attempts.

To do better than this classically we have to know more information:

- Is *S* sorted?
- Can I access any object directly as in "give me the fourth object"? This is called *random access*.
- Is *S* simply a linear collection of objects or does it have a more sophisticated data structure?

If *S* only has one entry, look at it. If *target* is there, we win.

If *S* is a sorted collection with random access, I can do a *binary search*.

Since the word "binary" is involved, this has something to do with the number 2.

I now show this with our previously sorted list of names. The problem we pose is seeing if *target* = Ruth is in *S*. There are 8 names in the list. If we do a linear search it would take 7 tries to find Ruth.

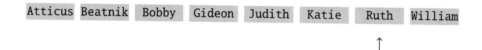

Let's take a different approach. Let $m = \frac{n}{2} = 4$ which is near the midpoint of the list of 8 names.

In case m is not a whole number, we round up. Examine the mth = fourth name in S. That name is Gideon. Since S is sorted and Gideon < Ruth, Ruth cannot be in the first half of the list. With this simple calculation we have already eliminated half the names in S. Now we need only consider

Judith Katie Ruth William

There are 4 names and so we divide this in half to get 2. The second name is Katie. Since Katie < Ruth, Ruth is again not in the first half of this list. We repeat with the second half.

Ruth William

The list has length 2 and we divide this in half and look at the first name. Ruth! Where have you been?

We found Ruth with only 3 searches. If by any chance that last split and compare had not found Ruth, the remaining sublist would have had only one entry and that had to be Ruth since we assumed that name was in S. We located our target with only $3 = \log_2(8)$ steps.

Binary search is $O(log(n))$ in the worst case, but remember than we imposed the conditions that S was sorted and had random access. As with sorting, there are many searching techniques and data structures that can make finding objects quite efficient.

As an example of a data structure, look at the binary tree of our example names as shown in Figure 2.1. The dashed lines show our root to Ruth. To implement a binary tree in a computer requires a lot more attention to memory layout and bookkeeping.

Question 2.8.3

I want to add two more names to this binary tree, Richard and Kristin. How would you insert the names and rearrange the tree? What about if I delete Ruth or Gideon from the original tree?

Question 2.8.4

For extra credit, look up how *hashing* works. Think about the combined performance of searching and what you must do to keep the underlying data structure of objects in a useful form.

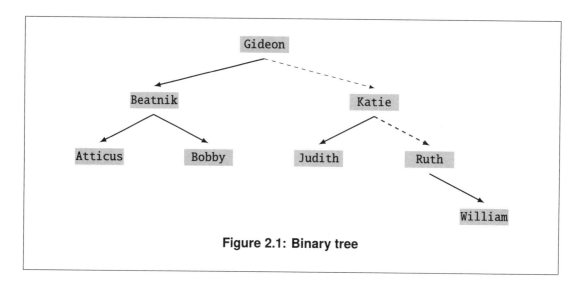

Figure 2.1: Binary tree

Entire books have been written on the related topics of sorting and searching. We return to this topic when we examine Grover's quantum algorithm for locating an item in an unsorted list without random access in only $O\left(\sqrt{n}\right)$ time in section 9.7.

> If we replace an $O(f(n))$ algorithm with an $O\left(\sqrt{f(n)}\right)$ one, we have made a *quadratic* improvement. If we replace it with a $O(\log(f(n)))$ algorithm, we have made an *exponential* improvement.

Suppose I have an algorithm that takes 1 million $= 10^6$ days to complete. That's almost 2,740 years! Forgetting about the constant in the $O()$ notation and using $\log_{10}$, a quadratic improvement would complete in $1,000 = 10^3$ days, which is about 2.74 years. An exponential improvement would give us a completion time of just 6 days.

To learn more

There are many kinds of sorting and searching algorithms and, indeed, algorithms for hundreds of computational use cases. Deciding which algorithm to use in which situation is very important to performance, as we have just seen. For some applications, there are algorithms to choose the algorithm to use! [5] [9]

2.9 Summary

Classical computers have been around since the 1940s and are based on using bits, 0s and 1s, to store and manipulate information. This is naturally connected to logic as we can think of a 1 or 0 as true or false, respectively, and vice versa. From logical operators like **and** we created real circuits that can perform higher-level operations like addition. Circuits implement portions of algorithms.

Since all algorithms to accomplish a goal are not equal, we saw that having some idea of measuring the time and memory complexity of what we are doing is important. By understanding the classical case we'll later be able to show where we can get a quantum improvement.

References

[1]   Thomas H. Cormen et al. *Introduction to Algorithms*. 3rd ed. The MIT Press, 2009.

[2]   R.W. Hamming. "Error Detecting and Error Correcting Codes". In: *Bell System Technical Journal* 29.2 (1950).

[3]   R. Hill. *A First Course in Coding Theory*. Oxford Applied Linguistics. Clarendon Press, 1986.

[4]   Institute for Advanced Study. *John von Neumann: Life, Work, and Legacy*. URL: https://www.ias.edu/von-neumann.

[5]   Donald E. Knuth. *The Art of Computer Programming, Volume 3: Sorting and Searching*. 2nd ed. Addison Wesley Longman Publishing Co., Inc., 1998.

[6]   Gordon E. Moore. "Cramming more components onto integrated circuits". In: *Electronics* 38.8 (1965), p. 114.

[7]   Ashok Muthukrishnan. *Classical and Quantum Logic Gates: An Introduction to Quantum Computing*. URL: http://www2.optics.rochester.edu/users/stroud/presentations/muthukrishnan991/LogicGates.pdf.

[8]   Oliver Pretzel. *Error-correcting Codes and Finite Fields*. Student Edition. Oxford University Press, Inc., 1996.

[9]   Robert Sedgewick and Kevin Wayne. *Algorithms*. 4th ed. Addison-Wesley Professional, 2011.

[10] The Unicode Consortium. *About the Unicode® Standard*. URL: http://www.unicode.org/standard/standard.html.

[11] B. Tuomanen. *Explore high-performance parallel computing with CUDA*. Packt Publishing, 2018.

3

More Numbers than You Can Imagine

*The methods of theoretical physics should be applicable
to all those branches of thought in which the
essential features are expressible with numbers.*

Paul Dirac
1933 Nobel Prize Banquet Speech

People use numbers for counting, percentages, ratios, prices, math homework, their taxes, and other practical applications.

$$1 \qquad 0 \qquad -1 \qquad 9.99999$$

$$-\sqrt{2}+1 \qquad \frac{22}{7} \qquad 3.14159265\ldots \qquad \pi$$

All these are examples of real numbers. In this chapter we look at the properties of real numbers and especially those of subsets like the integers. We extend those to other collections like the complex numbers that are core to understanding quantum computing.

For example, a quantum bit, or qubit, is defined as a pair of complex numbers with additional properties. Here we begin to lay the foundation for the algebraic side of quantum computing. In the next chapter we turn to geometry.

Topics covered in this chapter

3.1 Natural numbers

While there are special and famous numbers like π, the numbers we use for counting are much simpler: 1, 2, 3, ... I might say "Look, there is 1 puppy, 2 kittens, 3 cars, and 4 apples." If you give me 2 more apples, I will have 6. If I give my sister 1 of them, I will have 5. If I buy 2 more bags of 5 apples, I will have 15 in total, which is 3×5."

The set of natural numbers is the collection of increasing values

$$\{1, 2, 3, 4, 5, 6, 7, \dots\}$$

where we get from one number to the next by adding 1. 0 is not included. The braces "{" and "}" indicate we are talking about the entire set of these numbers.

When we want to refer to some arbitrary natural number but not any one specifically, we use a variable name like n and m.

The set of natural numbers is infinite. Suppose otherwise and that some specific number n is the largest natural number. But then $n + 1$ is larger and is a natural number, by definition. This *proof by contradiction* shows that the original premise, that there is a largest natural number, is false. Hence the set is infinite.

To avoid writing out "natural numbers" repeatedly, I sometimes abbreviate the collection of all natural numbers by using $\mathbb{N}$. What can we do if we restrict ourselves to working only with $\mathbb{N}$?

First, we can add them via the familiar arithmetic rules. $1 + 1 = 2$, $3 + 4 = 7$, $999998 + 2 = 1000000$, and so on.

Addition is so key to natural numbers that we consider it an essential part of the definition. In fact, we did: we described the values in the set, $\{1, 2, \dots\}$, but then also pointed out that we get from one value to the next by adding 1.

I've started with these basic numbers to make this point clear: we're not only concerned with a number here or a number there. We want to think about the entire collection and what we can do with them via operations like addition.

If we have two natural numbers and we add them together using "+", we always get another natural number. Hence $\mathbb{N}$ is closed under addition. The idea of closure or being closed with respect to doing something means that after we do it, the result is in the collection.

To choose something more exotic than basic arithmetic, consider the square root operation. The square root of 1 is still 1 and so a natural number, as is the square root of 4. But $\sqrt{2}$ is not a natural number and is, in fact, an irrational number.

$$\sqrt{2} = 1.41421356237\dots$$

$\mathbb{N}$ is not closed under the square root operation.

In $\mathbb{N}$ addition is commutative: $4 + 11 = 11 + 4$ and $n + m = m + n$ in general. No matter which order you count things, you always get the same answer.

What about subtraction, which is in some sense the complement of addition? Since $3 + 4 = 7$ then $3 = 7 - 4$ and $4 = 7 - 3$. We can also state the last as "4 is 7 minus 3, or 7 take away 3, or 7 subtract 3."

For all natural numbers n and m we always have $n + m$ in $\mathbb{N}$. However, we have $n - m$ in $\mathbb{N}$ only if $n > m$. $24 - 17 = 7$ is a natural number but $6 - 6$ is not a natural number because 0 is not in $\mathbb{N}$ by definition. $17 - 24$ is not a natural number since the smallest natural number is 1. $\mathbb{N}$ is not closed under subtraction.

We can use comparisons like "<" and ">" to tell if one natural number is less than or greater than another, respectively, in addition to testing for equality with "=". Because we can compare any two numbers in $\mathbb{N}$ in this way, we say that the natural numbers are ordered. Since we have comparison operations, we can sort collections of natural numbers in ascending or descending ways.

Given that we have addition, we can define multiplication "×" by saying that $n \times m$ is m added to itself n times. In particular, $1 \times n = n \times 1 = n$. Multiplication distributes over addition: $3 \times (8 + 11) = (3 \times 8) + (3 \times 11) = 57$.

Multiplication is commutative like addition: $n \times m = m \times n$. It's just a question of how you group things for counting:

$$\begin{aligned}
3 \times 7 &= 7 + 7 + 7 \\
&= (3 + 3 + 1) + (2 + 3 + 2) + (1 + 3 + 3) \\
&= 3 + 3 + (1 + 2) + 3 + (2 + 1) + 3 + 3 \\
&= 3 + 3 + 3 + 3 + 3 + 3 + 3 \\
&= 7 \times 3.
\end{aligned}$$

$\mathbb{N}$ is closed under multiplication but is not closed under division. $1/3$ is not a natural number, for example, even if $4/2$ is.

For natural numbers, the definition of multiplication follows directly from addition. For more sophisticated mathematical collections, multiplication can be much more complicated.

Let's start extending the collection to eliminate some of these problems regarding closure.

3.2 Whole numbers

If we append 0 to $\mathbb{N}$ as a new smallest value we get the whole numbers, denoted $\mathbb{W}$. The whole numbers are not used a lot by themselves in mathematics but let's see what we get with this additional value.

We are still closed under addition and multiplication and not closed under division. We do now have to watch out for division by zero. Expressions like $3 - 3$ or $n - n$ in general are in $\mathbb{W}$, so that's a little better for subtraction but this does not give us closure.

So far, there's not much that we've gained, it seems. Or have we?

> **0** is an *identity element* for addition, which is a new concept for us to consider. I've put it in bold to show how special it is. This is a unique (meaning there is one and only one) number such that for any whole number w we have $w + \mathbf{0} = \mathbf{0} + w = w$.
>
> Thus $14 + \mathbf{0} = \mathbf{0} + 14 = 14$. Also, $\mathbf{0} \times w = w \times \mathbf{0} = \mathbf{0}$.

For the whole numbers $\mathbb{W}$ we have a collection of values

$$\{\mathbf{0}, 1, 2, 3, \dots\},$$

an operation "+", and an identity element **0** for "+".

You may have realized by now that when we discussed the natural numbers we could have also noted that **1** is an identity element for multiplication. So let's restate everything we know about $\mathbb{N}$ and $\mathbb{W}$:

> The set of natural numbers $\mathbb{N}$ is the infinite ordered collection of values $\{\mathbf{1}, 2, 3, 4, \dots\}$ with a commutative operation "+" called *addition*. We get from one natural number to the next larger by adding **1**. $\mathbb{N}$ is closed under addition. We also have a commutative operation *multiplication* "$\times$" with identity element **1**. Multiplication distributes over addition. $\mathbb{N}$ is closed under multiplication. $\mathbb{N}$ is not closed under subtraction or division defined in the usual manner.

> The set of whole numbers $\mathbb{W}$ is the infinite ordered extension of $\mathbb{N}$ formed by adding a new smallest value **0**. **0** is an identity element for addition. $\mathbb{W}$ is closed under the commutative addition and multiplication operations but not under subtraction or division.

I hope you agree that we've come a long way from only considering the numbers 1, 2, 3, 4, 5, ... We've gone from thinking about counting and specific values to entire sets of numbers and their properties. Though I'll stop putting them in bold, 0 and 1 are not any random numbers; they play very special roles. Later on we'll see other "0-like" and "1-like" objects that are more than simple numbers.

Since we have addition and multiplication, we can define exponentiation. If a and w are whole numbers, w^a equals w multiplied by itself a times. Note the similarity with how we defined multiplication from addition:

$$3^7 = 3 \times 3 \times 3 \times 3 \times 3 \times 3 \times 3.$$

This means that $w^1 = w$. Less intuitively at the moment, $w^0 = 1$ even when $w = 0$.

What do we get when we multiply two expressions with exponents? When the base is the same (the thing we are raising to a power), we add the exponents:

$$2^3 \times 2^4 = (2 \times 2 \times 2) \times (2 \times 2 \times 2 \times 2)$$
$$= 2 \times 2 \times 2 \times 2 \times 2 \times 2 \times 2$$
$$= 2^7 = 2^{3+4}.$$

In general, $w^a \times w^b = w^{a+b}$. Also,

$$w^a \times w^0 = w^{a+0} = w^a = w^a \times 1$$

further showing why $w^0 = 1$ makes sense.

We omit the "$\times$" when it is clear that the context is multiplication: $2^3 \times 2^4 = 2^3 2^4 = 2^7$.

3.3 Integers

People are sometimes confused by negative numbers when they first encounter them. How can I have a negative amount of anything? I can't physically have fewer than no apples, can I?

To get around this, we introduce the idea that a positive number of things or amount of money means something that you yourself *have*. A negative number or amount means what you *owe* someone else.

If you have \$100 or €100 or ¥100 and you write a check or pay a bill electronically for 120, one of two things will likely happen. The first option is for the payment to fail and your bank may charge you a fee. The second is that the bank will pay the full amount, let you know that you are overdrawn, and charge you a fee. You will then need to pay the amount overdrawn quickly or have it paid from some other account.

Whatever currency you used, you started with 100 and ended up with -20 before repayment. You owed the bank 20. If you deposit 200 in your account immediately, your balance will be 180, which is $-20 + 200$.

The *integers*, denoted $\mathbb{Z}$, take care of the problem of the whole numbers not being closed under subtraction. We define an operation "$-$", called *negation*, for each whole number n such that $-0 = 0$ and $-n$ is a new value such that $n + -n = 0$. This new extended set of values with the operations and properties we will discuss is called the *integers*.

Integers like 1, 12, and 345 are said to be *positive*. More precisely, any integer that is also a natural number or a whole number greater that 0 is positive. Positive integers are greater than 0.

Integers like -4, -89, and -867253 are *negative*. Said differently, for n a natural number, any integer of the form $-n$ is negative. Negative integers are less than 0. 0 is neither positive nor negative.

Negation has the property that $--n = n$. Negation reverses the order of the values: since $4 < 7$ then $-4 > -7$, which is the same as $-7 < -4$. The set of ordered values in the integers looks like
$$\{\ldots, -4, -3, -2, -1, 0, 1, 2, 3, 4, 5, \ldots\}.$$

Negative signs cancel out in multiplication and division: $-1 \times -1 = 1$ and $-1/-1 = 1$. For any integers n and m we have relationships like $n \times -m = -n \times m = -(n \times m)$.

If n is a whole number then $(-1)^n$ is 1 if n is even and is -1 if n is odd.

Given an integer n, we define the *absolute value* of n to be 0 if n is 0, n if n is positive, and $-n$ if n is negative. We use vertical bars $|n|$ to denote the absolute value of n. Therefore

$$|n| = \begin{cases} n & \text{when } n > 0 \\ 0 & \text{when } n = 0 \\ -n & \text{when } n < 0. \end{cases}$$

Here are some examples:

$$|-87| = 87$$
$$|0| = 0$$
$$|231| = 231$$

Informally, you get the absolute value by discarding the negative sign in front of an integer in case it is present. We could also say that the absolute value of an integer is how far it is away from 0 where we don't worry whether it is less than or greater than 0.

For integers n and m, we always have $|nm| = |n| \times |m|$ and $|n + m| \le |n| + |m|$.

Absolute value is a measurement of size or length and it generalizes to other concepts in algebra and geometry. In fact, for a qubit, the absolute value is related to the probability of getting one answer or another in a computation.

Examples

$n + 3$ gives us n increased by 3 and $n + -3 = n - 3$ yields n decreased by 3.

- $7 + 3$ means we increase 7 by 3 to get 10.
- $7 + -3$ means we decrease 7 by 3 to get 4.
- $-7 + 3$ means we increase -7 by 3 to get -4.
- $-7 + -3$ means we decrease -7 by 3 to get -10.

Considering the usual rules for addition and the above, $n + -m = n - m$ and $n - -m = n + m$ for any two integers n and m.

With this you can see that the integers are closed under subtraction: if you subtract one integer from another you get another integer. Given addition and negation we can get away without subtraction, but we keep it for convenience and to reduce the complexity of expressions.

You may have been first introduced to these rules and properties when you were a pre-teen. I repeated them here so that you would think more generally about the operations of addition, subtraction, and negation as they applied to arbitrary integers versus simply doing some arithmetic.

Returning to negation, for any integer there is one and only one number you can add to it and get 0. If we have 33, we would add -33 to it to get 0. If we started with -74 then adding 74 yields 0. 0 is significant here because it is the identity element for addition.

Any number with absolute value 1 is called a *unit*. The integers have two: 1 and -1.

A *prime* is a positive integer greater than 1 whose only factors under multiplication are 1 and itself. These are primes:

$$2 = 1 \times 2 \qquad 3 = 1 \times 3 \qquad 37 = 1 \times 37$$

These are not primes:

$$0 \qquad\qquad 1 \qquad\qquad -25 = -5 \times 5$$
$$4 = 2 \times 2 \qquad 12 = 3 \times 4 = 2^2 \times 3 \qquad 500 = 2^2 \times 5^3$$

When one number divides another evenly we use "|" between them. Since 5 divides 500, we write $5|500$. An integer that is the product of two or more primes (which may be the same, such as 7×7) is composite.

3×2 and 2×3 are equivalent factorizations of 6. We usually display factorizations with the individual factors going from small to large.

You can uniquely factor a non-zero integer into zero or more primes times a unit, with some of the primes possibly repeated. In the factorization of 500 above, the prime 2 was repeated twice and the prime 5 was repeated three times.

There are an infinite number of primes. The study of primes and their generalizations is relevant to several areas of mathematics, especially number theory.

> The integers $\mathbb{Z}$ is the infinite ordered extension of $\mathbb{W}$ formed by appending the negative values $-1, -2, -3, -4, \ldots$. The integers have a unique identity element 0 such that for any integer n there exists a unique additive inverse $-n$ such that $n + -n = 0$. $\mathbb{Z}$ is closed under the commutative addition and multiplication operations, and subtraction, but not under division. Multiplication distributes over addition.

Taking a more geometric approach to the integers, we can draw the familiar number line with the negative values to the left of 0 and positive values to the right.

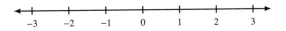

This is a visualization to help you think about the integers. The number line is *not* the set of integers but connects the algebra with a geometric aid.

Negating an integer corresponds to reflecting it to the other side of 0. Negating an integer twice means moving it across 0 and then back again to where it started. Hence double negation effectively does nothing. The absolute value means measuring how far to the left or right an integer is from 0.

Adding 0 does not move the position of a number on the line. Adding a positive integer means moving it that many units to the right, as does subtracting a negative integer. Adding a negative integer means moving it the absolute value of the integer units to the left, as does subtracting a positive integer.

> **Question 3.3.1**
>
> Using the number line, think about why negation reverses the order of two integers.

A line is one-dimensional. It takes only one number to exactly position a point anywhere on the line. Hence we can write (7) as the coordinate of the point exactly 7 units to the right of 0. With respect to this line, we call 0 the *origin*. Here is yet another special role for 0.

Mathematicians often work by taking a problem in one domain and translating it into another that they understand better or have better tools and techniques. Here I've shown some of the ways you can translate back and forth between the algebra and geometry of integers.

3.4   Rational numbers

The rational numbers, denoted $\mathbb{Q}$, take care of the problem of the integers not being closed under division by non-zero values.

3.4.1   Fractions

Let's start by talking about fractions, also known as the rational numbers, the way you may have been first introduced to them. This is elementary but useful to review to relate to what we have in the big picture with $\mathbb{Q}$.

Given a loaf of bread, if we cut it right down the middle, we say we have divided it into halves. Fraction-wise, one-half = 1/2. The two halves equal one whole loaf, so $1/2 + 1/2 = 2 \times 1/2 = 1$. Two halves is 2/2, which is 1. Four halves would make two loaves: 4/2 = 2.

Considering whole loaves, 1/1 is one loaf, 2/1 is two loaves, and 147/1 is one hundred and forty-seven loaves. We can represent any integer n as a fraction $n/1 = \frac{n}{1}$.

To multiply fractions, we multiply the tops (*numerators*) together and put those over the product (the result of multiplication) of the bottoms (*denominators*) and then simplify the result, a process I discuss below.

If we get another loaf and this time we cut it into three equal parts, or thirds, each part is 1/3 of the entire loaf. The three thirds equal the whole loaf, so $1/3 + 1/3 + 1/3 = 3 \times 1/3 = 1$.

If we had cut the loaf of bread so that there were two pieces but one was a third of the loaf and the other was two-thirds of the loaf, the equation would be

$$\frac{1}{3} + \frac{2}{3} = \frac{3}{3} = 1$$

If the thirds are each cut in half, we get six equal pieces, all of which add up to the original loaf. If we push two of them together we get back to a third. So $1/6 + 1/6 = 2 \times 1/6 = 1/3$. Written another way and in more detail:

$$2 \times \frac{1}{6} = \frac{2}{6} = \frac{2 \times 1}{2 \times 3} = \frac{2}{2} \times \frac{1}{3} = 1 \times \frac{1}{3} = \frac{1}{3}$$

Think of "×" as meaning "of." So half of a third being a sixth is

$$\frac{1}{2} \times \frac{1}{3} = \frac{1}{6}$$

The arithmetic of fractions, particularly addition and subtraction, is easy when we are only dealing with ones that have the same denominators such as halves (2), thirds (3), and sixths (6) as above. When the denominators are different we need to find the dreaded least common denominator (lcd).

What do we get if we push another one-sixth of a loaf of bread onto one-third of a loaf?

$$\frac{1}{6} + \frac{1}{3} = \frac{1}{6} + 2 \times \frac{1}{6} = 3 \times \frac{1}{6} = \frac{3}{6}$$

In this case, the least common denominator is simply 6 because we can always represent some number of thirds as twice that number of sixths.

What about $1/3 + 1/5$? We cannot easily represent thirds as fifths and so we have to subdivide each so that we have a common size. In this case, the smallest size is one-fifteenth: $1/3 = 5/15$ and $1/5 = 3/15$.

$$\frac{1}{3} + \frac{1}{5} = \frac{5}{15} + \frac{3}{15} = \frac{8}{15}$$

You might not think that fifteenths are as easy as halves, thirds, fourths, or fifths, but one fraction is as good as any other. In this example, 15 is the *least common multiple* (lcm) of 3 and 5: it is the smallest non-zero positive integer divisible by 3 and 5.

An example shows you how to compute the least common multiple of two integers. Let's work with -18 and 30. Since we care only about finding a positive integer, we can forget about the negative sign in front of 18.

Factor each number into primes: $18 = 2 \times 9 = 2 \times 3^2$ and $30 = 2 \times 3 \times 5$. We're going to gather these primes into a collection. For each prime contained in *either* factorization, put it in the collection with its exponent if it is not already there or replace what is in the collection if you find it with a larger exponent. Hence:

- The collection starts empty as { }.
- Process $18 = 2 \times 9 = 2 \times 3^2$ prime by prime.
    - 2 is not already in the collection, so insert it, yielding {2}.
    - 3 is not already in the collection with any exponent, so insert 3^2 yielding $\{2, 3^2\}$.
- Process $30 = 2 \times 3 \times 5$ prime by prime:

  - 2 is already in the collection, ignore it.
  - 3 is already in the collection with the larger exponent 2, so ignore it.
  - 5 is not in the collection, so insert it.
- The final collection is $\{2, 3^2, 5\}$.

Multiplying together all the numbers in the collection yields 90, the least common multiple. Our process ensures that each of the original numbers divides into 90 and that 90 is the smallest number for which this works.

When the numbers have no primes in common, the least common multiple is simply the absolute value of their product.

When the numerators are non-trivial (that is, not equal to 1) then we need to do some multiplication with them. As above, suppose we have found 15 to be the least common multiple of 3 and 5. Hence it is the least common denominator in $2/3 + 7/5$.

This is how we do the addition. Subtraction is similar.

$$\frac{2}{3} + \frac{7}{5} = \frac{5}{5} \times \frac{2}{3} + \frac{3}{3} \times \frac{7}{5}$$
$$= \frac{5 \times 2}{5 \times 3} + \frac{3 \times 7}{3 \times 5}$$
$$= \frac{10}{15} + \frac{21}{15}$$
$$= \frac{31}{15}$$

To raise a rational number to a whole number exponent, raise the numerator and denominator to that exponent.

$$\left(\frac{-3}{4}\right)^5 = \frac{(-3)^5}{4^5} = \frac{-243}{1024}$$

To simplify a fraction you express it in lowest terms, which means you have factored out common primes from the numerator and denominator. To further normalize it, make it contain at most one negative sign and, if it is present, put it in the numerator.

Examples

$$\frac{1}{-2} = \frac{-1}{2} \qquad\qquad \frac{5}{5} = \frac{5^1}{5^1} = \frac{5^0}{5^0} = \frac{1}{1} = 1$$

$$\frac{2}{8} = \frac{2^1}{2^3} = \frac{2^0}{2^2} = \frac{1}{4} \qquad\qquad \frac{12}{30} = \frac{2^2 \times 3^1}{2^1 \times 3^1 \times 5^1} = \frac{2^1}{5^1} = \frac{2}{5}$$

If a prime is present in the numerator and denominator then we have that prime divided by itself, which is 1. That means we can remove it from both. This is called *cancellation*.

There is no integer strictly between 3 and 4. Given two different rational numbers, you can always find a rational number between them. Just average them: $\frac{3+4}{2} = \frac{7}{2}$.

So, while we proceed from integer to integer by adding or subtracting one, we can't slip between an integer and its successor and find another integer.

Integers are infinite in each of the positive and negative directions, and so therefore are the rational numbers, but there are an infinite number of rational numbers between any two distinct rational numbers.

Greatest common divisor

There is a more direct way of calculating the least common multiple via the greatest common divisor.

> Let a and b be two non-zero integers. We can assume that they are positive. The *greatest common divisor* g is the largest positive integer such that $g|a$ and $g|b$. $g \leq a$ and $g \leq b$. We abbreviate the greatest common divisor to "gcd." One of the properties of the greatest common divisor g is that there are integers n and m so that $an + bm = g$.
>
> If $g = 1$ then we say that a and b are coprime.

Given this,

$$\text{lcm}(a, b) = \frac{a\,b}{\gcd(a, b)}.$$

If either a or b is negative, use its absolute value.

To calculate $\gcd(a, b)$ we use quotients and remainders in *Euclid's algorithm*. By the properties of division for positive integers a and b with $a \geq b$ there exist non-negative integers q and r such that

$$a = bq + r$$

with $0 \leq r < b$. q is called the *quotient* upon dividing a by b and r is the *remainder*. Because $r < b$, q is as large as it can be. If $r = 0$ then $b|a$ and $\gcd(a, b) = b$.

So let's suppose that n divides a and b. Then n divides $a - bq = r$. In particular, for $n = \gcd(a, b)$ then

$$\gcd(a, b) = \gcd(b, r).$$

We have replaced the calculation of the gcd of a and b with the calculation of the gcd of b and r, a smaller pair of numbers. We can keep repeating this process, getting smaller and smaller pairs. Since $r \geq 0$, we eventually stop.

Once we get an r that is 0, we go back and grab the previous remainder. That is the gcd.

Question 3.4.1

Compute $\gcd(15295, 38019)$. Factor the answer if you can.

Euclid originally used subtractions but with modern computers we can efficiently compute the quotients and remainders.

3.4.2 Getting formal again

Let's rewind now and look at the rational numbers and their operations in the ways we did with the natural numbers, whole numbers, and integers.

Just as we introduced $-w$ for a whole number w to be the unique integer value such that $w + -w = 0$, we set $1/w = \frac{1}{w}$ to be the unique value such that $1/w \times w = 1$ for non-zero w. $1/w$ is called the *inverse* of w. Technically, $1/w$ is the unique multiplicative inverse of w but that's quite a mouthful. If I don't state that w is non-zero, assume it is.

The rational numbers extend the integers with a multiplicative inverse and similarly extended rules for addition, subtraction, multiplication, and division.

For any two integers a and non-zero b we define $a/b = \frac{a}{b}$ to be $a \times \frac{1}{b}$. $\frac{b}{b} = 1$.

Two expressions of the form $\frac{c \times a}{c \times b}$ and $\frac{a}{b}$ for non-zero c and b refer to the very same rational number.

If a and b have no prime factors in common and b is positive, we say the rational number is shown in lowest terms. Simplifying a rational number expression means to write it in its lowest terms. We can now write out the rules for arithmetic.

Equality

Two expressions $\frac{a}{b}$ and $\frac{c}{d}$ with non-zero b and d represent the same rational number if $a \times d = c \times b$.

Addition

Process: Cross-multiply the numerators and denominators and add the results to get the new numerator, multiply the denominators to get the new denominator, simplify.

$$\frac{a}{b} + \frac{c}{d} = \frac{a \times d + c \times b}{b \times d}$$

for non-zero b and d. Alternatively, convert each fraction to have the same least common denominator, add the numerators, and simplify.

Subtraction

Process: Cross-multiply the numerators and denominators and subtract the results to get the new numerator, multiply the denominators to yield the new denominator, simplify.

$$\frac{a}{b} - \frac{c}{d} = \frac{a \times d - c \times b}{b \times d}$$

for non-zero b and d. Alternatively, convert each fraction to have the same least common denominator, subtract the numerators, and simplify.

Negation

Negating a value is the same as multiplying it by -1. Negative signs "cancel" across the numerator and denominator.

$$-\frac{a}{b} = \frac{-a}{b} = \frac{-1 \times a}{b} = \frac{a}{-1 \times b} = \frac{a}{-b}$$

for non-zero b.

Multiplication

Process: Multiply the numerators to yield the new numerator, multiply the denominators to yield the new denominator, simplify.

$$\frac{a}{b} \times \frac{c}{d} = \frac{a \times c}{b \times d}$$

for non-zero b and d.

Inversion

The inverse of a non-zero rational number is the rational number formed by swapping the numerator and denominator.

$$\frac{1}{\left(\frac{a}{b}\right)} = \left(\frac{a}{b}\right)^{-1} = \frac{b}{a}$$

for non-zero a and b. Raising a rational number to the -1 power means to compute its inverse.

Division

Divide two rational numbers by multiplying the first by the inverse of the second.

$$\frac{\left(\frac{a}{b}\right)}{\left(\frac{c}{d}\right)} = \frac{a}{b} \times \frac{1}{\left(\frac{c}{d}\right)} = \frac{a}{b} \times \frac{d}{c} = \frac{a \times d}{b \times c}$$

for non-zero b, c, and d.

Exponentiation

Similar to other numbers, raising a rational number to the 0th power yields 1. Raising it to a negative integer power means to swap the numerator and denominator and raise each to the absolute value of the exponent.

$$\left(\frac{a}{b}\right)^n = \begin{cases} \dfrac{a^n}{b^n} & \text{for integer } n > 0 \\ 1 & \text{for } n = 0 \\ \dfrac{b^{-n}}{a^{-n}} & \text{for integer } n < 0 \end{cases}$$

for non-zero b or if n is a negative integer, non-zero a.

The rational numbers $\mathbb{Q}$ is the infinite ordered extension of $\mathbb{Z}$ formed by appending the multiplicative inverses $\frac{1}{n}$ of all non-zero integers n, and then further extending by defining values $\frac{n}{m} = n \times \frac{1}{m}$ for integers n and non-zero m.

The rational numbers have a unique identity element **1** such that for any non-zero r there exists a unique multiplicative inverse $\frac{1}{r}$ such that $r \times \frac{1}{r} = \mathbf{1}$. $\mathbb{Q}$ is closed under the commutative addition and multiplication operations, subtraction, and under division by non-zero values.

The rational numbers seem to finally solve most of our problems about doing arithmetic and getting a valid answer. However, even though $\sqrt{4}$ and $\sqrt{\frac{1}{25}}$ are both rational numbers, neither $\sqrt{2}$ nor $\sqrt{\frac{1}{5}}$ is.

If $\sqrt{2}$ were rational, there would exist positive integers m and n such that $m/n = \sqrt{2}$, meaning that $\dfrac{m^2}{n^2} = 2$.

We can assume m and n have no factors in common. This is key!

Let's show this is not possible. Every even integer is of the form $2k$ for some other integer k. Similarly, all odd integers are of the form $2k + 1$. Therefore the square of a integer looks like $4k^2$, which is even, and the square of an odd integer looks like $4k^2 + 2k + 1$, which is odd.

This also shows that if an even integer is a square then it is the square of an even integer. If an odd integer is a square then it is the square of an odd integer.

If $\frac{m^2}{n^2} = 2$ then $m^2 = 2n^2$ and so m^2 and therefore m are even integers. So there is some integer j such that $m = 2j$ and $m^2 = 4j^2$.

We then have

$$m^2 = 4j^2 = 2n^2$$

or

$$2j^2 = n^2.$$

As before, this shows that n^2 and n are *even*. So both m and n are even and they share the factor 2.

But we assumed m and n have no factors in common, a contradiction! Thus there exist no such m and n and $\sqrt{2}$ *is not a rational number.*

Question 3.4.2

By similar methods, show $\sqrt{3}$ is not rational.

3.5   Real numbers

When we're were done looking at the real numbers we'll have finished analyzing the typical numbers most people encounter. Let's begin with decimals.

3.5.1 Decimals

A *decimal* expression for a real number looks like

- an optional minus sign,
- followed a finite number of digits 0, 1, 2, 3, 4, 5, 6, 7, 8, and 9,
- followed by a period, also called the *decimal point*,
- followed by a finite or infinite number of digits.

In many parts of the world, the decimal point is a comma instead of a period, but I use the United States and UK convention here.

If there are no digits after the decimal point then the decimal point may be omitted.

> Any trailing 0s on the right are usually omitted when you are using the number in a general mathematical context. They may be kept in situations where they indicate the precision of a measurement or a numeric representation in computer code.

Any leading 0s on the left are usually omitted. We have

$$0 = 0. = .0 = 000.00$$
$$1 = 1. = 1.0 = 000001$$
$$-3.27 = -03.27 = -3.27000000000$$

By convention it is common to have at least a single 0 before the decimal point and a single 0 after: 0.0 and −4.0, for example.

The integer 1327 is a shorthand way of writing

$$1 \times 10^3 + 3 \times 10^2 + 2 \times 10^1 + 7 \times 10^0.$$

Similarly, the integer −340 is

$$(-1)\left(3 \times 10^2 + 4 \times 10^1 + 0 \times 10^0\right).$$

We extend to the right of the decimal point by using negative powers of 10:

$$13.27 = 1 \times 10^1 + 3 \times 10^0 + 2 \times 10^{-1} + 7 \times 10^{-2}$$
$$-0.340 = (-1)\left(0 \times 10^0 + 3 \times 10^{-1} + 4 \times 10^{-2} + 0 \times 10^{-3}\right)$$

The decimal point is at the place where we move from 10^0 to 10^{-1}.

Since 10^{-1} is $\frac{1}{10}$ = one tenth, the digit immediately after the decimal point is said to be in the "tenths position." The one after that is the "hundredths position" because it corresponds to $10^{-2} = \frac{1}{100}$ = one hundreth. We continue this way to the thousandth, ten-thousanth, hundred-thousandth, millionth positions, and so on.

To convert a fraction like $1/2$ to decimal, we try to re-express it with denominator equal to some power of ten. In this case it is easy because $1/2 = 5/10$. So five-tenths is 0.5.

For $3/8$ we need to go all the way up to $375/1000$.

$$
\begin{aligned}
\frac{3}{8} &= \frac{3}{2^3} \times \frac{5^3}{5^3} \\
&= \frac{3}{2^3} \times \frac{125}{125} \\
&= \frac{375}{1000} \\
&= \frac{300}{1000} + \frac{70}{1000} + \frac{5}{1000} \\
&= \frac{3}{10} + \frac{7}{100} + \frac{5}{1000} \\
&= 3 \times 10^{-1} + 7 \times 10^{-2} + 5 \times 10^{-3} \\
&= .375
\end{aligned}
$$

Since $10 = 2 \times 5$, each power of 10 is the product of the same power of 2 times the same power of 5. That's why we chose 5^3 in the above example: $10^3 = 2^3 5^3$.

This method doesn't always work to convert a fraction to a decimal. The decimal expression of $1/7$ is

$$0.142857142857142857142857142857142857142857\ldots$$

Notice the section "142857" that repeats over and over.

$$0.\boxed{142857}\,142857\ 142857\ 142857\ 142857\ 142857\ 142857\ldots$$

It goes on repeating forever, block after adjacent block. This is an infinite decimal expansion. We write the repeating block with a line over it:

$$\frac{1}{7} = 0.\overline{142857}$$

> Any rational number has a finite decimal expression or an infinite one with a repeating block.

We showed above how to go from a finite decimal expansion to a fraction: express the decimal as a sum of powers of 10 and then do the rational number arithmetic.

$$2.13 = 2 \times 10^0 + 1 \times 10^{-1} + 3 \times 10^{-2}$$
$$= 2 + \frac{1}{10} + \frac{3}{100}$$
$$= \frac{213}{100}$$

This is already simplified but in general we need to do that at the end.

The general process is slightly more complicated. Let $r = 0.\overline{153846}$. The repeating block has **6** digits and it is immediately after the decimal point. Multiply both sides by $10^6 = 1000000$:

$$1000000r = 153846.\overline{153846}$$

and so

$$1000000r - r = 153846.\overline{153846} - 0.\overline{153846}$$

which gives

$$999999r = 153846$$
$$r = \frac{153846}{999999}$$
$$r = \frac{2}{13}$$

Question 3.5.1

How would you adjust this if the repeating block begins more to the right of the decimal point?

If the entire decimal expression does not repeat, you can separate it into a finite expansion plus the repeating one divided by the appropriate power of 10.

$$3.2\overline{153846} = 3.2 + 0.0\overline{153846}$$

$$= \frac{32}{10} + \frac{2}{13} \times 10^{-1}$$
$$= \frac{32}{10} + \frac{2}{130}$$
$$= \frac{32}{10} \times \frac{13}{13} + \frac{2}{130}$$
$$= \frac{416}{130} + \frac{2}{130}$$
$$= \frac{418}{130} = \frac{209}{65}$$

These calculations show the relationships between rational numbers and their decimal expansions.

Question 3.5.2

What is the rational number corresponding to $0.\overline{9}$?

If r is a real number then $\lfloor r \rfloor$, the *floor* of r, is the largest integer $\leq r$. Similarly, $\lceil r \rceil$, the *ceiling* of r, is the smallest integer $\geq r$.

3.5.2 Irrationals and limits

The case we have not considered is an infinite decimal expansion that does not have an infinitely repeating block. Not being rational, it is called an *irrational number*. The real numbers are the rational numbers in addition to all the irrational numbers. Since $\sqrt{2}$ is not rational, it must be irrational.

Let's consider approximation of a real number by a decimal.

$$\pi = 3.14159265358979323846264338327950\ldots$$

is an irrational number and so has no infinitely repeating blocks. It is not 22/7 and it is not 3.14. Those are rational and decimal approximations to π, and not even very good ones.

π exists as a number even though you cannot write it down as a fraction or write out the infinite number of digits that express it. π is in $\mathbb{R}$ but it is not in $\mathbb{Q}$.

Consider

$$3.1 \rightarrow 3.14 \rightarrow 3.141 \rightarrow 3.1415 \rightarrow 3.14159 \rightarrow 3.141592 \rightarrow 3.1415926 \rightarrow \cdots$$

This is a sequence of rational numbers (expressed as decimals) that get closer and closer to the actual value of π.

Want to be within one-millionth of the actual value? $\pi - 3.1415926 < 0.000001$. Within one-hundred-millionth? $\pi - 3.141592653 < 0.00000001$. We could keep going.

We have a sequence of rational numbers such that if we set a closeness threshold like one-millionth then one member of the sequence and all that follow it are at least that close to π. We say the irrational number π is the *limit* of the given sequence of rational numbers.

If you make the threshold smaller, we can find a possibly later sequence member so we will be at least that close from then on. We say the above sequence *converges* to π.

Think about this. All the members of the sequence are rational numbers but the limit is not. Informally, if we take the rational numbers and throw in all the limits of convergent sequences of them, we get the real numbers.

Of course, there are sequences of rational numbers that converge to rational numbers.

$$\frac{1}{1}, \frac{1}{2}, \frac{1}{3}, \dots, \frac{1}{n}, \dots$$

converges to the limit 0. Here we let n get larger and larger and we write

$$\lim_{n \to \infty} \frac{1}{n} = 0$$

Similarly,

$$-\frac{2}{1}, -\frac{3}{2}, -\frac{4}{3}, \dots, -\frac{n+1}{n}, \dots$$

converges to -1. For a non-obvious example, consider

$$\lim_{n \to \infty} \left(1 + \frac{1}{n}\right)^n$$

Even though n is getting bigger, the expression inside the parentheses is getting closer to 1. As n gets bigger the computed values appear to converge.

$$n = 1 \to 1.5$$
$$10 \to 2.5937424601\dots$$
$$10000 \to 2.71814592682\dots$$
$$100000 \to 2.71826823719\dots$$

This sequence converges to $e = 2.718281828459045235360\ldots$, the base of the natural logarithms and an irrational number. Like π, e is a special value in mathematics and shows up "naturally" in many contexts.

The sequence

$$1, 2, 3, 4, 5, \ldots, n, \ldots$$

does not converge to any finite rational number. We call it a *divergent sequence*.

> We define the real numbers to be the extension of $\mathbb{Q}$ that contains the limits of all convergent sequences of rational numbers. Furthermore, the real numbers are closed under taking the limits of convergent sequences of real numbers.

Closure, closure, closure. The concept really is fundamental to the kinds of numbers we use every day.

Limits are essential in calculus. The idea that we can have an infinite sequence of numbers that converges to a fixed and unique value is unlike anything most people have seen earlier in their mathematical studies. It can be a scary and daunting concept at first.

Remember above when I asked you to compute $0.\overline{9}$? This is the limit of the sequence

$$0.9$$
$$0.99$$
$$0.999$$
$$0.9999$$
$$0.99999$$

$$\ldots$$

Every time we move to the next member of the sequence we add another 9 on the far right. It appears the limit is 1. If you want to be within one quadrillionth of $1 = 10^{-15}$, then go out to 0.9999999999999999. However close you want to be to 1, I can add enough 9s to the end so I am at least that close and every member of the sequence after it is too. The sequence converges to 1, its limit.

We can also consider sequences where each member is a sum that builds on the previous member. This sequence

$$1$$

$$1 - \frac{1}{3}$$

$$1 - \frac{1}{3} + \frac{1}{5}$$

$$1 - \frac{1}{3} + \frac{1}{5} - \frac{1}{7}$$

$$1 - \frac{1}{3} + \frac{1}{5} - \frac{1}{7} + \frac{1}{9}$$

$$\cdots$$

converges to $\frac{\pi}{4}$ but it does so excruciatingly slowly. For use with computer calculations, it's critical to find sequences that converge quickly.

3.5.3 Binary forms

Just as we can write a whole number in base 10 form using the digits 0, 1, 2, 3, 4, 5, 6, 7, 8, and 9, we can also use only the bits 0 and 1 to represent it in binary form. We saw examples of this in section 2.2.

This algorithm converts from decimal to binary for w in $\mathbb{W}$:

1. If $w = 0$ then the result is also 0, and we are done.
2. Otherwise, let b be an initially empty placeholder where we put the bits.
3. If w is odd, put a 1 to the left of anything in b. Set w to $w - 1$. Otherwise, put a 0 to the left of anything in b. In both cases, now set w to $w/2$.
4. If $w = 0$, we are done and b is the answer. Otherwise, go back to step 3.

For example, let $w = 13$. Initially b is empty.

- w is odd, so b is now 1 and we set $w = (w - 1)/2 = 6$.
- w is even, so b is now 01 and we set $w = w/2 = 3$.
- w is odd, so b is now 101 and we set $w = (w - 1)/2 = 1$.
- w is odd, so b is now 1101 and we set $w = (w - 1)/2 = 0$.
- $w = 0$ and we are done. The representation of w in binary is $b = 1101_2$.

We put the subscript 2 at the end of the number to remind ourselves that we are in base 2.

Now suppose we start with a decimal r with $0 \leq r < 1$. We want to come up with a base 2 representation using just 0s and 1s to the right of the "binary point." We expand as we usually do but instead of using negative powers of 10, we use negative powers of 2.

$$.011_2 = 0 \times 2^{-1} + 1 \times 2^{-2} + 1 \times 2^{-3}$$

In base 10, this is $\frac{1}{4} + \frac{1}{8} = \frac{3}{8} = .375$.

Our algorithm for converting the fractional part of a real number to binary is reminiscent of the one above.

1. If $r = 0$ then the result is also 0, and we are done.
2. Otherwise, let b be a placeholder containing only "." where we put the bits.
3. Multiply r by 2 to get s. Since $0 \leq r < 1$, $0 \leq s < 2$. If $s \geq 1$, put a 1 to the right of anything in b and set $r = s - 1$. Otherwise, put a 0 to the right of anything in b and set $r = s$.
4. If $r = 0$, we are done and b is the answer. Otherwise, go back to step 3.

This sounds reasonable. Let's try it out with $r = .375_{10}$ to confirm our example above. r is not 0 and b starts with the binary point ".".

- Set $s = 2r = .75$, which is less than 1. We append a 0 to the right in b and set $r = s = .75$. b is now .0.
- Set $s = 2r = 1.5$, which is greater than or equal to 1. Append a 1 to the right in b and set $r = s - 1 = .5$. b is now .01.
- Set $s = 2r = 1$, which is greater than or equal to 1. Append a 1 to the right in b and set $r = s - 1 = 0$. b is now .011.
- Since $r = 0$, we are done and the answer is $.011_2$.

In a less verbose table form, this looks like

s	b	r
	.	.375
.75	.0	.75
1.5	.01	.5
1	.011	0

The first line holds the initial settings. The answer is b on the last line where $r = 0$.

The answer agrees with the previous example. Let's do another with $r = .2_{10}$.

s	b	r
	.	.2
.4	.0	.4
.8	.00	.8
1.6	.0001	.6
1.2	.00011	.2
.4	.000110	.4
⋮	⋮	⋮

I stopped because the process has started to repeat itself. Put another way, and using the notation we employed for repeating decimals,

$$.2_{10} = .\overline{00011}_2$$

There is no exact finite binary expansion for the decimal 0.2 but it repeats in a block.

> The following hold in parallel to the decimal case:
>
> - A base 10 rational number has either a finite binary expansion or it repeats in blocks.
> - A repeating block binary expansion is a base 10 rational number.
> - An irrational real number has an infinite, non-repeating-block binary expansion.
> - A binary expansion that has no repeating blocks is irrational.

Given a real value with whole number part w and decimal part $r < 1$, you create the full binary form by concatenating the binary forms of each. The full binary expansion of the decimal 5.125 is 110.001_2, for example.

Question 3.5.3

What is the binary expansion of 17.015625_{10}? Of $\frac{4}{3}$?

3.5.4   Continued fractions

There is yet another expansion for real numbers that is usually not taught in high school algebra classes. This is the *continued fraction* and two examples are on the right-hand sides in each of the following:

$$\tfrac{15}{11} = 1\tfrac{4}{11} = 1 + \cfrac{1}{2 + \cfrac{1}{1 + \cfrac{1}{3}}} \qquad \tfrac{11}{15} = 0 + \cfrac{1}{1 + \cfrac{1}{2 + \cfrac{1}{1 + \cfrac{1}{3}}}}$$

We write the integer portion out front and then construct a recurring sequence of fractions with 1 in the numerators.

Working through the first example shows you the algorithm. Begin with writing the integer portion out front.

First approximation: 1

What's left is $\tfrac{4}{11}$. Invert this to get $\tfrac{11}{4} = 2\tfrac{3}{4}$. Take the whole number part and use this as the second part of the expansion.

Second approximation: $1 + \cfrac{1}{2}$

Invert the remaining fractional part to get $\tfrac{4}{3} = 1\tfrac{1}{3}$. The whole number part goes into the expansion.

Third approximation: $1 + \cfrac{1}{2 + \cfrac{1}{1}}$

Invert $\tfrac{1}{3}$ to get 3 for the expansion. There is no non-zero fractional part and we are done.

Final expansion: $1 + \cfrac{1}{2 + \cfrac{1}{1 + \cfrac{1}{3}}}$

This is a finite continued fraction expansion since there are only a finite number of terms. Working through the fraction arithmetic, if you start with a finite continued fraction you end up with a rational number. What we are doing here is just a variation on Euclid's algorithm from subsection 3.4.1.

Even more interesting, every rational number terminates in this way when you do the expansion. We don't have to worry about repeating blocks for rational numbers converted to continued fractions as we do with decimal and binary expansions.

Question 3.5.4

What is the continued fraction expansion of $-\frac{97}{13}$? Of 0.375?

Using variable names, we can write a finite continued fraction as

$$b_0 + \cfrac{1}{b_1 + \cfrac{1}{b_2 + \cfrac{1}{\ddots + \cfrac{1}{b_n}}}}$$

Here all the b_j are in $\mathbb{Z}$ and $b_j > 0$ for $j > 0$. That is, b_0 can be negative but the rest must be positive integers. An alternative and much shorter notation for the above is

$$[b_0; b_1, b_2, \ldots, b_n].$$

It is also possible to represent a rational number using the form with one more term

$$[b_0; b_1, b_2, \ldots, b_{n-1}, b_n - 1, 1]$$

but I prefer the shorter version. If we decide that the last term cannot be 1, there is a unique representation for a rational number.

Question 3.5.5

Let $r > 0$ be in $\mathbb{R}$. Is

$$r[b_0; b_1, b_2, \ldots, b_n] = [rb_0; rb_1, rb_2, \ldots, rb_n]?$$

What if we are given an expansion that is infinite? It can't be a rational number and it does, in fact, converge to an irrational real number.

Every irrational real number has a unique infinite continued fraction expansion $f = [b_0; b_1, b_2, b_3, \ldots]$.

Infinite continued fractions can have repeating blocks or blocks that repeat via a formula. A line over a block means that it repeats, as usual.

The first expansion in the following table is the "golden ratio" while the last is for e, the base of the natural logarithms. In the expansion for e, note how the integer between the two 1s is incremented by 2 in every block.

Value	Expansion
$\frac{1+\sqrt{5}}{2}$	$[1; \overline{1}]$
$1 + \sqrt{2}$	$[2; \overline{2}]$
$\frac{3+\sqrt{13}}{2}$	$[3; \overline{3}]$
$\sqrt{3}$	$[1; \overline{1, 2}]$
$\sqrt{7}$	$[2; \overline{1, 1, 1, 4}]$
$\tan(1)$	$[1; 1, 1, 3, 1, 5, 1, 7, 1, 9, 1, 11, \dots]$
e	$[2; 1, 2, 1, 1, 4, 1, 1, 6, 1, 1, 8, 1, \dots]$

Question 3.5.6

Calculate the first 6 digits of the golden ratio from its continued fraction.

Let's revisit the first two examples, which I now write in short form:

$$\tfrac{15}{11} = [1; 2, 1, 3] \qquad \tfrac{11}{15} = [0; 1, 2, 1, 3].$$

What do you notice about these? First, the numbers are reciprocals and, second, they have the same expansion except the latter one has a 0 at the beginning. This is true in general.

Let r be in $\mathbb{Q}$ and positive. Suppose $r < 1$ and its continued fraction expansion is

$$[0; b_1, b_2, \dots, b_n].$$

The expansion for $\frac{1}{r}$ is

$$[b_1; b_2, \dots, b_n].$$

On the other hand, if $r \geq 1$ with expansion

$$[b_0; b_1, b_2, \ldots, b_n]$$

then the expansion for $\frac{1}{r}$ is

$$[0; b_0, b_1, b_2, \ldots, b_n]$$

Given an infinite continued fraction $f = [b_0; b_1, b_2, b_3, \ldots]$, it's natural to look at the sequence of finite fractions, the *convergents* of f,

$$f_0 = [b_0;] = \frac{x_0}{y_0}$$
$$f_1 = [b_0; b_1] = \frac{x_1}{y_1}$$
$$f_2 = [b_0; b_1, b_2] = \frac{x_2}{y_2}$$
$$\vdots$$
$$f_n = [b_0; b_1, b_2, b_3, \ldots, b_n] = \frac{x_n}{y_n}$$

and ask about the relationship between f and the f_j. Each x_j is an integer and each y_j is a positive integer. The f_j are expressed in reduced form. (That is, $\frac{1}{2}$ and not $\frac{3}{6}$.)

The convergents f_j have the following properties with respect to a specific convergent f_n:

- $f_1 > f_3 > f_5 > \cdots > f_n$ for all f_j with odd $j < n$.
- $f_2 < f_4 < f_6 < \cdots < f_n$ for all f_j with even $j < n$.
- If $j < k < n$ then $|f_n - f_k| < |f_n - f_j|$.

This means the convergents oscillate above and below f_n, always getting closer.

This example shows the rapid convergence to $\sqrt{3}$.

With f as above,

$$f_2 = [b_0; b_1, b_2] = b_0 + \cfrac{1}{b_1 + \cfrac{1}{b_2}} = \frac{b_0 b_1 b_2 + b_2 + b_0}{b_1 b_2 + 1} = \frac{b_2(b_0 b_1 + 1) + b_0}{b_1 b_2 + 1}.$$

Question 3.5.7

Compute f_1, x_1, and y_1, and f_3, x_3, and y_3. Establish a guess about how to compute the x_n and y_n given the values for $n - 1$ and $n - 2$. Confirm if this works for f_4, x_4, and y_4.

Convergence Properties of Continued Fractions [3]

Let r in $\mathbb{R}$ be the value of the infinite continued fraction $f = [b_0; b_1, b_2, b_3, \dots]$. Let $f_j = \frac{x_j}{y_j}$ be the convergents.

- Each convergent is a reduced fraction. That is, $\gcd(x_j, y_j) = 1$.
- If $k > j$ then $y_k > y_j$.
- The denominators y_j are increasing exponentially:

$$y_j \geq 2^{\frac{j-1}{2}}$$

- We can approximate r as closely as we wish by computing a divergent with large enough j.

$$\left| r - f_j \right| = \left| r - \frac{x_j}{y_j} \right| < \frac{1}{y_j y_{j+1}}$$

Question 3.5.8

What do you need to modify in the preceding statements so they hold for a finite continued fraction?

To learn more

Continued fractions are a fascinating but somewhat specialized area of mathematics. They're not difficult but they are not used in every field. The topic is often covered briefly in algebra and number theory texts but there are only a few dedicated books about them. [7, Chapter 10] [3] [8]

3.6 Structure

I took some time to show the operations and the properties of the real numbers and its subsets like the integers and rational numbers because these are very common in other parts of mathematics when properly abstracted. This structure allows us to learn and prove things and then apply them to new mathematical collections as we encounter them. We start with three: *groups*, *rings*, and *fields*.

3.6.1 Groups

Consider a collection of objects which we call **G**. For example, **G** might be $\mathbb{Z}$, $\mathbb{Q}$, or $\mathbb{R}$ as above. We also have some pairwise operation between elements of **G** we denote by "∘". It's a placeholder for an action that operates on two objects.

This "∘" operation could be addition "+" or multiplication "×" for numbers, but might be something entirely different. Use your intuition with numbers, but understand that the general case is, well, more general. We call the collection together with its operation (**G**, ∘).

We write the use of "∘" the same as we would normally with addition or multiplication, between the elements. We write $a \circ b$ for a and b in **G**.

This is called *infix* notation. Negation like -7 uses *prefix* notation. The factorial operation $n! = 1 \times 2 \times \cdots \times (n-1) \times n$ uses *postfix* notation.

We say (**G**, ∘) is a group if the following conditions are met

- If a and b are in **G** then $a \circ b$ is in **G**. This is closure.
- If a, b, and c are in **G** then $(a \circ b) \circ c = a \circ (b \circ c)$ is in **G**. This is associativity.
- There exists a unique element id in **G** such that $a \circ id = id \circ a = a$ for every a in **G**. This is the existence of a unique identity element.
- For every a in **G**, there is an element denoted a^{-1} such that $a^{-1} \circ a = a \circ a^{-1} = id$. This is the existence of a inverse.

The inverse is unique. Suppose there are two elements b and c such $b \circ a = a \circ b = id$ and $c \circ a = a \circ c = id$. Then $b \circ a \circ c = id \circ c$ by applying "∘c" on the right. So $b \circ (a \circ c) = c$. Since c is an inverse of a, $b = c$.

We do not require the "∘" operation to be commutative: $a \circ b$ need not equal $b \circ a$. When this is true for all a and b, we call **G** a commutative group.

In the mathematical literature, commutative groups are called *abelian groups* in honor of the early nineteenth century mathematician Niels Henrik Abel, but we stick to the descriptive name.

While you are likely aware of the quadratic formula for finding the roots of a polynomial like $x^2 + x - 6$, you may not know that there are also (very messy) formulas for third and fourth degree polynomials. Despite others working on this for hundreds of years, Abel finally proved that there is no corresponding formula for polynomials of degree 5.

A subset of **G** that uses the same operation "∘" and is closed under it, contains id, and is closed under inversion is called a *subgroup* of **G**.

Examples

- The natural numbers are not a group under addition because of the lack of 0 and negative numbers.
- The whole numbers are not a group under addition because all positive numbers lack their negative counterparts.
- The integers $\mathbb{Z}$, rational numbers $\mathbb{Q}$, and real numbers $\mathbb{R}$ are each a group under addition with identity element 0. $\mathbb{Z}$ is a subgroup of $\mathbb{Q}$, which is a subgroup of $\mathbb{R}$.
- The even integers are a group under addition with identity element 0. They are a subgroup of $\mathbb{Z}$.

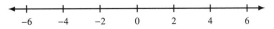

- The odd integers are not a group under addition.
- The integers $\mathbb{Z}$ are not a group under multiplication because most integers lack multiplicative inverses.
- The rational numbers $\mathbb{Q}$ are not a group under multiplication because there is no multiplicative inverse for 0.
- The rational numbers without 0 are a group under multiplication but not under addition.
- Similarly, the non-zero real numbers are a group under multiplication but not under addition.

Much of the fuss we made when we proceeded from $\mathbb{W}$ to $\mathbb{N}$ to $\mathbb{Z}$ was to systematically examine their properties to ultimately show $\mathbb{Z}$ is a group under "+".

In all the examples above where we have groups, they are commutative groups. The study of groups, finite and infinite, commutative and non-commutative, is a fascinating and essential topic at the core of much of mathematics and physics.

For another example, imagine you live on a world that is one long straight infinite street.

For our group, take movements of the form "walk 12 meters to the right" and "walk 4 meters to the left." The group operation is composition "∘" thought of as "and then." We can write

$$a = \text{"walk 12 meters to the left"}$$
$$b = \text{"walk 4 meters to the right"}$$
$$a \circ b = \text{"walk 12 meters to the left"} \; \textit{and then} \; \text{"walk 4 meters to the right"}$$
$$= \text{"walk 8 meters to the left"}$$

Note I didn't specify where on the street to start: all movements are relative. The inverse of a "walk to the right" element is the corresponding "walk to the left" element. The identity element *id* is "walk to the right 0 meters" which we take to be the same as "walk to the left 0 meters."

Verify for yourself that "∘" is associative and commutative.

You can extend this group in two dimensions by adding similar "walk forward" and "walk backward" elements. For three dimensions you would indicate up and down movements. Think about associativity and commutativity in each case.

Question 3.6.1

Is this a finite or infinite group? What are some subgroups?

Rather than being on a straight line, consider being on a circle of diameter 4 meters. You are only able to move in natural number meter increments clockwise or counterclockwise.

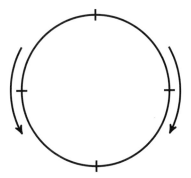

We can equate moving 4 meters in either direction with moving 0 meters since we move all the way around the circle. Moving 5 meters counterclockwise is the same element as moving 1 meter in that direction or 3 meters in the other.

Question 3.6.2

Is this a finite or infinite group? If we allow moving clockwise or counterclockwise in any non-negative real number increments, is it a finite or infinite group?

3.6.2 Rings

When we have more than one operation, we get a more sophisticated structure called a *ring* if certain requirements are met. For convenience we call these two operations "+" and "×" but remember they could behave in very different ways from the addition and multiplication we have for numbers.

We say $(\mathbf{R}, +, \times)$ is a *ring* if the following conditions are met

- $\mathbf{R}$ is a commutative group under "+" with identity element 0.
- If a, b, and c are in $\mathbf{R}$ then $(a \times b) \times c = a \times (b \times c)$ is in $\mathbf{R}$. This is associativity for "$\times$".
- There exists an element 1 in $\mathbf{R}$ such that $a \times 1 = 1 \times a = a$ for every a in $\mathbf{R}$. This is the existence of a multiplicative identity element.
- Multiplication "$\times$" distributes over addition "+". If a, b, and c are in $\mathbf{R}$ then $a \times (b + c) = (a \times b) + (a \times c)$ and $(b + c) \times a = (b \times a) + (c \times a)$.

Note that $0 \neq 1$.

While addition is commutative in a ring, multiplication need not be. As you might guess, a *commutative ring* is one with commutative multiplication. A *non-commutative ring* has the obvious definition.

A subgroup of $\mathbf{R}$ under "+" is a *subring* if it also shares the same "$\times$", contains 1, and is closed under "$\times$".

Examples

- The integers $\mathbb{Z}$, rational numbers $\mathbb{Q}$, and real numbers $\mathbb{R}$ are each a ring under addition and multiplication with identity elements 0 and 1, respectively. $\mathbb{Z}$ is a subring of $\mathbb{Q}$ which is a subring of $\mathbb{R}$. When we want to consider $\mathbb{Z}$ as an additive group, we write $\mathbb{Z}^+$.
- The even integers are not a subring of $\mathbb{Z}$ because they do not contain 1.
- Consider all elements of $\mathbb{R}$ of the form $a + b\sqrt{2}$ for a and b integers. We have

$$\mathbf{0} = 0 + 0\sqrt{2}$$

$$\mathbf{1} = 1 + 0\sqrt{2}$$

$$-\left(a + b\sqrt{2}\right) = -a - b\sqrt{2}$$

$$\left(a + b\sqrt{2}\right) + \left(c + d\sqrt{2}\right) = (a + c) + (b + d)\sqrt{2}$$

$$\left(a + b\sqrt{2}\right) \times \left(c + d\sqrt{2}\right) = (ac + 2bd) + (ad + bc)\sqrt{2}$$

We call this $\mathbb{Z}\left[\sqrt{2}\right]$ and it is a commutative ring that extends $\mathbb{Z}$. Other than $\mathbb{Q}$, this is the first ring we have seen that is larger than $\mathbb{Z}$ but smaller than $\mathbb{R}$ itself.

In some commutative rings it is possible for $a \times b = 0$ with neither a nor b being 0. **R** is called an *integral domain* if this is not possible. Said otherwise, we must have a or b being 0 for the product to be 0. All the rings we have seen so far are integral domains.

3.6.3 Fields

> A *field* **F** is a commutative ring where every non-zero element has a multiplicative inverse. A field is closed under division by non-zero elements.

$\mathbb{Q}$ and $\mathbb{R}$ are fields but $\mathbb{Z}$ is not. $\mathbb{Q}$ is a *subfield* of $\mathbb{R}$. Viewed from the opposite direction, $\mathbb{R}$ is an *extension* of $\mathbb{Q}$. For example, if we look at all numbers of the form $r + \sqrt{2}s$ with r and s in $\mathbb{Q}$, and perform the arithmetic operations in the usual way, we have an extension field of $\mathbb{Q}$ that is a subfield of $\mathbb{R}$. We denote this field by $\mathbb{Q}[\sqrt{2}]$.

All fields are integral domains. Suppose otherwise and $a \times b = 0$ with neither a nor b being 0. Then there exists a^{-1} such that $a^{-1} \times a = 1$. Hence $a^{-1} \times a \times b = a^{-1} \times 0$ means $1 \times b = 0$. But we said b is not 0! We have a contradiction and there can be no such b. So the field is an integral domain.

3.6.4 Even greater abstraction

Though we do not need them in the rest of this book, I want to name two additional algebraic structures since we have already seen examples of them. If $\mathbb{Z}$ is a group under addition, what is $\mathbb{W}$? It doesn't contain the additive inverses like -2 and so it can't be a group.

Even worse, what is $\mathbb{N}$? Here we don't even have 0, the additive identity element.

> We say $(\mathbf{G}, \circ)$ is a *semigroup* if the following conditions are met:
>
> - If a and b are in **G** then $a \circ b$ is in **G**. This is closure.
> - If a, b, and c are in **G** then $(a \circ b) \circ c = a \circ (b \circ c)$ is in **G**. This is associativity.
>
> $(\mathbf{G}, \circ)$ is a *monoid* if we also have:
>
> - There exists a unique element *id* in **G** such that $a \circ id = id \circ a = a$ for every a in **G**. This is the existence of a unique identity element.

With this, $\mathbb{N}$ is a semigroup and $\mathbb{W}$ is a monoid. All groups are monoids and all monoids are semigroups.

In the summary in section 3.10, I've provided a table and chart showing how these algebraic structures are related to each other and the collections of numbers we are working through in this chapter.

To learn more

Group theory is ubiquitous across many areas of mathematics and physics. [10] Rings, fields, and other structures are fundamental to areas of mathematics like algebra, algebraic number theory, commutative algebra, and algebraic geometry. [1] [5]

3.7 Modular arithmetic

There are an infinite number of integers and hence rationals and real numbers. Are there sets of numbers that behave somewhat like them but are finite?

Consider the integers modulo 6: $\{0, 1, 2, 3, 4, 5\}$. We write 3 mod 6 when we consider the 3 in this collection. Given any integer n, we can map it into this collection by computing the remainder modulo 6. Arithmetic can be done in the same way:

$$7 \equiv 1 \bmod 6 \qquad\qquad (4-5) \equiv 5 \bmod 6$$
$$-2 \equiv 4 \bmod 6 \qquad\qquad (3 \times 7) \equiv 3 \bmod 6$$
$$(5+4) \bmod 6 \equiv 3 \bmod 6 \qquad\qquad (2+4) \equiv 0 \bmod 6$$

Instead of using "=", we write "$\equiv$" and say that a is *congruent* to b mod 6 when we see $a \equiv b \bmod 6$. This means that $a - b$ is evenly divisible by 6: $6 | (a - b)$.

This is a group under addition with identity 0. In the last example, 2 is the additive inverse of 4. We denote this $\mathbb{Z}/6\mathbb{Z}$.

Question 3.7.1

What is $-1 \bmod 6$? For n a natural number greater than 1, what is $-1 \bmod n$?

Let's consider the same collection but without the 0. Instead of addition, use multiplication with identity 1.

Is this a group? Is it closed under multiplication? Does every element have an inverse?

Since $2 \times 3 = 6 \equiv 0$ and 0 is not in the collection, it is not closed under multiplication!

The elements that **do not** have multiplicative inverses are 2, 3, and 4 because each of these share a factor with 6.

The inverse of 1 is itself. $(5 \times 5) \bmod 6 \equiv 5^2 \bmod 6 \equiv 25 \bmod 6 \equiv 1 \bmod 6$, so it too is its own inverse.

If we restrict ourselves to the elements that do not have a factor in common with 6 then we do get a group $\{1, 5\}$, though it is not a very big one. The fancy mathematical way of writing this group is $(\mathbb{Z}/6\mathbb{Z})^\times$.

If instead of 6 we had chosen 15, then the elements in $(\mathbb{Z}/15\mathbb{Z})^\times$ would be

$$\{1, 2, 4, 7, 8, 11, 13, 14\}.$$

We have a better was of expressing "does not have a factor in common" and that is via the greatest common divisor. The integers a and b do not share a non-trivial factor if and only if $\gcd(a, b) = 1$. That is, a and b are coprime. If this is the case, there are integers n and m such that $an + bm = 1$. If we look at this modulo b then

$$1 \equiv an + bm \bmod b \equiv an \bmod b.$$

Hence n is equal to a^{-1} modulo b!

The phase "X if and only if Y" means "if X is true, then Y is true, and if Y is true, then X is true." We cannot have one of them true and the other false.

The integers a that are in $(\mathbb{Z}/15\mathbb{Z})^\times$ have $0 < a < 15$ and $\gcd(a, 15) = 1$. If we use 7 instead, the elements in $(\mathbb{Z}/7\mathbb{Z})^\times$ are $\{1, 2, 3, 4, 5, 6\}$.

Well this case is interesting! We got all the non-zero elements *because 7 is prime*. If p is a prime number then none of the numbers $1, 2, 3, \ldots, p - 1$ share a factor with p. That is, they are coprime with p.

The elements $1, 2, \ldots, p - 1$ form a multiplicative group with identity element 1 if and only if p is prime. The group has $p - 1$ elements.

The elements $0, 1, 2, \ldots, p - 1$ form a field under "+" and "$\times$" with identity elements 0 and 1, respectively, if and only if p is prime. The field has p elements and is denoted $\mathbb{F}_p$.

There are more finite fields than these but any other finite field that is not formed this way is an extension of one of these. The number of elements in any finite field is a power of a prime number p. Any two finite fields with the same number of elements are isomorphic.

One way mathematicians differentiate among the kinds of fields is via their *characteristic*. For p prime, a field that is either an $\mathbb{F}_p$ or an extension of it has characteristic p. Otherwise the field has characteristic 0. $\mathbb{Q}$ and $\mathbb{R}$ are examples of fields with characteristic 0.

The smallest field is $\mathbb{F}_2$ and it has only two elements, 0 and 1. Fields with characteristic 2 often require special handling because 2 is the only prime that is an even number.

> **To learn more**
>
> Finite fields have many applications in pure and applied mathematics, including computer error correction and cryptography. [2] [4] [6] [9]

3.8 Doubling down

So far we've seen finite and infinite groups, rings, and fields, with some of them being extensions of others. In this section we look at combining them.

Consider the collection of all pairs of integers (a, b) where we define addition and multiplication component-wise. This means

$$\mathbf{0} = (0, 0)$$
$$\mathbf{1} = (1, 1)$$
$$-(a, b) = (-a, -b)$$
$$(a, b) + (c, d) = (a + c, b + d)$$
$$(a, b) \times (c, d) = (ac, bd)$$

This is a ring, denoted $\mathbb{Z}^2$, but it is not an integral domain.

For example, $(1, 0) \times (0, 1) = (0, 0)$ but neither of the factors is $\mathbf{0}$.

For the same reason, neither $\mathbb{Q}^2$ nor $\mathbb{R}^2$ can be an integral domain. In particular, they are not fields with these operations.

Let's change the definition of multiplication for $\mathbb{R}^2$:

$$(a, b) \times (c, d) = (ac - bd, ad + bc)$$

For (a, b) not $\mathbf{0}$, define

$$(a, b)^{-1} = \left(\frac{a}{a^2 + b^2}, -\frac{b}{a^2 + b^2} \right)$$

With this unusual definition for multiplication, we now have a field. Where did this come from?

3.9 Complex numbers, algebraically

In subsection 3.6.2 I gave an example of how we can extend the integers by considering elements of the form $a + b\sqrt{2}$. We can similarly extend $\mathbb{R}$.

The real numbers $\mathbb{R}$ do not contain square roots of negative numbers. We *define* the value i to be $\sqrt{-1}$, which means $i^2 = -1$.

For a and b in $\mathbb{R}$, consider all elements of the form $z = a + bi$. This is the field of *complex numbers* $\mathbb{C}$ formed as $\mathbb{R}[i] = \mathbb{R}\left[\sqrt{-1}\right]$.

3.9.1 Arithmetic

We call a the *real part* of z and denote it by $\mathrm{Re}(z)$. b is the *imaginary part* $\mathrm{Im}(z)$. a and b are real numbers. Every real number is also a complex number with a zero imaginary part.

While we can always determine if $x < y$ for two real numbers, there is no equivalent ordering for arbitrary complex ones that extends what works for the reals.

The equations for arithmetic are

$$\mathbf{0} = 0 + 0i$$
$$\mathbf{1} = 1 + 0i$$
$$-(a + bi) = -a - bi$$
$$(a + bi) + (c + di) = (a + c) + (b + d)i$$
$$(a + bi) - (c + di) = (a - c) + (b - d)i$$

Multiplication is more complicated.

$$(a + bi) \times (c + di) = (ac - bd) + (ad + bc)i$$
$$\mathrm{Re}((a + bi) \times (c + di)) = ac - bd$$
$$\mathrm{Im}((a + bi) \times (c + di)) = ad + bc$$

Let's work this out.

$$
\begin{aligned}
(a + bi) \times (c + di) &= a \times (c + di) + bi\,(c + di) \\
&= ac + adi + bic + (bi)(di) \\
&= ac + (ad + bc)i + bd(ii) \\
&= ac + (ad + bc)i + bd(-1) \\
&= ac + (ad + bc)i - bd \\
&= (ac - bd) + (ad + bc)i
\end{aligned}
$$

For example,

$$(2 + 3i) \times (4 + 6i) = 2(4 + 6i) + 3i(4 + 6i)$$
$$= 2 \times 4 + 2 \times 6i + 3i \times 4 + 3i \times 6i$$
$$= 8 + 12i + 12i + 18 \times i \times i$$
$$= 8 + 24i + 18 \times -1$$
$$= -10 + 24i$$

It's conventional to use the variable z when referring to a complex number perhaps because "Zahl" is the German word for "number." w is our frequent choice if we need another.

3.9.2 Conjugation

Complex numbers have one operation we have not seen before: *conjugation*. For $z = a + bi$, the conjugate of z, denoted $\bar{z}$, is $a - bi$.

- The product of a complex number $z = a + bi$ with its conjugate is a *non-negative* real number: $z\bar{z} = a^2 + b^2$.
- $z = 0$ if and only if $\bar{z} = 0$.
- $\mathrm{Re}(z) = \mathrm{Re}(\bar{z}) = a$.
- $\mathrm{Im}(z) = -\mathrm{Im}(\bar{z}) = -b$.
- When $\mathrm{Im}(z) = 0$, z is real and conjugation does nothing: $z = \bar{z}$.

It's hard to overstate the importance of conjugation in the algebra of complex numbers. We return to it later to get a geometric interpretation.

We use the conjugate to compute the inverse of a non-zero complex number z.

$$z^{-1} = \frac{1}{z} = \frac{\bar{z}}{\bar{z}z} = \frac{a - bi}{a^2 + b^2} = \frac{a}{a^2 + b^2} - \frac{b}{a^2 + b^2}i$$

Question 3.9.1

Confirm that

$$(a + bi)\left(\frac{a}{a^2 + b^2} - \frac{b}{a^2 + b^2}i\right) = 1.$$

This is the division formula for complex numbers.

Conjugation behaves very nicely regarding all the standard operations on complex numbers. For complex numbers z and w,

$$z = \overline{\overline{z}}$$

$$\overline{z^{-1}} = \overline{\left(\frac{1}{z}\right)} = \frac{1}{\overline{z}}$$

$$\overline{z+w} = \overline{z}+\overline{w}$$

$$\overline{\left(\frac{z}{w}\right)} = \frac{\overline{z}}{\overline{w}} \text{ for } w \neq 0$$

$$\overline{z-w} = \overline{z}-\overline{w}$$

$$\overline{z^n} = \overline{z}^n \text{ for } n \text{ an integer}$$

$$\overline{z \times w} = \overline{z} \times \overline{w}$$

The *absolute value* of a complex number z is $|z| = \sqrt{z\overline{z}}$. For $z = a + bi$, this is $\sqrt{a^2+b^2}$. Any complex number with absolute value 1 is called a *unit*.

Question 3.9.2

Can you prove $|z| = |\overline{z}|$?

3.9.3 Units

A unit is a number with absolute value 1. The complex numbers have an infinite number of units unlike the real numbers, which only have 1 and -1.

As complex numbers, 1, -1, i, and $-i$ are all units but so are all $a + bi$ with $a^2 + b^2 = 1$. Thus $\frac{\sqrt{2}}{2} \pm \frac{\sqrt{2}}{2}i$ and $\frac{\sqrt{3}}{2} \pm \frac{1}{2}i$ are units.

Question 3.9.3

Where have you seen numbers like those in the real and imaginary parts before?

Here's a hint: for any real number x, $\sin^2(x) + \cos^2(x) = 1$. That means if we look at complex numbers of the form $z = \cos(x) + \sin(x)i$ then $|z| = 1$. Numbers in this form are all units in $\mathbb{C}$. Even better, these are the *only* units in $\mathbb{C}$.

Seemingly out of nowhere, we have connected $\sqrt{-1}$, the real numbers extended by it, and trigonometry. These are key tools we need when we look at a qubit, the fundamental information object of quantum computing.

When we look at the geometry of complex numbers, I explain *Euler's formula* $e^{xi} = \cos(x) + \sin(x)i$. From this comes *Euler's identity*

$$e^{\pi i} = -1$$

where e is the base of the natural logarithms $= 2.171828\ldots$.

Many people consider this the most beautiful equation in mathematics, connecting the integers (-1), an irrational number basic to calculus (e), another irrational central to trigonometry (π), and the complex numbers (i).

3.9.4 Polynomials and roots

We can't always use variable names like a, b, c, x, y, and z when we have to use many names. Therefore I use subscripted variables whenever necessary. In x_i, i is the subscript. expressions that involve all the n constants $a_1, a_2, \ldots, a_n$. Eventually, we'll encounter expressions involving multiple subscripts like $a_{2,3}$.

Let $p(z)$ be a polynomial such that

$$p(z) = a_n z^n + a_{n-1} z^{n-1} + \cdots + a_2 z^2 + a_1 z + a_0$$

where at least one of the complex numbers $a_1, \ldots, a_n$ is not 0. Then there exists at least one complex number s such that $p(s) = 0$. s is a *root* of p and $z - s$ divides $p(z)$ exactly.

The a_i are called the *coefficients* of p. Insisting that one of $a_1, \ldots, a_n$ is not 0 means p is not a constant polynomial. For example,

$$p(z) = 2 - \frac{4}{5}i$$

is a constant polynomial. There is no complex number s you can substitute for z that makes $p(s) = 0$. If $a_n \neq 0$ then n is called the *degree* of the polynomial.

Let's back up and see what this means. For $p(z) = z^2 - 1$, both $s = 1$ and $s = -1$ work. That is, $p(1) = 0$ and $p(-1) = 0$. In this case $p(z)$ has the real number coefficients $a_2 = 1$ and $a_0 = -1$. The roots are also real. The roots are related to the factorization of p via

$$p(z) = (z - (1))\,(z - (-1))$$

where the roots are shown in parentheses. Written more simply,

$$p(z) = (z - 1)\,(z + 1).$$

Consider another polynomial with real coefficients, $p(z) = z^2 + 1$. This is as simple as the previous one and here $a_2 = 1$ and $a_0 = 1$. Stating $p(z) = 0$ means $z^2 + 1 = 0$ and so $z^2 = -1$. So any s that works must be a square root of -1! There is no such real number. There are two and only two complex numbers that work: i and $-i$.

$$p(i) = p(-i) = 0$$
$$p(z) = (z - i)(z + i)$$

When $p(z)$ has real coefficients *it is not automatically the case that it has real roots*. Just as we saw way back when we considered that $\mathbb{N}$ was not closed under subtraction, $\mathbb{R}$ is not closed under finding real roots for non-constant polynomials with real coefficients.

Just as $\mathbb{Z}$ fixed the subtraction closure problem for $\mathbb{N}$, $\mathbb{C}$ fixes the non-constant polynomial root finding problem for $\mathbb{R}$. The statement at the beginning of this section says even more: $\mathbb{C}$ is closed under finding complex roots for non-constant polynomials with complex coefficients.

This is a very strong and important property for a field to have. The phrase we use is that the complex numbers are *algebraically closed*. $\mathbb{Q}$ and $\mathbb{R}$ are not algebraically closed.

Consider the quadratic polynomial equation $ax^2 + bx + c = 0$ with a, b and c real numbers and $a \neq 0$. We solve for x by the technique of completing the square. In particular, we use

$$\left(x + \frac{b}{2a}\right)^2 = x^2 + \frac{b}{a}x + \left(\frac{b}{2a}\right)^2 = x^2 + \frac{b}{a}x + \frac{b^2}{4a^2}$$

This may seem like an arbitrary thing to do, but you learn to solve equations by looking for patterns and understanding what you can do to each side to isolate the term or variable you want.

The symbol "$\Rightarrow$" means "implies" in the following.

$$ax^2 + bx + c = 0 \Rightarrow x^2 + \frac{b}{a}x + \frac{c}{a} = 0 \text{ because } a \neq 0$$

$$\Rightarrow \left(x^2 + \frac{b}{a}x + \frac{c}{a}\right) + \frac{b^2}{4a^2} = \frac{b^2}{4a^2}$$

$$\Rightarrow \left(x^2 + \frac{b}{a}x + \frac{b^2}{4a^2}\right) + \frac{c}{a} = \frac{b^2}{4a^2}$$

$$\Rightarrow x^2 + \frac{b}{a}x + \frac{b^2}{4a^2} = \frac{b^2}{4a^2} - \frac{c}{a}$$

$$\Rightarrow \left(x + \frac{b}{2a}\right)^2 = \frac{b^2}{4a^2} - \frac{c}{a}$$

$$\Rightarrow x + \frac{b}{2a} = \pm\sqrt{\frac{b^2}{4a^2} - \frac{c}{a}}$$

$$\Rightarrow x = \pm\sqrt{\frac{b^2}{4a^2} - \frac{c}{a}} - \frac{b}{2a}$$

$$\Rightarrow x = \frac{-b \pm \sqrt{b^2 - 4ac}}{2a}$$

This is the *quadratic formula*, and now you know (or remember) how it is derived!

The significance of it, and why I went into such detail to calculate it, is that there are two possible values for x that you can plug into $ax^2 + bx + c$ to get 0. One is

$$\frac{-b + \sqrt{b^2 - 4ac}}{2a}$$

and the other is

$$\frac{-b - \sqrt{b^2 - 4ac}}{2a}.$$

These are the roots of the polynomial.

When $b^2 - 4ac = 0$, we have one root $-\frac{b}{2a}$ but it is repeated twice. For example, $x^2 - 4x + 4$ has the repeated root 2.

When $b^2 - 4ac > 0$, we get two different real roots. The polynomial $x^2 + x - 6$ has the real roots 2 and -3.

When $b^2 - 4ac < 0$, we get two non-real complex roots *and they are conjugate.*

$$b^2 - 4ac < 0 \Rightarrow 4ac - b^2 > 0$$

So

$$\frac{-b + \sqrt{b^2 - 4ac}}{2a} = \frac{-b + \sqrt{4ac - b^2}i}{2a} = \frac{-b}{2a} + \sqrt{\frac{4ac - b^2}{4a^2}}i$$

and

$$\frac{-b - \sqrt{b^2 - 4ac}}{2a} = \frac{-b - \sqrt{4ac - b^2}i}{2a} = \frac{-b}{2a} - \sqrt{\frac{4ac - b^2}{4a^2}}i$$

Since a real number is its own conjugate, this statement holds in all cases.

> If s is a complex root of the polynomial $ax^2 + bx + c$ with a, b, and c **real numbers** and $a \neq 0$, meaning that $as^2 + bs + c = 0$, then $\bar{s}$ is a root. If s is real then it does not give us an additional root.

It also holds for polynomials of all degrees greater than 0 with real coefficients.

> If s is a complex root of the nth degree polynomial
>
> $$a_n x^n + a_{n-1} x^{n-1} + \cdots + a_1 x + a_0$$
>
> where all coefficients are **real**, $n > 0$, and $a_n \neq 0$, then $\bar{s}$ is also a root.

This is called the *complex conjugate root theorem*, though you sometimes see the word "conjugate" omitted in the title.

3.10 Summary

There's more to numbers than you might have thought as you've used them in your daily life. Starting with the simplest kind, the natural numbers $\mathbb{N}$, we systematically added operations and properties to gain functionality. The idea of "closure" was important in driving us to understand the value of extending to larger collections of numbers that could handle the problems we wanted to solve.

We briefly delved into abstract algebra to look at groups, rings, and fields and to see the common structure they provide. The complex numbers are key to working with quantum computing and we began to look at their algebraic properties. Though they involve the imaginary i, they are very real in describing the way the universe evidently works.

The following table and diagram bring together all the forms of numbers we have seen and some of their properties. The diagram shows the inclusion relationships among the collections of numbers. The expressions

$$A \subset B \quad \text{or} \quad \begin{matrix} B \\ \cup \\ A \end{matrix}$$

mean that A is included in B. For example, $\mathbb{Z}$ is included in $\mathbb{Q}$, which is included in $\mathbb{Q}[\sqrt{2}]$.

Number Collection	Symbol	Closure	Structure
natural numbers	$\mathbb{N}$	addition, multiplication	semigroup
whole numbers	$\mathbb{W}$	addition, multiplication	monoid
integers (additive subgroup)	$\mathbb{Z}^+$	addition, subtraction	group
integers	$\mathbb{Z}$	addition, subtraction, multiplication	ring
rational numbers	$\mathbb{Q}$	addition, subtraction, multiplication, division	field
a rational number field extension	$\mathbb{Q}\left[\sqrt{2}\right]$	addition, subtraction, multiplication, division	field
real numbers	$\mathbb{R}$	addition, subtraction, multiplication, division, limits	field
complex numbers	$\mathbb{C}$	addition, subtraction, multiplication, division, polynomial factorization, $\sqrt{-1}$	field
modular integers	$\mathbb{Z}/n\mathbb{Z}$	addition, subtraction, multiplication	ring
modular integers (multiplicative subgroup)	$(\mathbb{Z}/n\mathbb{Z})^{\times}$	multiplication, division	group
finite fields	$\mathbb{F}_p$	addition, subtraction, multiplication, division	field

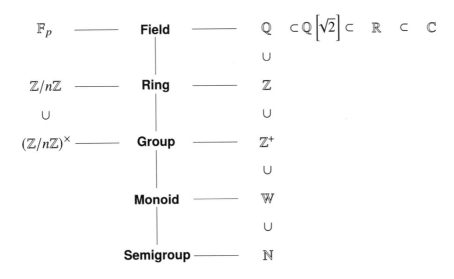

References

[1]   D. S. Dummit and R. M. Foote. *Abstract Algebra*. 3rd ed. Wiley, 2004.

[2]   Kenneth Ireland and Michael Rosen. *A Classical Introduction to Modern Number Theory*. 2nd ed. Graduate Texts in Mathematics 84. Springer-Verlag New York, 1990.

[3]   A. Ya. Khinchin. *Continued Fractions*. Revised. Dover Books on Mathematics. Dover Publications, 1997.

[4]   Neal Koblitz. *A Course in Number Theory and Cryptography*. 2nd ed. Graduate Texts in Mathematics 114. Springer-Verlag, 1994.

[5]   S. Lang. *Algebra*. 3rd ed. Graduate Texts in Mathematics 211. Springer-Verlag, 2002.

[6]   Robert J. McEliece. *Finite Fields for Computer Scientists and Engineers*. 10th ed. The Springer International Series in Engineering and Computer Science 23. Springer US, 1987.

[7]   S.J. Miller et al. *An Invitation to Modern Number Theory*. Princeton University Press, 2006.

[8]   C. D. Olds. *Continued Fractions*. Mathematical Association of America, 1963.

[9]   Oliver Pretzel. *Error-correcting Codes and Finite Fields*. Student Edition. Oxford University Press, Inc., 1996.

[10] J. Rotman. *An Introduction to the Theory of Groups*. Graduate Texts in Mathematics 148. Springer New York, 1999.

4

Planes and Circles and Spheres, Oh My

No employment can be managed without arithmetic,
no mechanical invention without geometry.

Benjamin Franklin

In the last chapter we focused on the algebra of numbers and collections of objects that behave like numbers. Here we turn our attention to geometry and look at two and three dimensions. When we start working with qubits in chapter 7, we represent a single qubit as a sphere in three dimensions. Therefore, it's necessary to get comfortable with the geometric side of the mathematics before we tackle the quantum computing aspect.

Topics covered in this chapter

4.1   Functions

A function is one of the concepts in math that sounds pretty abstract but is really straightforward once you get experience with it. Thought of in terms of numbers, a function takes a value and returns one and only one value.

For example, for any real number, we can square it. That process is a function. For any non-negative real number, if we take the positive square root of it then we get another function. If we were to say we got both the positive and negative square roots, we would not have a function.

We use the notation $f(x)$ for a function, meaning that we start with some value x, do something to it indicated by the definition of f, and the result is $f(x)$. The f can be any letter or word, but we use f because the word "function" starts with it and we are not being especially creative. It's common to see g and h but, really, we can use anything.

We write a function definition like $f(x) = x^2$ or $g(x) = \sqrt{x}$ or $h(x) = 2x - 3$.

The set of values that we can use for x is called the *domain* of the function. The domain of $f(x) = x^2$ could be the integers, the real numbers, the complex numbers, or any other set that makes sense. When considering real numbers, the domain of $g(x) = \sqrt{x}$ is the set of non-negative real numbers, for example.

The *range* of a function is the set of values it produces. So if the domain of $f(x) = x^2$ is $\mathbb{R}$ then the range is the set of all non-negative real numbers. The same is true for the absolute value $|x|$. I didn't mention $f(x)$ or $g(x)$ or something similar when I said that $|x|$ is a function.

It's not mandatory to use the $f(x)$ notation when the context makes it clear we have a function. We can also use the arrow notation $x \mapsto f(x)$ to show the input is x and the result is $f(x)$.

When the range of function is some subset of $\mathbb{R}$, we say that "f is real-valued." Similarly we might say it is integer-valued or complex-valued.

Functions don't have to work on or produce numeric inputs and outputs only. Consider the function $c(n)$ on the integers that produces the color white if n is even and the color black if n is odd. Each integer is mapped to one of white or black, though an infinite number of integers are mapped to each color.

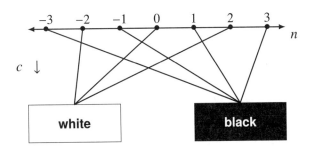

We can use another variation of the arrow notation to show the domain and range of a function.

$$c : \mathbb{Z} \rightarrow \{\text{white}, \text{black}\}$$

shows that the domain of c is the integers and the range is the set $\{\text{white}, \text{black}\}$. A common range that we will see later is $\{0, 1\}$.

A function that has only one value in its range is called *constant*. The function $zero(x) = 0$ is constant. So is the function $\sin^2(x) + \cos^2(x)$, though you need to know the trigonometric identity to realize that.

I use x and n as the domain variables in the above examples. Just like f is only one possibility for the function name, we can use any letter or word for the domain variable. t is another common choice, especially when we think of the domain as real numbers representing time. Don't use the same letter or word as the name of the function!

Question 4.1.1

What would be your personal function if we defined $height(t)$ where t is your age in days?

Functions don't have to be defined in a simple way with only one formula. They may be stated for portions of their domains. For example,

$$f(x) = \begin{cases} -x^2 & \text{when } x < 0 \\ x^2 & \text{when } x \geq 0. \end{cases}$$

There are many kinds of functions. In section 2.7 we saw exponential, logarithmic, and quadratic functions when we talked about growth. Polynomials define functions, as do sine and cosine among the trigonometric functions. Of special importance are the linear functions, which are essential to quantum computing.

A real-valued function f on the real numbers is called *linear* if the following conditions hold:

- if x and y are real numbers then $f(x+y) = f(x) + f(y)$, and
- if a is a real number then $f(ax) = af(x)$.

This implies that $f(ax + by) = af(x) + bf(y)$ for real numbers a, b, x, and y.

The only linear functions that map real numbers to real numbers look like $f(x) = cx$, for c a fixed value. We call such a c a *constant*. For example, $f(x) = 7x$ or $f(x) = -x$. When we consider functions applied to collections that have more structure, the linear functions will be significantly less trivial.

What are the real-valued linear functions on the real numbers such that $|f(x)| = |x|$ for any x? Since $f(x) = cx$, then

$$|f(x)| = |cx| = |c| \times |x| = |x|$$

So $|c| = 1$ and $c = 1$ or $c = -1$. When $c = 1$ we have *identity function*: $\text{id}(x) = x$.

Can you reverse the effect of a function? If $f(x) = 3x - 1$ then if we apply $g(x) = \frac{1}{3}x + \frac{1}{3}$ we get

$$
\begin{aligned}
g(f(x)) &= \frac{1}{3}f(x) + \frac{1}{3} \\
&= \frac{1}{3}(3x - 1) + \frac{1}{3} \\
&= x - \frac{1}{3} + \frac{1}{3} \\
&= x = \text{id}(x)
\end{aligned}
$$

g is the inverse of f because $g(f(x)) = x$. $\text{id}(x)$ is its own inverse.

Question 4.1.2

Does $f(x) = 5$ have an inverse?

Question 4.1.3

What conditions do you need to impose on the domain of $f(x) = x^2$ so that the positive $\sqrt{x}$ is the inverse?

If f and g are functions such that $g(f(x)) = x$ and $f(g(y)) = y$ for all x and y in the domains of f and g, respectively, then f and g are *invertible* and one is the inverse of the other.

4.2 The real plane

When we were building up the structure of the integers, we showed the traditional number line

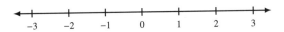

with the negative integers to the left of 0 and the positive ones to the right. Really, though, this was just part of the real number line

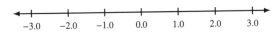

This is one-dimensional in that we need only one value, or coordinate, to locate a point uniquely on the line. For a real number x, we represent the point on the line by (x). For example, the point (-2.6) is between the markings -3 and -2. We use or omit the parentheses when it is clear from context whether we are referring to the point or the real number that gives its relative position from 0.

I drop the decimal points on the labels now that it is clear we have real numbers.

4.2.1 Moving to two dimensions

Now suppose the number line sits in two dimensions so that we extend upwards and downwards.

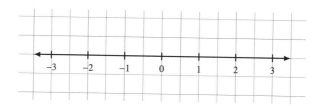

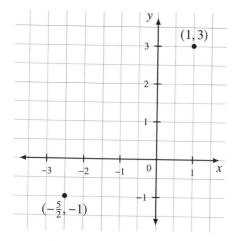

Since we need two coordinates for two dimensions, we position points with two real number x and y coordinates and label the point (x, y). These are called *Cartesian coordinates* after the mathematician René Descartes.

We positioned the point $(1, 3)$ in the upper right side of the graph and $(-\frac{5}{2}, -1)$ in the lower left. The horizontal line is called the x axis and the vertical line is the y axis. Cartesian coordinates are also called *rectangular coordinates*.

It may help you visualize this better if you think of the axes and points laid out on a table in front of you. This is the graph of the *real plane*, denoted $\mathbb{R}^2$.

One last bit of information to help us navigate: the axes divide the plane into four areas, or *quadrants*. The first quadrant, the one in the upper right, has both x and y positive.

Moving counterclockwise, the second quadrant has negative x and positive y. The third and fourth quadrants have $x < 0$ and $y < 0$, and $x > 0$ and $y < 0$, respectively.

Portrait of René Descartes (1596-1650) by Franz Hals. Painting is in the public domain. ⓟⓓ

4.2.2 Distance and length

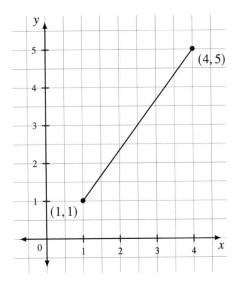

Given two points in the real plane, how far apart are they?

What is the distance from $(1,1)$ to $(4,5)$? This is equivalent to asking for the length of the line segment between the two points.

Are we talking about miles, kilometers, feet, centimeters, or what? Distance is usually mentioned in some particular unit. In math, we often don't say which unless we are translating from a particular problem. We refer to "units" though we might note "x is measured in meters." Unless necessary, we omit the particular kind of units.

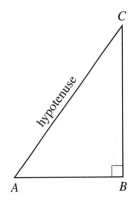

The Pythagorean theorem states that in a right triangle, the square of the length of one side plus the square of the length of the other is equal to the square of the length of the hypotenuse.

We labeled the points on the corners of the triangle A, B, and C. One side of the triangle is the line segment $\overline{AB}$, the other is $\overline{BC}$, and the hypotenuse is $\overline{AC}$.

If we use the notation $\left|\overline{AB}\right|$ to mean the length of the line segment $\overline{AB}$, the theorem tells us

$$\left|\overline{AC}\right|^2 = \left|\overline{AB}\right|^2 + \left|\overline{BC}\right|^2$$

or

$$\left|\overline{AC}\right| = \sqrt{\left|\overline{AB}\right|^2 + \left|\overline{BC}\right|^2}$$

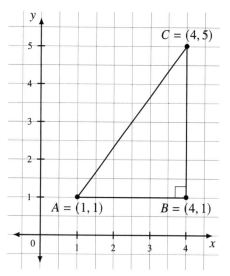

Expanding the example above, we have the graph on the left.

Let $A = (1,1)$, $B = (4,1)$, and $C = (4,5)$. By counting units on the axes we have $\left|\overline{AB}\right| = 3$ and $\left|\overline{BC}\right| = 4$. This means

$$\left|\overline{AC}\right|^2 = 3^2 + 4^2 = 25$$

and so

$$\left|\overline{AC}\right| = \sqrt{25} = 5.$$

"Counting units" along the x axis means to take the absolute value of the difference of the x coordinates. The similar statement holds for the y axis.

It's rare for a triangle to have all its sides being natural numbers, but this is an example of a $3 : 4 : 5$ triangle where things work out nicely. If you multiply or divide each side by the same natural number, the relationship still holds.

Question 4.2.1

A group of three natural numbers $n : m : p$ is called a Pythagorean triple if $n^2 + m^2 = p^2$. Find two more such triples that are not simple multiples of $3 : 4 : 5$.

If the two sides each had length 1, the length of the hypotenuse would be $\sqrt{2}$.

Given two points in the real plane with coordinates (a, b) and (c, d), the *distance* between the two points is

$$\sqrt{(a - c)^2 + (b - d)^2}.$$

In particular, if the second point is the origin $(0, 0)$, the distance is

$$\sqrt{a^2 + b^2}.$$

4.2.3 Geometric figures in the real plane

Before we look at three dimensions and qubits, you must get comfortable with finding your way around $\mathbb{R}^2$ with the standard geometric figures like lines and circles. Let's look at them and review general plotting of functions such as exponentials and logarithms.

Lines

> Given two different points with coordinates (a, b) and (c, d), there is only one line you can draw between those points. The essential notion of a line is that when you move from one point to another, the ratio of the change in distance between the y coordinates and the change in distance between the x coordinates is *constant*. This ratio is called the *slope* of the line.

If $(1, 1)$ is a point on the line and the slope is 2, for every unit we move in the x direction, we change twice that much in the y direction. Move 1 unit in x (that is, to the right), move 2 units in the y (up). Move -3 unit in x (that is, to the left), move -6 in the y (down).

For (x, y) to be on a line with slope m and given point (c, d), it must obey this equation

$$\frac{y - d}{x - c} = m$$

How do we find m? If (a, b) is another given point on the line, it must also obey

$$\frac{b - d}{a - c} = m.$$

Example: find the equation of the line that passes between the points $(-1, -2)$ and $(2, 3)$.

We begin by calculating the slope

$$m = \frac{-2 - 3}{-1 - 2} = \frac{-5}{-3} = \frac{5}{3}.$$

Then we have

$$\frac{y - -2}{x - -1} = \frac{y + 2}{x + 1} = \frac{5}{3}.$$

We rewrite this equation so we can compute y if we are given x:

$$y = \frac{5}{3}(x+1) - 2 = \frac{5}{3}x - \frac{1}{3}$$

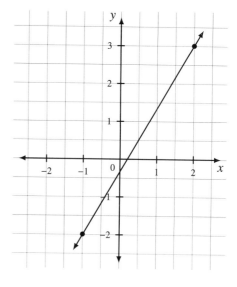

This is called the *slope-intercept* form $y = mx + b$. The slope m is the number in front of x, $\frac{5}{3}$. When $x = 0$, the place where the line crosses the y axis is $b = -\frac{1}{3}$. b is called the y-intercept.

This expresses y as a function of x. By the way, no one seems to know exactly why m is the standard variable name used for slope.

Plotting functions

When we have a real-valued function on all or part of $\mathbb{R}$, we can plot it as we just saw for lines where $y = f(x)$. If you were to do this by hand, you would choose enough values of x, compute $f(x)$ for each, plot $(x, f(x))$ for each, and then connect the points. This may or may not be a good rendition of what it really looks like.

Let's take $y = (x, f(x)) = \frac{1}{3}x^2$. We let x have the integer values between -2 and 2.

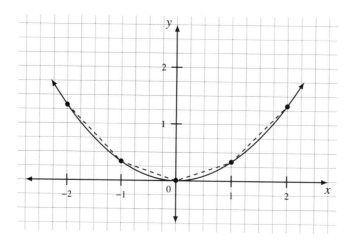

The outer smooth curve is the correct plot while the inner dashed one has straight lines connecting the few points we calculated.

It's also possible to make mistakes while plotting because you inadvertently include a value that is not in the domain. Let $y = f(x) = \left|\frac{1}{x}\right|$ when $x \neq 0$. If we plot for $x = -2, -1, 2, 1$ and are careless we might see the dashed line plot below. The function is not defined at 0 and the plot looks more like the smooth solid curve.

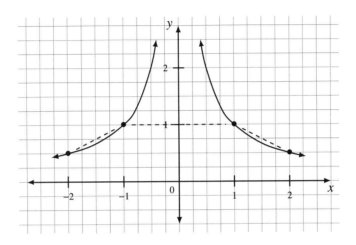

We also say that $f(x) = \left|\frac{1}{x}\right|$ is not *continuous* at $x = 0$.

Functions can be more sophisticated than mapping a number to a number. Let's instead think of mapping a number to a point in the plane. If the input variable is t, we can think of the formula for the point as $(g(t), h(t))$. This includes the cases above because we can define $g(t)$ to be t. This is called a *parametrized function*.

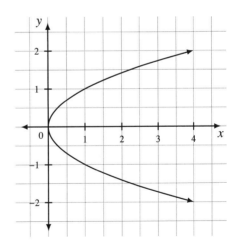

That is, rather than considering the function t^2 from $\mathbb{R}$ to $\mathbb{R}$ and then plotting the points (t, t^2), we instead directly plot the function $t \mapsto (t, t^2)$ from $\mathbb{R}$ to $\mathbb{R}^2$. The plot of the function $t \mapsto (t^2, t)$ looks like a parabola oriented horizontally.

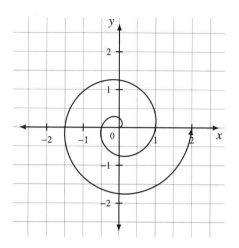

Plotting parametrized functions can generate many beautiful graphs such as spirals and multi-petal flowers. On the left is the plot of

$$t \mapsto \left(\frac{t \cos(t)}{2\pi}, \frac{t \sin(t)}{2\pi} \right)$$

for $0 \le t \le 4\pi$.

Circles

A circle is defined by choosing one point as the center and then taking all points at a given fixed positive distance from that center. This distance is called the *radius* of the circle.

Let r be a positive real number that we use as the radius of the circle. If the coordinates of the center are (c, d) and a point (x, y) is distance r from the center, it satisfies the formula

$$\sqrt{(x - c)^2 + (y - d)^2} = r$$

or

$$(x - c)^2 + (y - d)^2 = r^2.$$

When $r = 1$ this is a *unit circle*.

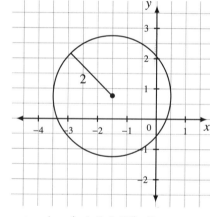

To the above right is the plot of the circle of radius 2 centered at $(-1.5, 0.75)$. Its equation is $(x + 1.5)^2 + (y - 0.75)^2 = r^2$.

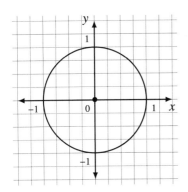

This is the plot of the unit circle of radius 1 centered at $(0,0)$, the origin: Its equation is $x^2 + y^2 = 1$.

Question 4.2.2

How would you write y as a function of x in the equation of the unit circle? What are the domain and range of the function?

4.2.4 Exponentials and logarithms

An exponential function has the form

$$y = f(x) = c\,a^x$$

for constant real numbers $a > 0$, $a \neq 1$, and $c \neq 0$. Common choices for a are 2, 10, and e. For real x, $f(x)$ is positive or negative if c is. $f(0) = c$.

When c is positive and $0 < a < 1$, we have exponential decay.

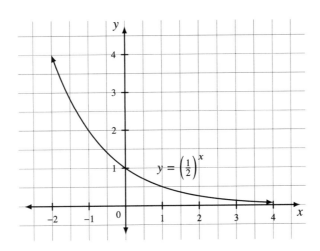

$$y = \left(\tfrac{1}{2}\right)^x$$

Radioactive decay follows this pattern.

If $a > 1$ we get exponential growth. Here are two examples of growth:

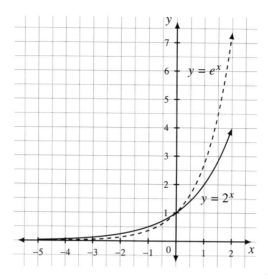

Moore's Law, which was always more of an observation than a legally or physically binding statement, said that roughly every two years the computational power of a computer processor would double, its size would halve, and its energy requirements would also be cut in half. While this no longer seems to be the case, it was an example of exponential growth of power and decay of size and energy requirements. [5]

Compare the growth the exponential $f(x) = 10^x$ and linear $g(x) = x$ functions.

$$
\begin{aligned}
f(0) &= 1 & g(0) &= 0 \\
f(1) &= 10 & g(1) &= 1 \\
f(6) &= 1000000 & g(6) &= 6 \\
f(15) &= 1000000000000000 & g(15) &= 15
\end{aligned}
$$

Generally, exponential growth is something to be controlled or avoided in computation unless it somehow involves your money.

If exponential functions can show fast growth, logarithmic ones do the opposite. A *logarithm* is an inverse function of an exponential one.

The function $\log_2(x)$ returns the answer to the question "to what power should I raise 2 to get x?" It is only defined for positive real numbers x.

$$\log_2(1) = 0 \text{ because } 2^0 = 1$$
$$\log_2(2) = 1 \text{ because } 2^1 = 2$$
$$\log_2(4) = 2 \text{ because } 2^2 = 4$$
$$\log_2(64) = 6 \text{ because } 2^6 = 64$$
$$\log_2(2048) = 11 \text{ because } 2^{11} = 2048$$

If you work with computers long enough, you learn your powers of 2. $\log_2$ is called the *binary logarithm*.

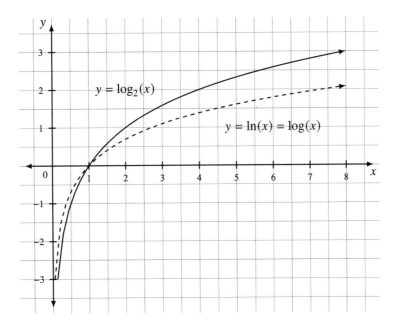

The "inverse" behavior in the above means that $x = 2^{\log_2(x)}$ and $x = \log_2(2^x)$.

The natural logarithm is the inverse function to exponentiation by e. As a function name and shorthand for "$\log_e$", some authors use "ln" and others use "log". I use the latter.

For any positive base a, real number t, and positive real numbers x and y,

$$\log_a (xy) = \log_a (x) + \log_a (y)$$

$$\log_a \left(\frac{x}{y}\right) = \log_a (x) - \log_a (y)$$

$$\log_a \left(x^t\right) = t \log_a (x)$$

We can convert from one logarithmic base to another. For another base b,

$$\log_a (x) = \frac{\log_b (x)}{\log_b (a)}.$$

To derive this, if we set $t = \log_a (x)$ then $a^t = x$. Apply $\log_b$ to both sides and solve for t.

Question 4.2.3

Work out the details of deriving this formula. Show your work.

4.3 Trigonometry

Trigonometry is the study of triangles, angles, and length. The Greek word *trigōnon* means "triangle" and *metron* means "measure." For the most part we restrict ourselves to triangles with hypotenuse equal to 1, and so within the unit circle.

4.3.1 The fundamental functions

Many people have heard that a circle has 360 degrees, also written 360°. Why 360? If you look around the web you'll find stories about ancient Mesopotamians, Egyptians, and base 60 number systems. Whatever the reason, 360 is a great number because it is divisible by so many other numbers like 2, 3, 4, 5, 6, 8, 10, 12, 15, and so on. That is, it's easy to work with portions of 360° that are nice round whole numbers.

Degrees don't really have a natural meaning in mathematics though. They are simply a convenient unit of measurement. Instead, we use *radians*.

A circle of radius r has circumference $2\pi r$. If we go half way around the circle we cover a distance of πr. One quarter of the way cover $\frac{\pi}{2} r$.

When $r = 1$ we use fractions of 2π to determine how far around the circle we have gone. This is the *radian measure* of the corresponding angle. Half way around is π radians. If we go around three times we have 6π radians. This has real geometric meaning compared to degrees. When we move counterclockwise the radian measure is positive and it is negative when we move clockwise.

With either degrees or radians, we begin measuring the angle from the positive x axis.

Because we are dealing with a circle, an angle of $\frac{3\pi}{4}$ radians lands us on the same point as does one of $-\frac{5\pi}{4}$ radians.

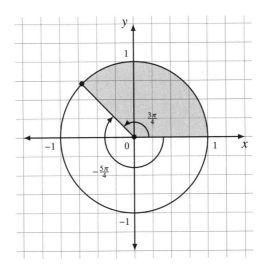

It's customary to use Greek letters as variable names when we work with radians or degrees. The two most commonly used are θ (theta) and φ (phi).

Using θ as the measurement of the angle, consider the plot where P is a point on the unit circle with $\theta = \frac{\pi}{3}$.

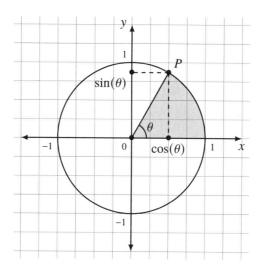

We define the function $\cos(\theta)$, the *cosine* of θ, to be the x coordinate of the point P. Similarly, the function $\sin(\theta)$, *sine* of θ is the y coordinate of P. Though we drew P in the first quadrant, we extend the definition to all points on the unit circle in all quadrants and on the axes.

The cosine is just an x coordinate and the sine is just a y coordinate. As functions they have many elegant properties, but it all comes down to the geometry of a radius 1 circle around the origin.

Because P is on the unit circle, it's distance from the origin $(0,0)$ is 1. This means

$$1 = \sqrt{(\cos(\theta) - 0)^2 + (\sin(\theta) - 0)^2} = \sqrt{\cos^2(\theta) + \sin^2(\theta)}.$$

Squaring, we get

$$1 = \cos^2(\theta) + \sin^2(\theta).$$

This is the *fundamental identity of trigonometry.* It follows naturally from the definition of length and the Pythagorean theorem.

Instead of writing the longer $(\cos(\theta))^2$ in the above, I used the common shorthand $\cos^2(\theta)$.

If θ is negative, we consider $|\theta|$ but rotated clockwise from the positive x axis. Look at the above plot to convince yourself that $\cos(\theta) = \cos(-\theta)$ and $\sin(\theta) = -\sin(-\theta)$. While this is by no means a rigorous mathematical proof, developing your geometric intuition is more important here.

A function f like cos that has the property $f(x) = f(-x)$ is called an *even* function. If $f(-x) = -f(x)$ then the function is *odd.* sin is an odd function.

θ	$\cos(\theta)$	$\sin(\theta)$	$\tan(\theta)$	$\sec(\theta)$	$\csc(\theta)$	$\cot(\theta)$
0	1	0	0	1	undefined	undefined
$\frac{\pi}{6}$	$\frac{\sqrt{3}}{2}$	$\frac{1}{2}$	$\frac{1}{\sqrt{3}}=\frac{\sqrt{3}}{3}$	$\frac{2}{\sqrt{3}}=\frac{2\sqrt{3}}{3}$	2	$\sqrt{3}$
$\frac{\pi}{4}$	$\frac{\sqrt{2}}{2}$	$\frac{\sqrt{2}}{2}$	1	$\frac{2}{\sqrt{2}}=\sqrt{2}$	$\frac{2}{\sqrt{2}}=\sqrt{2}$	1
$\frac{\pi}{3}$	$\frac{1}{2}$	$\frac{\sqrt{3}}{2}$	$\sqrt{3}$	2	$\frac{2}{\sqrt{3}}=\frac{2\sqrt{3}}{3}$	$\frac{1}{\sqrt{3}}=\frac{\sqrt{3}}{3}$
$\frac{\pi}{2}$	0	1	undefined	undefined	1	0

Figure 4.1: Common values of the trigonometric functions in the first quadrant

Another essential trigonometric function is the tangent, written $\tan(\theta)$ and defined by

$$\tan(\theta) = \frac{\sin(\theta)}{\cos(\theta)} \text{ for } \cos(\theta) \neq 0.$$

When does $\cos(\theta) = 0$? This happens whenever the x coordinate is 0 and so θ corresponds to a point on the y axis. These values are $\theta = \frac{\pi}{2}, \frac{3\pi}{2}, -\frac{\pi}{2}, -\frac{3\pi}{2}$, and integer multiples of each of these by 2π. These multiples correspond to going around the circle more than one time in either direction

The tangent is the slope of the line segment from the origin $(0,0)$ to P.

The three remaining standard trigonometric functions are the secant, cosecant, and cotangent. These are defined, respectively, by

$$\sec(\theta) = \frac{1}{\cos(\theta)} = (\cos(\theta))^{-1} \text{ for } \cos(\theta) \neq 0$$

$$\csc(\theta) = \frac{1}{\sin(\theta)} = (\sin(\theta))^{-1} \text{ for } \sin(\theta) \neq 0$$

$$\cot(\theta) = \frac{\cos(\theta)}{\sin(\theta)} = (\tan(\theta))^{-1} \text{ for } \sin(\theta) \neq 0$$

These are connected to the fundamental identity by dividing by $\cos^2(\theta)$ or $\sin^2(\theta)$.

$$1 = \cos^2(\theta) + \sin^2(\theta)$$

$$\frac{1}{\cos^2(\theta)} = \frac{\cos^2(\theta)}{\cos^2(\theta)} + \frac{\sin^2(\theta)}{\cos^2(\theta)} \Rightarrow \sec^2(\theta) = 1 + \tan^2(\theta)$$

$$\frac{1}{\sin^2(\theta)} = \frac{\cos^2(\theta)}{\sin^2(\theta)} + \frac{\sin^2(\theta)}{\sin^2(\theta)} \Rightarrow \csc^2(\theta) = \cot^2(\theta) + 1$$

These identities hold when the denominators are not 0.

4.3.2 The inverse functions

Do the trigonometric functions have inverses? Yes, but we need to be very careful at stating the domains and ranges. We also have to decide what to call them. Three notations are common for the inverse tangent:

$$\tan^{-1}(x) \qquad \arctan(x) \qquad \mathrm{atan}(x)$$

The first is consistent with calling f^{-1} the inverse of the function f, but is confusing because of our shorthand with forms like $\tan^2(x)$ meaning $\tan(x)^2$.

The third version is common in programming languages like Python, Swift, C++, and Java. We use the middle one as it is traditional and relates to the arc that spans the angle in which we are interested.

The sine function gives its name to the archetypal example of a wave, a *sinusoidal wave*.

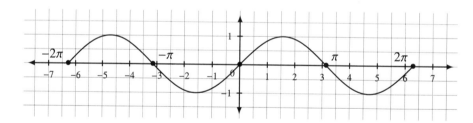

In the portion of the graph I've shown here, the plot crosses the x axis five times. On this graph $\sin(x) = 0$ when $x = -2\pi, -\pi, 0, \pi$, and 2π. So what is $\arcsin(0)$? We need to choose a range that ensures arcsin is single-valued and therefore a function.

arcsin(x) is a function with domain $-1 \leq x \leq 1$ and range $\frac{\pi}{2} \leq \text{arcsin}(x) \leq \frac{\pi}{2}$.

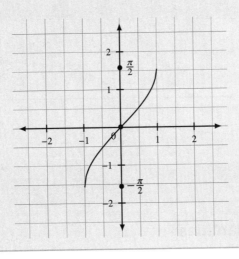

The inverse cosine arccos has a different range.

arccos(x) is a function with domain $-1 \leq x \leq 1$ and range $0 \leq \text{arccos}(x) \leq \pi$.

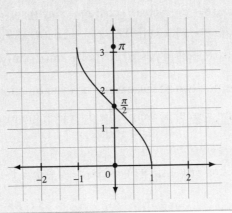

The inverse tangent arctan requires closer consideration because in the definition

$$\tan(x) = \frac{\sin(x)}{\cos(x)}$$

the denominator can be 0 when $\cos(x)$ is. The values of x we need to look out for are $\frac{\pi}{2}$ and $-\frac{\pi}{2}$.

arctan(x) is a function with domain all of $\mathbb{R}$ and range $\frac{\pi}{2} < $ arctan(x) $ < \frac{\pi}{2}$.

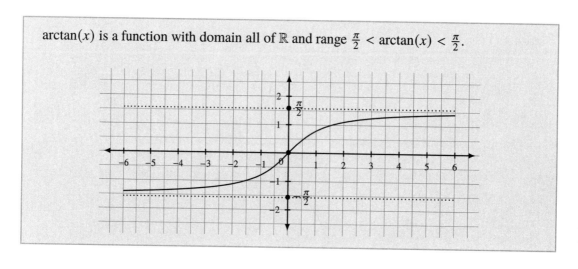

4.3.3 Additional identities

There are several other trigonometric formulas which are useful. The sum and difference formulas express sin and cos when you have addition and subtraction of angles in the arguments.

$$\sin(\theta + \varphi) = \sin(\theta)\cos(\varphi) + \cos(\theta)\sin(\varphi)$$
$$\cos(\theta + \varphi) = \cos(\theta)\cos(\varphi) - \sin(\theta)\sin(\varphi)$$
$$\sin(\theta - \varphi) = \sin(\theta)\cos(\varphi) - \cos(\theta)\sin(\varphi)$$
$$\cos(\theta - \varphi) = \cos(\theta)\cos(\varphi) + \sin(\theta)\sin(\varphi)$$

The last two follow from the first two because cos is an even function and sin is an odd function.

If we let $\theta = \varphi$ in the sum formulas, we get the double angle formulas.

$$\sin(2\theta) = 2\sin(\theta)\cos(\theta)$$
$$\cos(2\theta) = \cos^2(\theta) - \sin^2(\theta) = 2\cos^2(\theta) - 1 = 1 - 2\sin^2(\theta)$$

4.4   From Cartesian to polar coordinates

Each point on the unit circle is uniquely determined by an angle φ given in radians such that $0 \leq \varphi < 2\pi$.

Even though a point on the unit circle is in $\mathbb{R}^2$, which is two-dimensional, it takes only one value, φ, to determine it. We lost the need for a second value by insisting that the point has distance 1 from the origin.

More generally, let $P = (a, b)$ be a non-zero point (that is, a point which is not the origin) in $\mathbb{R}^2$. Let $r = \sqrt{a^2 + b^2}$ be the distance from P to the origin. Then the point

$$Q = \left(\frac{a}{r}, \frac{b}{r} \right)$$

is on the unit circle. There is a unique angle φ such that $0 \leq \varphi < 2\pi$ that corresponds to Q. With r we can uniquely identify

$$P = (r \cos(\varphi), r \sin(\varphi)).$$

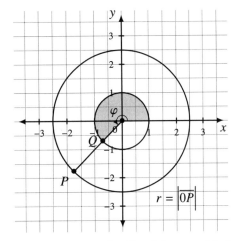

(r, φ) are called the *polar coordinates* of P. You may sometimes see the Greek letter ρ (rho) used instead of r.

Each non-zero point in $\mathbb{R}^2$ is uniquely determined by an angle φ given in radians such that $0 \leq \varphi < 2\pi$, and a positive real number r.

To uniquely identify P in this plot, $r = 2.5$ and $\varphi = \frac{5\pi}{4}$.

4.5   The complex "plane"

In the last chapter we discussed the algebraic properties of $\mathbb{C}$, the complex numbers. We return to them again here to look at their geometry. For any point (a, b) in the real plane, consider the corresponding complex number $a + bi$.

In the graph of the complex numbers, the horizontal axis is the real part of the complex variable z and the vertical axis is the imaginary part. These replace the x and y axes, respectively.

The plot to the right shows several complex values. Despite appearances and some authors' use of the terminology, that is not a *complex* plane. A plane has two dimensions. We visualized $\mathbb{C}$, which is one-dimensional, in the two-dimensional *real* plane. We return to these issues about dimensions with respect to a field in the next chapter when we look at vector spaces.

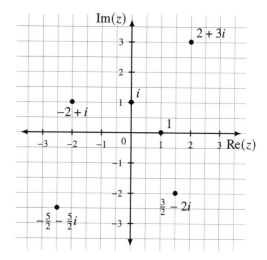

Conjugation

Conjugation reflects a complex number across the horizontal $\mathrm{Re}(z)$ axis. If the number has 0 imaginary part, it is already on the $\mathrm{Re}(z)$ axis and so nothing happens.

If we conjugate a complex number z and then conjugate the result, we end up at z, the same place at which we started. Conjugation is its own inverse.

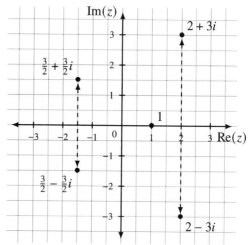

Question 4.5.1

What does conjugation do to 0? i?

Polar coordinates

As we saw above for the Cartesian plane $\mathbb{R}^2$, we can use polar coordinates for a point and so also for any non-zero complex number.

Each non-zero point $z = a + bi$ in $\mathbb{C}$ is uniquely determined by an angle φ given in radians such that $0 \leq \varphi < 2\pi$, and a positive real number $r = |z|$. The arg function associates φ with z: $\varphi = \arg(z)$.

$$z = r\cos(\varphi) + r\sin(\varphi)\,i$$
$$= |z|\cos(\arg(z)) + |z|\sin(\arg(z))\,i$$

The angle φ is called the *phase* of z and $|z|$ is its *magnitude*.

The term "phase" for the angle φ of a complex number is used in physics more than it is in mathematics. I introduce it here because we need it when we describe the representation of a qubit.

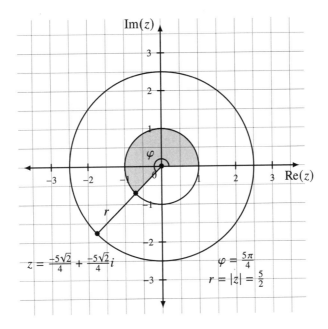

$$z = \frac{-5\sqrt{2}}{4} + \frac{-5\sqrt{2}}{4}i$$

$$\varphi = \frac{5\pi}{4}$$
$$r = |z| = \frac{5}{2}$$

Any φ which is a non-zero integer multiple of 2π plus φ lands us on the same point, but we normalize our choice to be between 0 and 2π.

Instead of using the (r, φ) notation or the longer $r\cos(\varphi) + r\sin(\varphi)\,i$ expression using polar coordinates, we adopt an exponential form incorporating both r and φ via Euler's formula.

Euler's formula

Leonhard Euler was a prolific eighteenth century mathematician who made many contributions in multiple fields of science. He was the first to prove the connection between the extension of the exponential function to the complex numbers and the $a + bi$ notation involving polar coordinates. That is, he showed

$$re^{\varphi i} = r\cos(\varphi) + r\sin(\varphi)i.$$

This is the notation we use for the polar form of a complex number.

> **To learn more**
>
> The proof of this is beyond the scope of this book. It falls with the branch of mathematics called *complex analysis*, which you may think of as the extension of calculus from the real numbers to the complex ones. [1] [3]

If z_1 and z_2 are non-zero complex numbers then $|z_1 z_2| = |z_1| \times |z_2|$ and $\arg(z_1 z_2) = \arg(z_1) + \arg(z_2)$. Expressed in long form, if

$$z_1 = r_1\cos(\varphi_1) + r_1\sin(\varphi_1)i$$
$$z_2 = r_2\cos(\varphi_2) + r_2\sin(\varphi_2)i$$

then

$$z_1 z_2 = r_1 r_2\cos(\varphi_1 + \varphi_2) + r_1 r_2\sin(\varphi_1 + \varphi_2)i$$
$$z_1/z_2 = \frac{r_1}{r_2}\cos(\varphi_1 - \varphi_2) + \frac{r_1}{r_2}\sin(\varphi_1 - \varphi_2)i$$

> Using Euler's formula, if $z_1 = r_1 e^{\varphi_1 i}$ and $z_2 = r_2 e^{\varphi_2 i}$ then
>
> $$z_1 z_2 = r_1 r_2 e^{(\varphi_1 + \varphi_2)i}$$
> $$z_1/z_2 = \frac{r_1}{r_2} e^{(\varphi_1 - \varphi_2)i}$$

This is considerably simpler to use when we are multiplying or dividing complex numbers, but much less so when we are adding or subtracting. It also means that multiplying a complex number by $e^{\varphi i}$ translates geometrically to rotating the complex number φ radians around 0. If φ is positive, the rotation is counterclockwise; for negative values it is clockwise.

If $z = re^{\varphi i}$ then $\bar{z} = re^{-\varphi i}$.

Whether you think of a complex number as a real and imaginary part or as a phase and a magnitude, it holds two independent pieces of data represented as real numbers. This is very powerful and is one of the reasons why complex numbers show up in some unexpected places in physics and engineering.

4.6 Real three dimensions

When plotting in three dimensions, we need either three Cartesian coordinates (x_0, y_0, z_0) or a radius r and two angles φ and θ.

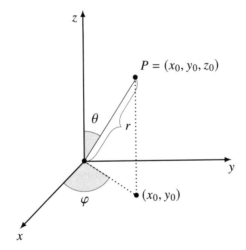

The radius $r = |P| = \sqrt{x_0^2 + y_0^2 + z_0^2}$. φ is the angle from the positive x axis to the dotted line from $(0,0)$ to the projection (x_0, y_0) of P into the xy-plane. θ is the angle from the positive z axis to the line segment $\overline{OP}$.

That's a lot to absorb, but it builds up systematically from what we saw in $\mathbb{R}^2$. When $r = 1$ we
get the *unit sphere* in $\mathbb{R}^3$. It's the set of all points
(x_0, y_0, z_0) in $\mathbb{R}^3$ where $x_0{}^2 + y_0{}^2 + z_0{}^2 = 1$.
The *unit ball* is the set of all points where
$x_0{}^2 + y_0{}^2 + z_0{}^2 \leq 1$.

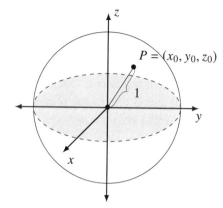

We return to this graphic frequently when
we consider the Bloch sphere representation of
a qubit.

If we were in four real dimensions we would need 4 coordinates (x, y, z, w). The *unit
hypersphere* is the set of all points such that $1 = x^2 + y^2 + z^2 + w^2$.

4.7 Summary

After handling algebra in the last chapter, we tackled geometry here. The concept of "function"
is core to most of mathematics and its application areas like physics. Functions allow us to
connect one or more inputs with some kind of useful output. Plotting functions is a good way
to visualize their behavior.

Two- and three-dimensional spaces are familiar to us and we learned and reviewed the common tools to allow us to effectively use them. Trigonometry demonstrates beautiful connections
between algebra and geometry and falls out naturally from relationships like the Pythagorean
theorem.

The complex "plane" is like the real plane $\mathbb{R}^2$ but the algebra and geometry gives more
structure than points alone can provide. Euler's formula nicely ties together complex numbers
and trigonometry in an easy-to-use notation that is the basis for how we define many quantum
operations in chapter 7 and chapter 8.

To learn more

Many readers were introduced to geometry in high school or its equivalent via theorems
and proofs. If that part of the subject interests you, more advanced treatments explore
axiomatic and geometric-algebraic approaches. [2] [4]

References

[1]   L. V. Ahlfors. *Complex Analysis. An introduction to the theory of analytic functions of one complex variable*. 3rd ed. International Series in Pure and Applied Mathematics 7. McGraw-Hill Education, 1979.

[2]   R. Artzy. *Linear Geometry*. Addison-Wesley series in mathematics. Addison-Wesley Publishing Company, 1974.

[3]   J. B. Conway. *Functions of One Complex Variable*. 2nd ed. Graduate Texts in Mathematics 11. Springer-Verlag, 1978.

[4]   H. S. M. Coxeter. *Introduction to geometry*. 2nd ed. Wiley classics library. John Wiley and Sons, 1999.

[5]   Gordon E. Moore. "Cramming more components onto integrated circuits". In: *Electronics* 38.8 (1965), p. 114.

5

Dimensions

... from a purely mathematical point of view
it's just as easy to think in 11 dimensions,
as it is to think in three or four.

Stephen Hawking [1]

We are familiar with many properties of objects like lines and circles in two dimensions, and cubes and spheres in three dimensions. If I ask you how long something is, you might take out a ruler or a tape measure. When you take a photo, you rotate your camera or phone in three dimensions without thinking too much about it.

Alas, there is math behind all these actions. The notion of something existing in one or more dimensions, indeed even the idea of what a dimension is, must be made more formal if we are to perform calculations. This is the concept of *vector spaces*. The study of what they are and what you can do with them is called *linear algebra*.

Linear algebra is essential to pure and applied mathematics, physics, engineering, and the parts of computer science and software engineering that deal with graphics. It's also a valuable tool in many parts of AI like machine learning.

Although many books have been written about linear algebra I'm only going to cover the bare minimum we need for quantum computing, plus 10%. Why the extra 10%? It's because some properties and connections are just too interesting to ignore! They also give you jumping off points to explore the topics in greater detail.

Topics covered in this chapter

5.1 $\mathbb{R}^2$ and $\mathbb{C}^1$

We looked earlier at the real plane as a set of standard Cartesian coordinate pairs (x, y) with x and y in $\mathbb{R}$ representing points we can plot. Now we give these pairs an algebraic structure so that if u and v are in $\mathbb{R}^2$ then so is $u + v$. Also, if r is in $\mathbb{R}$, then rv is in $\mathbb{R}^2$ as well. We carry out the addition coordinate by coordinate. The multiplication by r, called *scalar multiplication*, is also done that way.

If $u = (u_1, u_2)$ and $v = (v_1, v_2)$,

$$u + v = (u_1 + v_1, u_2 + v_2)$$
$$ru = (ru_1, ru_2)$$

Using the origin $O = (0,0)$ as the identity element, $\mathbb{R}^2$ is a commutative group under addition. $\mathbb{R}^2$ is a two-dimensional *vector space* over $\mathbb{R}$. This is possible because $\mathbb{R}$ is a field.

Rather than considering them as pairs or points, we now call **u** and **v** *vectors*. I use **bold** to indicate a variable or a "point" is a vector. When we plot a vector we draw it as an arrow from the origin $(0, 0)$ to the point represented by the Cartesian coordinates.

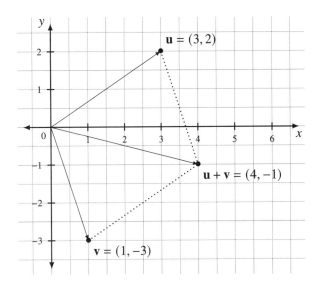

There are two special vectors: $\mathbf{e_1}$, associated with the Cartesian coordinates $(1, 0)$, and $\mathbf{e_2}$, associated with $(0, 1)$. Any vector associated with the coordinates (a, b) may be written as $a\,\mathbf{e_1} + b\,\mathbf{e_2}$. If this sum equals $\mathbf{O} = (0, 0)$ then a and b must be zero themselves, meaning $\mathbf{e_1}$ and $\mathbf{e_2}$ are *linearly independent*.

Because they have these properties, $\mathbf{e}_1$ and $\mathbf{e}_2$ are called *basis vectors*, and these two are the *standard basis vectors*. Together they constitute a *basis* for $\mathbb{R}^2$, the *standard basis*.

> The whole idea of Cartesian coordinates is that we are using $\mathbf{e}_1$ and $\mathbf{e}_2$. When we mention a point with coordinates (a, b), we really mean the point corresponding to the vector $a\,\mathbf{e}_1 + b\,\mathbf{e}_2$. When we refer to a point (a, b), assume we are using Cartesian coordinates unless stated otherwise.

Other pairs have this property as well. Let

$$\mathbf{h}_1 = \tfrac{\sqrt{2}}{2}\mathbf{e}_1 + \tfrac{\sqrt{2}}{2}\mathbf{e}_2 \quad \text{and} \quad \mathbf{h}_2 = \tfrac{\sqrt{2}}{2}\mathbf{e}_1 - \tfrac{\sqrt{2}}{2}\mathbf{e}_2$$

then

$$\mathbf{h}_1 + \mathbf{h}_2 = \sqrt{2}\,\mathbf{e}_1$$

and

$$\mathbf{h}_1 - \mathbf{h}_2 = \sqrt{2}\,\mathbf{e}_2.$$

Given the new vector

$$3\mathbf{e}_1 + 0\mathbf{e}_2 = \frac{3}{\sqrt{2}}\,(\mathbf{h}_1 + \mathbf{h}_2) + \frac{0}{\sqrt{2}}\,(\mathbf{h}_1 - \mathbf{h}_2)$$

$$= \frac{3\sqrt{2}}{2}\,(\mathbf{h}_1 + \mathbf{h}_2)$$

$$= \frac{3\sqrt{2}}{2}\mathbf{h}_1 + \frac{3\sqrt{2}}{2}\mathbf{h}_2.$$

This means the point with Cartesian coordinates $(3, 0)$ has the coordinates $\left(\frac{3\sqrt{2}}{2}, \frac{3\sqrt{2}}{2}\right)$ when using the basis $\mathbf{h}_1$ and $\mathbf{h}_2$.

> Coordinates are relative to the basis you are using. There are times when a change of basis makes the algebra and geometry less complex and easier to understand.

For $\mathbf{u} = \mathbf{e}_1 + 2\mathbf{e}_2$, $\mathbf{v} = 2\mathbf{e}_1 - 3\mathbf{e}_2$, and $r = -1$, we have $\mathbf{u} + \mathbf{v} = 3\mathbf{e}_1 - \mathbf{e}_2$ and $r\,\mathbf{u} = -\mathbf{e}_1 - 2\mathbf{e}_2$. Here's the plot.

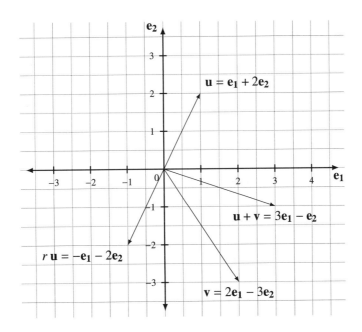

Observe that $\mathbf{u}$ and $r\,\mathbf{u}$ fall on the same line and this means together they are not a basis: $\mathbf{O} = r\,\mathbf{u} - (r\,\mathbf{u})$.

The length of a vector $a\,\mathbf{e_1} + b\,\mathbf{e_2}$ in $\mathbb{R}^2$ is $\sqrt{a^2 + b^2}$. It is denoted $\|a\,\mathbf{e_1} + b\,\mathbf{e_2}\|$.

A vector of length 1 is a *unit vector*. $\mathbf{e_1}$ and $\mathbf{e_2}$ are unit basis vectors. So are $\mathbf{h_1}$ and $\mathbf{h_2}$.

For $\mathbb{R}^1$, $\mathbf{e_1}$, corresponding to (1) in its Cartesian coordinate, is the standard unit basis. For a vector $a\,\mathbf{e_1}$, its length is $\|a\,\mathbf{e_1}\| = |a|$. The $|a|$ is the usual absolute value for real numbers.

For $\mathbb{R}^3$, we use $\mathbf{e_1}$ corresponding to $(1,0,0)$ in Cartesian coordinates, $\mathbf{e_2}$ to $(0,1,0)$, and $\mathbf{e_3}$ to $(0,0,1)$. If $\mathbf{u} = u_1\mathbf{e_1} + u_2\mathbf{e_2} + u_3\mathbf{e_3}$ is in $\mathbb{R}^3$ then the length of $\mathbf{u}$ is

$$\|\mathbf{u}\| = \sqrt{u_1{}^2 + u_2{}^2 + u_3{}^2}$$

As vectors spaces over $\mathbb{R}$, $\mathbb{R}^1$, $\mathbb{R}^2$, and $\mathbb{R}^3$ have dimensions 1, 2, and 3 respectively.

If you are driving at 65 miles per hour or 100 kilometers per hour, those numbers are the *speed* at which you are traveling. If you then add in a direction, such as "I'm driving north at 60 kph," you are talking about *velocity*.

Velocity is a vector. If we represent north by $\mathbf{e_2}$ in $\mathbb{R}^2$, the above becomes $60\,\mathbf{e_2}$. Velocity does not say where you are, just how fast you are going in some direction.

Momentum is also a vector and is equal to the mass of an object times its velocity. At the quantum particle level, the *Heisenberg Uncertainty Principle* says that the more precisely we know either the particle's position or momentum, the less precisely we can know the other.

$\mathbb{C}$ is a two-dimensional vector space over $\mathbb{R}$ with standard unit basis vectors $\mathbf{e_1} = \mathbf{1} = (1, 0)$ and $\mathbf{e_2} = \mathbf{i} = (0, 1)$. $\mathbb{C}^1$ is a one-dimensional vector space over $\mathbb{C}$ with standard unit basis vector $\mathbf{e_1} = (1)$.

Since we are considering multiple fields with $\mathbb{R}$ contained in $\mathbb{C}$, let's fine tune how we define a linear function beyond what we said in section 4.1.

A complex-valued function f on $\mathbb{C}$ is called *linear* (more precisely, *complex linear*) if the following conditions hold:

- if z and w are in $\mathbb{C}$ then $f(z + w) = f(z) + f(w)$, and
- if a is in $\mathbb{C}$ then $f(az) = af(z)$.

This implies $f(az + bw) = af(z) + bf(w)$ for $a, b, z,$ and w in $\mathbb{C}$.

A complex-valued function f on $\mathbb{C}$ is called *real linear* if the following conditions hold:

- if z and w are in $\mathbb{C}$ then $f(z + w) = f(z) + f(w)$, and
- if a is in $\mathbb{R}$ then $f(az) = af(z)$.

The first definition is useful when we think of $\mathbb{C} = \mathbb{C}^1$ as a one-dimensional vector space over itself. The second applies when we consider it to be a two-dimensional vector space over $\mathbb{R}$. We are primarily be concerned with vector spaces over $\mathbb{C}$ for quantum computing.

If f is a function on $\mathbb{C}$ such that $|f(x)| = |x|$, then f preserves length. Such functions are called *isometries*. If f is also linear then it is called a linear isometry.

The word stem *iso* comes from the Greek and means "equal" or "identical." "Isometry" then means something that has equal measure before and after being applied.

Isothermal means that the temperature remains the same. An *isobaric* curve on a weather map connects places that have the same air pressure.

An isometry maps complex values on the unit circle to the unit circle. Any function $f(z) = e^{\theta i} z$ for $0 \le \theta < 2\pi$ is a linear isometry. Note $f(z)\overline{f(z)} = 1$.

What about conjugation? If $z = a + bi$ then

$$|z| = \sqrt{a^2 + b^2} = |\overline{z}|.$$

So conjugation is an isometry but is it linear? For $f(z) = \overline{z}$ and a in $\mathbb{C}$,

$$f(az) = \overline{az} = \overline{a}\,\overline{z} = \overline{a}f(z).$$

This only happens when a is real. So conjugation is not complex linear, but it's close! In fact, it is real linear. As we look at higher dimensional complex vector spaces, we need to take conjugation into account frequently.

With this grounding in low dimensional real vector spaces and the algebraic and geometric relationships between $\mathbb{R}^2$ and $\mathbb{C}$, we're ready to tackle generalizations of what we've been discussing and higher dimensions.

This is necessary because in quantum computing the vector spaces get very large. For example, IBM first put a 5-qubit quantum computer on the cloud in 2016. The mathematics for even such a small system requires 32 complex dimensions! A 20-qubit system involves $2^{20} = 1,048,576$ complex dimensions.

The terminology and notation we develop simplifies how we understand and manipulate such large vector spaces. Although we used coordinates like (x, y, z) in three dimensions, we do **not** have to resort to anything like

$$(x_1, x_2, x_3, x_4, x_5, x_6, x_7, x_8, x_9, x_{10}, x_{11}, x_{12}, x_{13}, x_{14}, x_{15}, x_{16}, x_{17},$$
$$x_{18}, x_{19}, x_{20}, x_{21}, x_{22}, x_{23}, x_{24}, x_{25}, x_{26}, x_{27}, x_{28}, x_{29}, x_{30}, x_{31}, x_{32})$$

Until this point I have distinguished between a point and a vector to which it corresponds. The coordinates of the point have been the coefficients when we write out the basis expression for the vector.

I now stop using language like "the vector associated with the point" since we will primarily be speaking about vectors. I leave it to you to go back and forth as necessary between points and vectors as the content dictates.

When I do give a list of coordinates like $\mathbf{v} = (v_1, v_2, v_3, v_4)$ to describe a vector, remember these are all coefficients with respect to some particular basis. By default it is the standard basis but the discussion will make it clear if another basis is involved.

5.2   Vector spaces

The last section introduced several ideas about vector spaces using familiar notions from $\mathbb{R}^2$ and $\mathbb{C}$. It's time to generalize.

Let $\mathbb{F}$ be a field, for example $\mathbb{R}$ or $\mathbb{C}$, and let V be a set of objects. These objects are called vectors and are shown in bold such as $\mathbf{v}$. We are interested in defining a special kind of multiplication, called scalar multiplication, and addition.

If s is in $\mathbb{F}$ then we insist $s\mathbf{v}$ is in V for all $\mathbf{v}$ in V. This means the set V is closed under multiplication by scalars from the field $\mathbb{F}$. While V may have some kind of multiplication defined between its elements, we do not need to consider it here.

For any $\mathbf{v_1}$ and $\mathbf{v_2}$ in V, we also insist $\mathbf{v_1} + \mathbf{v_2}$ is in V and that the addition is commutative. Thus V is closed under addition. In fact, we demand V has an element $\mathbf{O}$ and additive inverses so that V is a commutative additive group.

V is almost a vector space over $\mathbb{F}$ but we have to insist on a few more conditions related to scalar multiplication. They concern the usual arithmetic properties we first saw with numbers. Let s_1 and s_2 be in $\mathbb{F}$. All the following must hold:

$$1\,\mathbf{v_1} = \mathbf{v_1} \text{ for } 1, \text{ the multiplicative identity of } \mathbb{F}$$
$$s_1\,(\mathbf{v_1} + \mathbf{v_2}) = s_1\mathbf{v_1} + s_1\mathbf{v_2}$$
$$(s_1 + s_2)\,\mathbf{v_1} = s_1\mathbf{v_1} + s_2\mathbf{v_1}$$
$$(s_1 s_2)\,\mathbf{v_1} = s_1\,(s_2\mathbf{v_1})$$

When all these requirements are met, we say V is a vector space over $\mathbb{F}$.

A vector space may have finite or infinite dimensions. The number is related to the size of a *basis*.

Let X be a possibly infinite subset of the vectors of V. Notationally, we refer to the elements of X as $\mathbf{x_1}$, $\mathbf{x_2}$ or $\mathbf{x_n}$ in general. If any vector $\mathbf{v}$ can be represented as a finite sum of elements of X multiplied by values s_i from $\mathbb{F}$, then X *spans V*.

For example, a given $\mathbf{v}$ might be given by

$$\mathbf{v} = s_3\mathbf{x_3} + s_7\mathbf{x_7} + s_{31}\mathbf{x_{31}}$$

and another $\mathbf{u}$ in V might be

$$\mathbf{u} = s_1\mathbf{x_3} + s_{70}\mathbf{x_{70}} + s_{397}\mathbf{x_{397}} + s_{7243}\mathbf{x_{7243}}.$$

$\mathbf{u}$ and $\mathbf{v}$ are linear combinations of elements of X.

X may contain more vectors than are strictly needed. With spanning we are only making sure that finite linear combinations of vectors from X given us all vectors in V.

Consider $\mathbb{R}^2$ and the vector collection X comprising $\mathbf{x_1} = \mathbf{e_1}$, $\mathbf{x_2} = \mathbf{e_2}$, and $\mathbf{x_3} = (1, 1)$. Then X spans V, but we really only need two of the three vectors in it to do so. In fact, $\mathbf{x_3} = \mathbf{x_1} + \mathbf{x_2}$.

If we rewrite the last equation as $\mathbf{x_3} - \mathbf{x_1} - \mathbf{x_2} = \mathbf{O} = (0,0)$ then we have a linear combination of vectors of X with non-zero coefficients, yet their sum is $\mathbf{O}$. When this happens, the vectors are *linearly dependent*. If it cannot happen, then the vectors in X are *linearly independent*.

> Let X be a possibly infinite subset of the vectors of a vector space V over a field $\mathbb{F}$. If the vectors of X span V and are linearly independent, then X is a *basis* for V. If the number of vectors in X is finite, then that number is the *dimension* of V.

The linear independence also implies there is a unique way to write a given vector as a linear combination of basis vectors. If

$$\mathbf{v} = a_1\mathbf{x_1} + a_2\mathbf{x_2} + \cdots + a_n\mathbf{x_n}$$
$$= b_1\mathbf{x_1} + b_2\mathbf{x_2} + \cdots + b_n\mathbf{x_n}$$

then $a_1 = b_1$, $a_2 = b_2$, and it continues like this all the way to $a_n = b_n$. When we say "unique" remember we mean "one and only one."

The standard basis, denoted E, is the set of vectors $\mathbf{e_i}$ which has 1 in the ith coordinate position and 0s elsewhere. For example,

$$\mathbf{e_1} = (1, 0, 0, \ldots, 0)$$
$$\mathbf{e_2} = (0, 1, 0, \ldots, 0) \text{ and so on.}$$

For a given V there can be different sets X that form bases. They are all finite or all infinite. The dimension is independent of your choice of basis. In some situations it may make sense to change from one basis to another.

It's a little confusing when you use "basis" or "bases." *Basis* refers to an entire collection of vectors that span the vector space and are linearly independent. These are called *basis vectors*. If I have more than one basis then they are *bases*.

I usually pronounce "bases" in this context as "bay-seas" versus "base-is."

5.3   Linear maps

We've looked at linear functions several times now to get a concrete idea of how they work. We must generalize this idea for vector spaces.

Let U and V be vector spaces over the same field $\mathbb{F}$. The function $L : U \to V$ is a *linear map*

- if $\mathbf{u_1}$ and $\mathbf{u_2}$ are in U then $L(\mathbf{u_1} + \mathbf{u_2}) = L(\mathbf{u_1}) + L(\mathbf{u_2})$, and
- if a_1 is in $\mathbb{F}$ then $L(a_1\mathbf{u_1}) = a_1 L(\mathbf{u_1})$.

In particular, for a_2 also in $\mathbb{F}$ we have

$$L(a_1\mathbf{u_1} + a_2\mathbf{u_2}) = a_1 L(\mathbf{u_1}) + a_2 L(\mathbf{u_2}) .$$

When $U = V$ we also say L is a *linear transformation* of U or a linear operator on U.

All linear transformations on $\mathbb{R}^2$ look like

$$(x, y) \mapsto (ax + by, cx + dy)$$

using Cartesian coordinates and a, b, c, d, x, and y in $\mathbb{R}$. This is interesting because the linear transformations on $\mathbb{R}^1$ all look like the somewhat trivial $(x) \mapsto (ax)$. To take it further, the linear transformations on $\mathbb{R}^3$ look like the messy

$$(x, y, z) \mapsto (ax + by + cz, dx + fy + gz, hx + jy + kz).$$

I skipped using e and i because they are special values. All numbers are real.

Look again at the forms for two- and three-dimensional linear transformations. Note

- If we think of x, y, and z as variables then they each appear with exponent 1 in each coordinate position.
- There are no constant terms like "+7" or "−1" in any position.
- Any or all of the coefficients can be 0.

For $\mathbb{R}^2$, then the map $(x, y) \mapsto (1x + 0y, 0x + 1y) = (x, y)$ is the identity I. It is its own inverse. The map $(x, y) \mapsto (0x + 0y, 0x + 0y) = (0, 0)$ is the zero transformation Z. It is not invertible as it discards essential information. I think of it as kind of the black hole of linear maps.

Question 5.3.1

What are the identity and zero transformations for $\mathbb{R}^3$?

In $\mathbb{R}^1$ the transformation $(x) \mapsto (ax)$ is invertible if and only if $a \neq 0$. For $\mathbb{R}^2$, it is invertible if and only if $ad - bc \neq 0$. Clearly this idea of linearity and invertibility is less simple than you might initially think.

A special kind of linear transformation is when we map to the field $\mathbb{F}$ itself, thought of as a one-dimensional vector space.

Let $L : V \to \mathbb{F}$ be a linear map from a vector space V to its field of scalars. Then L is called a *linear form* or *linear functional*. All linear forms look like

$$(v_1, v_2, \ldots, v_n) \mapsto a_1 v_1 + a_2 v_2 \cdots a_n v_n$$

with all a_i in $\mathbb{F}$.

Once we get above one dimension the structure of linear maps and transformations becomes more sophisticated.

5.3.1 Algebraic structure of linear transformations

Let V be a vector space over a field $\mathbb{F}$. If L and M are linear transformations on V then we can compose them by applying one after another

$$(L \circ M)(\mathbf{v}) = L(M(\mathbf{v}))$$

for $\mathbf{v}$ in V. It is very often the case that $L \circ M \neq M \circ L$. This means the order in which you apply linear transformations on a vector space of dimension 2 or greater makes a difference. Composition is not commutative in general.

The identity transformation I maps a vector to itself $I(\mathbf{v}) = \mathbf{v}$.

> The collection of invertible linear transformations on a vector space V over a field $\mathbb{F}$ form a non-commutative group under composition "$\circ$" with identity element I.

You can see that by understanding the concept of a group we can compress a lot of information about linear transformations into a simple statement.

We can add and subtract linear transformations, and take advantage of scalar multiplication in a vector space:

$$(L + M)(\mathbf{v}) = L(\mathbf{v}) + M(\mathbf{v})$$
$$-L(\mathbf{v}) = (-1)L(\mathbf{v}) = L(-\mathbf{v})$$
$$(L - M)(\mathbf{v}) = (L + -M)(\mathbf{v}) = L(\mathbf{v}) + (-1)M(\mathbf{v}) = L(\mathbf{v}) + M(-\mathbf{v})$$

Because addition is commutative in V,

$$(L + M)(\mathbf{v}) = L(\mathbf{v}) + M(\mathbf{v}) = M(\mathbf{v}) + L(\mathbf{v}) = (M + L)(\mathbf{v}).$$

The zero transformation Z maps every vector $\mathbf{v}$ to the zero vector $\mathbf{O}$.

> The collection of linear transformations on a vector space V over a field $\mathbb{F}$ form a commutative group under addition "$+$" with identity element the zero transformation Z.
>
> More than that, the collection of linear transformations on a vector space V over a field $\mathbb{F}$ form a non-commutative ring with the addition operation "$+$" and multiplication being the composition operation "$\circ$".

5.3.2 Example linear transformations on $\mathbb{R}^2$

Stretches and compressions

A stretch transformation moves points in some direction by a non-zero multiple. For example, it might double the y coordinate. Included in this category are the transformations that compress, or shrink, by a non-zero amount. An example is replacing the x coordinate by one-third of its value. These two transformations are

$$(x, y) \mapsto (x, 2y)$$

$$(x, y) \mapsto (\frac{1}{3}x, y)$$

Here you can see the effect of the two stretches on points on the unit circle.

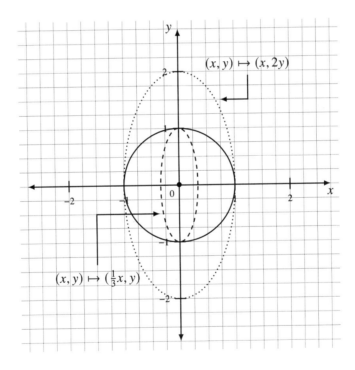

A stretch can move in both x and y directions at the same time by either the same or different amounts: $(x, y) \mapsto (\pi x, \frac{5}{7}y)$, for example.

Stretches and compressions are reversible, that is, invertible. For non-zero a and b, the inverse of $(x, y) \mapsto (ax, by)$ is $(x, y) \mapsto (\frac{x}{a}, \frac{y}{b})$.

If either a or b is 0 then we cannot reverse the effect. For this reason, if a or b is zero we prefer to consider this to be an example of a projection followed by a stretch, or vice versa.

What if we want to stretch in other directions such as the direction of the line $y = x$? You can accomplish this by a stretch in the x direction and then a $\frac{\pi}{4}$ rotation (that is, $45°$).

Reflections

A reflection moves points from one side of a line to the other. If a point lies on the line, nothing happens. The reflected image is the same distance from the line as the original point. The line segment connecting the original and reflected image is perpendicular is to the reflection line.

Here we reflect across the line $y = x$. The linear transformation is $(x, y) \mapsto (y, x)$.

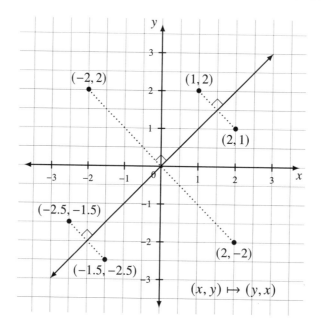

The linear transformations $(x, y) \mapsto (-x, y)$ and $(x, y) \mapsto (x, -y)$ reflect across the y and x axes, respectively. The latter corresponds to complex conjugation as a real linear map.

Reflections are their own inverses.

Rotations

To rotate any point (x, y) by φ radians around the origin, apply the linear transformation

$$(x, y) \mapsto (x \cos(\varphi) - y \sin(\varphi), x \sin(\varphi) + y \cos(\varphi))$$

But then we knew this already! Why?

Think of (x, y) as the number $x + yi$ in $\mathbb{C}$. From our discussion of Euler's formula we know multiplying by $e^{\varphi i}$ rotates a complex number by φ radians around 0. Since $e^{\varphi i} = \cos(\varphi) + \sin(\varphi)i$

$$e^{\varphi i} \times (x + yi) = (\cos(\varphi) + \sin(\varphi)i) \times (x + yi)$$
$$= x \cos(\varphi) + x \sin(\varphi)i + yi \cos(\varphi) + yi \sin(\varphi)i$$
$$= (x \cos(\varphi) - y \sin(\varphi)) + (x \sin(\varphi) + y \cos(\varphi))i$$

which corresponds back to the point $(x \cos(\varphi) - y \sin(\varphi), x \sin(\varphi) + y \cos(\varphi))$.

If we want to rotate by $\frac{\pi}{6}$ ($= 30°$), then we first note

$$\cos\left(\tfrac{\pi}{6}\right) = \tfrac{\sqrt{3}}{2} \quad \text{and} \quad \sin\left(\tfrac{\pi}{6}\right) = \tfrac{1}{2} \;.$$

The linear transformation is

$$(x, y) \mapsto \left(\frac{\sqrt{3}}{2}x - \frac{1}{2}y, \frac{1}{2}x + \frac{\sqrt{3}}{2}y\right)$$

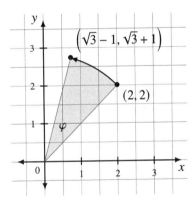

When φ is positive or negative, the rotation is counterclockwise or clockwise, respectively. If $|\varphi| \geq 2\pi$ we can replace it with one where $0 \leq |\varphi| < 2\pi$. A negative rotation φ can be replaced by a positive one and we can always choose $0 \leq \varphi < 2\pi$.

A rotation by φ is invertible: rotate by $-\varphi$ to get back to where you started.

Shears

Consider

$$f : (x, y) \mapsto (x, y + 3x) \quad \text{and} \quad g : (x, y) \mapsto (x + 3y, y) \;.$$

They are linear transformations. The first is a *vertical shear* and the second is a *horizontal shear*.

The first keeps the x coordinate constant while shifting the y coordinate by a multiple of the x coordinate. When x is large, the shift is large but the change is linear.

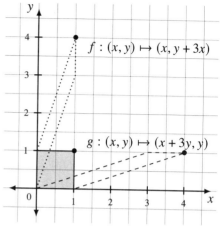

The second example leaves y constant but shifts x by an amount proportional to y. In the following you can see the effect of each transformation on the coordinates $(0, 0)$, $(1, 0)$, $(1, 1)$, and $(0, 1)$, the corners of the *unit square*.

I've marked $(1, 1)$ and where it goes under the two shear transformations.

The horizontal shear looks as if you took a deck of playing cards and smoothly pushed the upper cards by varying amounts to create a slanted rhombus.

What is the inverse of f? By examination, we need to do nothing to the x coordinate since f does not change it. For the second coordinate, $y - 3x$ undoes the shear. Vertical and horizontal shears are invertible linear transformations.

Projections

The previous linear transformations we considered in this section were all invertible. Here we see some that clearly are not.

Choose a line A and a point. If the point is on the line, we are done as it is the value of the projection onto A. If not, draw a line through the point that is perpendicular to A. The intersection of the line and A is the value of the projection.

The map $(x, y) \mapsto (0, y)$ is linear and it throws away the x coordinate, replacing it with 0. It's not invertible because the point $(0, 3)$ is the image of an infinite number of points including $(4, 3)$, $(-3636, 0)$, and $(e, 3)$. Anything of the form $(a, 3)$ maps to it, for a a real number. If we apply this transformation we have no idea which a we started with, and so cannot reverse it. It's a projection onto the y axis.

Geometrically, let's start with a circle of radius 0.5 centered at $(1, 1)$. This transformation maps all points onto the y axis between 0.5 and 1.5.

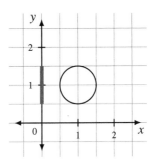

Since all the points are on the y axis, we can think of this as a linear form, from $\mathbb{R}^2$ to $\mathbb{R}$: $(x, y) \mapsto (y)$.

Question 5.3.2

What does the projection onto the x axis look like?

If you go back and look at the reflection across the $y = x$ line, we can also compute the projection onto that line. Essentially, a simple projection is a halfway reflection. For that particular reflection, all we had to do was swap the two coordinates. Convince yourself that

$$(x, y) \mapsto \left(\frac{x + y}{2}, \frac{x + y}{2} \right)$$

is the projection. Anything along the line $y = -x$ ends up at the origin.

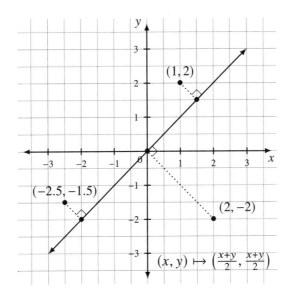

Injections

A linear map f is injective if $f(x) = 0$ implies $x = 0$. Put another way, if $f(x) = f(y)$ then $x = y$. A linear transformation is injective if it is invertible.

An *injection* is an injective linear map. Since we are not covering all of linear algebra, I'm going to use injections in a much less general way. I want you to imagine taking a vector space like $\mathbb{R}^1$ and putting it inside something bigger.

Indeed, we first showed this when we went from $\mathbb{R}^1$, the "number line," to the real plane, which is $\mathbb{R}^2$.

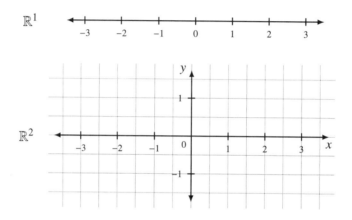

If (t) is in $\mathbb{R}^1$ then we can inject $\mathbb{R}^1$ into $\mathbb{R}^2$ by making the number line become the x axis: $(t) \mapsto (t, 0)$.

We can inject $\mathbb{R}^1$ onto the line $y = x$ in $\mathbb{R}^2$ via $(t) \mapsto (t, t)$. The projection $(x, y) \mapsto (x)$ undoes the injection.

5.4 Matrices

If we write out the details of a linear transformation of a five-dimensional vector space over a field $\mathbb{F}$ it looks like this:

$$(v_1, v_2, v_3, v_4, v_5) \mapsto (a_1 v_1 + a_2 v_2 + a_3 v_3 + a_4 v_4 + a_5 v_5,$$
$$b_1 v_1 + b_2 v_2 + b_3 v_3 + b_4 v_4 + b_5 v_5,$$
$$c_1 v_1 + c_2 v_2 + c_3 v_3 + c_4 v_4 + c_5 v_5,$$
$$d_1 v_1 + d_2 v_2 + d_3 v_3 + d_4 v_4 + d_5 v_5,$$
$$f_1 v_1 + f_2 v_2 + f_3 v_3 + f_4 v_4 + f_5 v_5).$$

This is not practical, especially if we start looking at the formulas for compositions of linear maps. Hence we introduce matrices.

The plural of "matrix" is "matrices."

We begin with notation and then move on to the algebra.

5.4.1 Notation and terminology

This array of values with entries in the field $\mathbb{F}$

$$\begin{bmatrix} 2 & 5 & -1 & 0 \\ 1 & -9 & 2 & -4 \\ -3 & 0 & 0 & 6 \end{bmatrix}$$

is a 3 by 4 *matrix*: it has three rows, four columns, and $12 = 3 \times 4$ entries. The *dimension* of this matrix is 3 by 4.

When we need to use subscripts for the entries in a matrix, they look like this:

$$\begin{bmatrix} a_{1,1} & a_{1,2} & a_{1,3} \\ a_{2,1} & a_{2,2} & a_{2,3} \\ a_{3,1} & a_{3,2} & a_{3,3} \end{bmatrix}$$

for a 3 by 3 matrix. The first subscript is the *row index* and the second is the *column index*. The dimension of this matrix is 3 by 3.

In mathematics and physics we use 1 as the first row or column index. In computer science and software coding we use 0. Remember this if you are translating from a formula to code or vice versa.

The general case for m rows and n columns is

$$\begin{bmatrix} a_{1,1} & a_{1,2} & \cdots & a_{1,n} \\ a_{2,1} & a_{2,2} & \cdots & a_{2,n} \\ \vdots & \vdots & \ddots & \vdots \\ a_{m,1} & a_{m,2} & \cdots & a_{m,n} \end{bmatrix}.$$

When the number of rows equals the number of columns, $m = n$, we have a square matrix. Most of the matrices we consider are square because these correspond to linear transformations.

Matrix entries like $a_{1,1}$, $a_{2,2}$, and $a_{3,3}$ lie on the diagonal of the matrix.

$$\begin{bmatrix} \boxed{a_{1,1}} & a_{1,2} & a_{1,3} \\ a_{2,1} & \boxed{a_{2,2}} & a_{2,3} \\ a_{3,1} & a_{3,2} & \boxed{a_{3,3}} \end{bmatrix}.$$

I put the diagonal entries in boxes for visual emphasis only.

If all the off-diagonal entries are 0, then we have a *diagonal matrix*. These are the easiest kinds of matrices with which to work.

A matrix with all 0 entries is a *zero matrix*, not surprisingly. A square matrix with 1s along the diagonal and 0s elsewhere is an *identity matrix*. We denote by I_n the n by n identity matrix. Here is I_4:

$$\begin{bmatrix} 1 & 0 & 0 & 0 \\ 0 & 1 & 0 & 0 \\ 0 & 0 & 1 & 0 \\ 0 & 0 & 0 & 1 \end{bmatrix}.$$

A matrix is *sparse* if almost all its entries are 0.

The *transpose* of a matrix is the new matrix we get by reflecting the elements across the diagonal. In the 2 by 2 case:

$$\begin{bmatrix} a & b \\ c & d \end{bmatrix} \xrightarrow{\text{transpose}} \begin{bmatrix} a & c \\ b & d \end{bmatrix}.$$

If A is a matrix, then A^T denotes its transpose. If

$$A = \begin{bmatrix} a_{1,1} & a_{1,2} & a_{1,3} \\ a_{2,1} & a_{2,2} & a_{2,3} \\ a_{3,1} & a_{3,2} & a_{3,3} \end{bmatrix} \quad \text{then} \quad A^\mathsf{T} = \begin{bmatrix} a_{1,1} & a_{2,1} & a_{3,1} \\ a_{1,2} & a_{2,2} & a_{3,2} \\ a_{1,3} & a_{2,3} & a_{3,3} \end{bmatrix}.$$

Taking the transpose does not change any of the diagonal entries. Taking the transpose of a square diagonal matrix does not change the matrix.

An n by n square matrix A is *symmetric* if $A = A^\mathsf{T}$. If such a symmetric matrix A has all entries in $\mathbb{R}$, it is *positive definite* if for any non-zero real vector $\mathbf{v}$ of length n, $\mathbf{v}^\mathsf{T} A \mathbf{v} > 0$. If instead we only have $\mathbf{v}^\mathsf{T} A \mathbf{v} \geq 0$, then A is *positive semi-definite*.

When the matrix has complex entries, that is, $\mathbb{F} = \mathbb{C}$, we call it a complex matrix. The conjugate of a complex matrix is the new matrix formed by taking the conjugate of each of the entries.

$$A = \begin{bmatrix} 0 & 1+i & 3-2i \\ 7+i & -i & \frac{\sqrt{2}}{2} - \frac{\sqrt{2}}{2}i \end{bmatrix} \qquad \overline{A} = \begin{bmatrix} 0 & 1-i & 3+2i \\ 7-i & i & \frac{\sqrt{2}}{2} + \frac{\sqrt{2}}{2}i \end{bmatrix}.$$

> The *adjoint* of a complex matrix is its conjugate transpose $\overline{A}^{\mathsf{T}}$ and is denoted $A^{\dagger}$.
>
> In either order, you take the complex conjugates of the entries and transpose the matrix.

Conjugation and taking adjoints have different notation between mathematics and physics and sometimes even between subfields of the subjects. We use the "line over" notation for conjugation such as $\overline{z}$, while some people use an asterisk superscript like z^{*}. This is further confusing since mathematicians use A^{*} for the adjoint. Physicists use a dagger superscript for the same thing: $A^{\dagger}$.

We use $\overline{z}$ and $A^{\dagger}$.

> An n by n square complex matrix A is *Hermitian* if $A = A^{\dagger}$. A Hermitian matrix is *self-adjoint*. An n by n Hermitian matrix is positive definite if for any non-zero complex vector $\mathbf{v}$ of length n, $\mathbf{v}^{\dagger} A \mathbf{v} > 0$. It is positive semi-definite if for any non-zero complex vector $\mathbf{v}$ of length n, $\mathbf{v}^{\dagger} A \mathbf{v} \geq 0$.

Hermitian matrices are named after the nineteenth century mathematician Charles Hermite. Being Hermitian for a complex matrix is the extension of the concept of a real matrix being symmetric.

Question 5.4.1

Why are all the diagonal entries of a Hermitian matrix in $\mathbb{R}$?

Charles Hermite, circa 1901. Photo is in the public domain. ⓟ

Question 5.4.2

Is the n by n identity matrix I_n positive definite as a real matrix? As a Hermitian matrix?

5.4.2 Matrices and linear maps

Vector notation

Before we get into the topic of this section, we need to have a talk about how we write vectors. So far we have used the "bold text" notation like $\mathbf{v}$ and $\mathbf{e_1}$, and the "I'm going to write this like the coordinates of a point but you know it is really a vector" notation like $(2, 3)$ and $(v_1, v_2, \ldots, v_n)$.

While we continue to use these, we have to add two additional forms so we can work with matrices.

For the vector $\mathbf{v} = (v_1, v_2, \ldots, v_n)$,

$$\begin{bmatrix} v_1 & v_2 & \cdots & v_n \end{bmatrix} \quad \text{and} \quad \begin{bmatrix} v_1 \\ v_2 \\ \vdots \\ v_n \end{bmatrix}$$

are the row and column vector forms of $\mathbf{v}$. It is no coincidence they look like 1 by n and n by 1 matrices. Whenever it is convenient, we think of them as such.

A vector in column form is a *column vector*. You can guess what a *row vector* is.

The coordinates are always with respect to some basis X, which may be the standard basis E. If we need to indicate what basis we are using, we use its name as a subscript as in

$$(v_1, v_2, \ldots, v_n)_X \text{ or } \begin{bmatrix} v_1 & v_2 & \cdots & v_n \end{bmatrix}_X \text{ or } \begin{bmatrix} v_1 \\ v_2 \\ \vdots \\ v_n \end{bmatrix}_X .$$

There are yet two more vector notations we will use, Dirac's *bra-ket* forms. We define these in section 7.2 when we consider the mathematics of qubits.

Vector operations

The transpose of a row vector is a column vector and vice versa.

$$\begin{bmatrix} v_1 & v_2 & \cdots & v_n \end{bmatrix}^\mathsf{T} = \begin{bmatrix} v_1 \\ v_2 \\ \vdots \\ v_n \end{bmatrix} \quad \text{and} \quad \begin{bmatrix} v_1 \\ v_2 \\ \vdots \\ v_n \end{bmatrix}^\mathsf{T} = \begin{bmatrix} v_1 & v_2 & \cdots & v_n \end{bmatrix} .$$

The conjugate of a complex vector is the new vector we get by taking the complex conjugate of each entry:

$$\overline{\mathbf{v}} = (\overline{v_1}, \overline{v_2}, \ldots, \overline{v_n})$$

We have similar relationships with the adjoint as we do with conjugation of row and column vectors:

$$\begin{bmatrix} v_1 & v_2 & \cdots & v_n \end{bmatrix}^\dagger = \begin{bmatrix} \overline{v_1} \\ \overline{v_2} \\ \vdots \\ \overline{v_n} \end{bmatrix} \quad \text{and} \quad \begin{bmatrix} v_1 \\ v_2 \\ \vdots \\ v_n \end{bmatrix}^\dagger = \begin{bmatrix} \overline{v_1} & \overline{v_2} & \cdots & \overline{v_n} \end{bmatrix}.$$

The *adjoint* of a complex vector $\mathbf{v}$ is its conjugate transpose $\overline{\mathbf{v}^\mathsf{T}}$ and is denoted $\mathbf{v}^\dagger$.

Applying the matrix of a linear transformation to a vector

When we apply a linear transformation to a vector, we get a different vector unless we are using the identity transformation I. To do the arithmetic we must use coordinates.

Apply a matrix to a column vector as shown in this 2 by 2 example:

$$\begin{bmatrix} a & b \\ c & d \end{bmatrix} \begin{bmatrix} x \\ y \end{bmatrix} = \begin{bmatrix} ax + by \\ cx + dy \end{bmatrix}$$

We start with a vector and end up with another: $(x, y) \mapsto (ax+by, cx+dy)$. This is the definition of a linear map from a two-dimensional vector space to a two-dimensional vector space.

To apply a linear transformation represented as a matrix to a vector, the number of columns in the matrix must equal the number of entries in the vector. The number of rows is the size of the target vector space.

Application is done row by row. To start, the first matrix entry in the first row is multiplied by the first vector entry. This is then added to the second matrix entry in the first row multiplied by the second vector entry. We continue through the row and the sum becomes the first coordinate of the result vector. We then move on to the second row and so forth.

$$\begin{bmatrix} a_{1,1} & a_{1,2} & \cdots & a_{1,n} \\ a_{2,1} & a_{2,2} & \cdots & a_{2,n} \\ \vdots & \vdots & \ddots & \vdots \\ a_{m,1} & a_{m,2} & \cdots & a_{m,n} \end{bmatrix} \begin{bmatrix} v_1 \\ v_2 \\ \vdots \\ v_n \end{bmatrix} = \begin{bmatrix} a_{1,1}v_1 + a_{1,2}v_2 + \cdots + a_{1,n}v_n \\ a_{2,1}v_1 + a_{2,2}v_2 + \cdots + a_{2,n}v_n \\ \vdots \\ a_{m,1}v_1 + a_{m,2}v_2 + \cdots + a_{m,n}v_n \end{bmatrix}$$

We are mapping an n-dimensional vector to an m-dimensional one via an n by m matrix.

If A is a 1 by n matrix or a row vector of length n, it defines a linear form to the field of scalars $\mathbb{F}$. For a vector $\mathbf{v}$ in column form,

$$\begin{bmatrix} a_{1,1} & a_{2,1} & \cdots & a_{n,1} \end{bmatrix} \begin{bmatrix} v_1 \\ v_2 \\ \vdots \\ v_n \end{bmatrix} = a_{1,1}v_1 + a_{2,1}v_2 + \cdots + a_{n,1}v_n$$

Coordinates for a vector are with respect to a particular basis. If we were to use a different basis, the coordinates for the vector would be different but *it would still refer to the same vector*.

A matrix is always represented with respect to a particular basis. If we change to a different basis, the matrix changes but both correspond to **the same linear map**.

This means we really are talking about a (matrix, basis) pair when making a linear transformation concrete for computation. If I do not state the basis, assume it is the standard basis.

The rank of a matrix A is the maximum number of linearly independent columns of A, thought of as vectors. Equivalently, the rank is the maximum number of linearly independent rows of A. The rank is a property of the linear map to which the matrix corresponds in the given basis.

Example: Stretches and compressions

A stretch or compression in the direction of a basis vector is a diagonal matrix.

$$(x, y) \mapsto (2x, 0.25y) \qquad \begin{bmatrix} 2 & 0 \\ 0 & 0.25 \end{bmatrix} \begin{bmatrix} x \\ y \end{bmatrix} = \begin{bmatrix} 2x \\ 0.25y \end{bmatrix}$$

This doubles the first coordinate in a stretch but compresses the second to one-quarter its previous size. You can see how it affects the points on the unit circle.

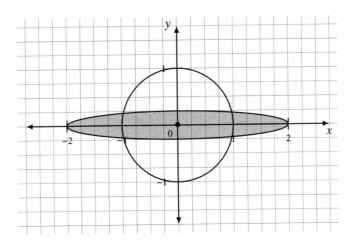

Example: Reflections

The "conjugation reflection" negates the second coordinate:

$$(x, y) \mapsto (x, -y) \qquad \begin{bmatrix} 1 & 0 \\ 0 & -1 \end{bmatrix} \begin{bmatrix} x \\ y \end{bmatrix} = \begin{bmatrix} x \\ -y \end{bmatrix}$$

In three dimensions, negating the y coordinate reflects across the xz-plane.

$$C : (x, y, z) \mapsto (x, -y, z) \qquad \begin{bmatrix} 1 & 0 & 0 \\ 0 & -1 & 0 \\ 0 & 0 & 1 \end{bmatrix} \begin{bmatrix} x \\ y \\ z \end{bmatrix} = \begin{bmatrix} x \\ -y \\ z \end{bmatrix}$$

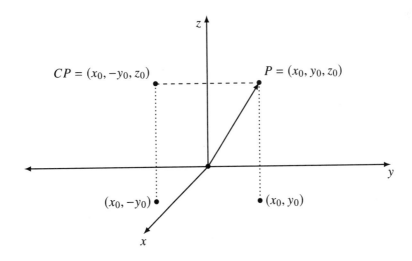

The reflection across the line $y = x$ swaps the two coordinates:

$$(x, y) \mapsto (y, x) \qquad \begin{bmatrix} 0 & 1 \\ 1 & 0 \end{bmatrix} \begin{bmatrix} x \\ y \end{bmatrix} = \begin{bmatrix} y \\ x \end{bmatrix}$$

The main diagonal is 0 and all the "work" is done by the off-diagonal elements.

Example: Rotations

As we previously developed, the linear transformation

$$(x, y) \mapsto (x \cos(\theta) - y \sin(\theta), x \sin(\theta) + y \cos(\theta))$$

$$\begin{bmatrix} \cos(\theta) & -\sin(\theta) \\ \sin(\theta) & \cos(\theta) \end{bmatrix} \begin{bmatrix} x \\ y \end{bmatrix} = \begin{bmatrix} x \cos(\theta) - y \sin(\theta) \\ x \sin(\theta) + y \cos(\theta) \end{bmatrix}$$

rotates any point (x, y) by θ radians around the origin. This matrix has rank 2.

A $\frac{\pi}{2}$ $(= 90°)$ rotation brings $(1, 0)$ to $(0, 1)$. This looks like it might be a reflection across $y = x$ but $(0, 1)$ itself is brought to $(-1, 0)$.

Question 5.4.3

What is the matrix for this linear transformation? What happens if you apply it four times?

In three dimensions,

$$(x, y, z) \mapsto (x \cos(\theta) - y \sin(\theta), x \sin(\theta) + y \cos(\theta), z)$$

$$\begin{bmatrix} \cos(\varphi) & -\sin(\varphi) & 0 \\ \sin(\varphi) & \cos(\varphi) & 0 \\ 0 & 0 & 1 \end{bmatrix} \begin{bmatrix} x \\ y \\ z \end{bmatrix} = \begin{bmatrix} x \cos(\varphi) - y \sin(\varphi) \\ x \sin(\varphi) + y \cos(\varphi) \\ z \end{bmatrix}$$

rotates any point (x, y, z) by φ radians around the z axis. Similarly,

$$(x, y, z) \mapsto (x, y \cos(\theta) - z \sin(\theta), y \sin(\theta) + z \cos(\theta))$$

$$\begin{bmatrix} 1 & 0 & 0 \\ 0 & \cos(\theta) & -\sin(\theta) \\ 0 & \sin(\theta) & \cos(\theta) \end{bmatrix} \begin{bmatrix} x \\ y \\ z \end{bmatrix} = \begin{bmatrix} x \\ y \cos(\theta) - z \sin(\theta) \\ y \sin(\theta) + z \cos(\theta) \end{bmatrix}$$

rotates any point (x, y, z) by θ radians around the x axis.

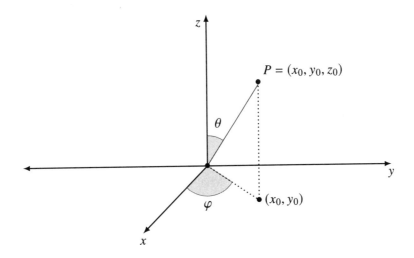

These matrix and trigonometry equations allow you to move points around when doing computer graphics in video games.

Example: Shears

The vertical shear and horizontal shears

$$(x, y) \mapsto (x, y + 3x) \quad \text{and} \quad (x, y) \mapsto (x + 3y, y)$$

have the matrices

$$\begin{bmatrix} 1 & 0 \\ 3 & 1 \end{bmatrix} \quad \text{and} \quad \begin{bmatrix} 1 & 3 \\ 0 & 1 \end{bmatrix},$$

respectively.

At this point you should be getting comfortable with the interplay of the matrix entries with the original coordinates.

Example: Projections

The transformation

$$(x, y, z) \mapsto (x, y) \qquad \begin{bmatrix} 1 & 0 & 0 \\ 0 & 1 & 0 \end{bmatrix} \begin{bmatrix} x \\ y \\ z \end{bmatrix} = \begin{bmatrix} x \\ y \end{bmatrix}$$

projects from 3 dimensions into 2. In terms of $\mathbb{R}^3$, it throws away the z coordinate. The image is $\mathbb{R}^2$.

If you want to stay in $\mathbb{R}^3$, this version projects all points to the xy-plane, which is equivalent to $\mathbb{R}^2$.

$$(x, y, z) \mapsto (x, y, 0) \qquad \begin{bmatrix} 1 & 0 & 0 \\ 0 & 1 & 0 \\ 0 & 0 & 0 \end{bmatrix} \begin{bmatrix} x \\ y \\ z \end{bmatrix} = \begin{bmatrix} x \\ y \\ 0 \end{bmatrix}$$

The matrices for this projection have rank 2.

Question 5.4.4

What is this matrix multiplied by itself? What does this result mean geometrically?

The $\mathbb{R}^2$ projection onto the line $y = x$ is

$$(x, y) \mapsto \left(\frac{x + y}{2}, \frac{x + y}{2} \right) \qquad \begin{bmatrix} \frac{1}{2} & \frac{1}{2} \\ \frac{1}{2} & \frac{1}{2} \end{bmatrix} \begin{bmatrix} x \\ y \end{bmatrix} = \begin{bmatrix} \frac{x+y}{2} \\ \frac{x+y}{2} \end{bmatrix}$$

Question 5.4.5

What is this matrix multiplied by itself? What does this result mean geometrically? What is the rank of this matrix?

Example: Injections

Casually, an injection is when you map a vector space into another without sending multiple vectors to the same target vector. For these examples I'm going to use it in a more limited sense of embedding a vector space inside another with a minimum number of changes.

$\mathbb{R}^1$ sits inside $\mathbb{R}^2$ in two different ways using the standard bases. The first maps it to the x axis:

$$(t) \mapsto (t, 0) \qquad \begin{bmatrix} 1 \\ 0 \end{bmatrix} \begin{bmatrix} t \end{bmatrix} = \begin{bmatrix} t & 0 \end{bmatrix}$$

and the second to the y axis:

$$(t) \mapsto (0, t) \qquad \begin{bmatrix} 0 \\ 1 \end{bmatrix} \begin{bmatrix} t \end{bmatrix} = \begin{bmatrix} 0 & t \end{bmatrix}$$

To inject $\mathbb{R}^2$ into $\mathbb{R}^3$ using the standard bases you can send it to the xy-plane, the yz-plane, or the xz-plane. This last case is

$$(x, y) \mapsto (x, 0, y) \qquad \begin{bmatrix} 1 & 0 \\ 0 & 0 \\ 0 & 1 \end{bmatrix} \begin{bmatrix} x \\ y \end{bmatrix} = \begin{bmatrix} x & 0 & y \end{bmatrix}$$

Example: Compositions

You compose transformations by applying one after another. Let's look at the $\frac{\pi}{4}$ rotation and the reflection across the x axis.

Per the above examples

$$\begin{bmatrix} x \\ y \end{bmatrix} \xrightarrow{\text{rotate by } \frac{\pi}{4}} \begin{bmatrix} x\cos(\frac{\pi}{4}) - y\sin(\frac{\pi}{4}) \\ x\sin(\frac{\pi}{4}) + y\cos(\frac{\pi}{4}) \end{bmatrix} = \begin{bmatrix} -y \\ x \end{bmatrix} \xrightarrow{\text{reflect across } x \text{ axis}} \begin{bmatrix} -y \\ -x \end{bmatrix}$$

and

$$\begin{bmatrix} x \\ y \end{bmatrix} \xrightarrow{\text{reflect across } x \text{ axis}} \begin{bmatrix} x \\ -y \end{bmatrix} \xrightarrow{\text{rotate by } \frac{\pi}{4}} \begin{bmatrix} x\cos(\frac{\pi}{4}) - (-y)\sin(\frac{\pi}{4}) \\ x\sin(\frac{\pi}{4}) + (-y)\cos(\frac{\pi}{4}) \end{bmatrix} = \begin{bmatrix} y \\ x \end{bmatrix}$$

In the first case, the point $(1, 0)$ is mapped to $(0, 1)$ and then $(0, -1)$. In the second, the same point is mapped to itself because it is on the x axis and then is rotated to get $(0, 1)$. The formulas are not the same and we demonstrated how the results vary for a specific vector.

In general, the composition of linear transformations is not commutative.

There is something else worth noting in this example. In the second computation we began with (x, y) and ended with (y, x).

(Rotation by $\frac{\pi}{4}$) $\circ$ (Reflection across the x axis) = (Reflection across the line $y = x$)

Composition is done from right to left: $(f \circ g)(x) = f(g(x))$.

If you are trying to see how functions like linear maps work, apply them to simple examples like $(1, 0)$ and $(0, 1)$, or higher dimensional analogs. That is, *see what the maps do to the standard basis vectors*.

We've now seen many of the common kinds of linear transformations. Composition gives us even more of them as well as some we have seen before. There is math going on here, of course, so it is time to look at matrices as algebraic structures.

5.5 Matrix algebra

So far we have looked at matrices and their relationships to linear maps. We now investigate operations that are relevant when two or more matrices are involved. We cover the general case first of matrices which may have different numbers of rows and columns and then move on to square matrices.

All matrices are over fields in this section and when we are manipulating multiple matrices they all have entries in the same field. We can consider matrices over rings such as the integers but we have no need to make this restriction for quantum computing.

5.5.1 Arithmetic of general matrices

Matrices that are the same size, meaning they have the same number of rows and columns, can be added together entry by entry. For example,

$$\begin{bmatrix} a & b & c \\ d & f & g \end{bmatrix} + \begin{bmatrix} 1 & 3 & -\frac{4}{7} \\ -2 & \pi & 0 \end{bmatrix} = \begin{bmatrix} a+1 & b+3 & c-\frac{4}{7} \\ d-2 & f+\pi & g \end{bmatrix}.$$

The same is true for subtraction and negation.

$$-\begin{bmatrix} 1 & 3 & -\frac{4}{7} \\ -2 & \pi & 0 \end{bmatrix} = \begin{bmatrix} -1 & -3 & \frac{4}{7} \\ 2 & -\pi & 0 \end{bmatrix}$$

Scalar multiplication is done entry by entry.

$$5\begin{bmatrix} a & b \\ c & d \end{bmatrix} = \begin{bmatrix} 5a & 5b \\ 5c & 5d \end{bmatrix}$$

The set of n by m matrices over a field $\mathbb{F}$ for given integers n and $m \geq 1$ is a vector space of dimension $n \times m$.

Everything we learned about the algebra of vector spaces now applies to such collections of matrices.

Question 5.5.1

Verify that this is a vector space based on the definition in section 5.2. What do you think the "standard basis" is?

The transpose of a sum or difference is the sum or difference of the transposes.

$$(A + B)^\mathsf{T} = A^\mathsf{T} + B^\mathsf{T}$$
$$(A - B)^\mathsf{T} = A^\mathsf{T} - B^\mathsf{T}$$
$$(-A)^\mathsf{T} = -A^\mathsf{T}$$

The same holds for adjoints for complex matrices.

$$(A + B)^\dagger = A^\dagger + B^\dagger$$
$$(A - B)^\dagger = A^\dagger - B^\dagger$$
$$(-A)^\dagger = -A^\dagger$$

Now that we can add, subtract, and invert matrices, it's time to look at matrix multiplication. Its definition is more complicated and requires an interesting prerequisite on some of the dimensions.

> To be able to multiply matrices A and B, the number of columns of A must equal the number of rows of B. Said another way, if A is an n by m matrix and B is a m by p matrix, then we can multiply them.

When you see a matrix multiplication from this point on in this book, this condition on the rows and columns is met.

Before we look at the general formula, here is how it works in a 2 by 2 case.

$$AB = \begin{bmatrix} a & b \\ c & d \end{bmatrix} \begin{bmatrix} u & v \\ w & x \end{bmatrix} = \begin{bmatrix} au + bw & av + bx \\ cu + dw & cv + dx \end{bmatrix} = C$$

Taking the first matrix A row by row, we successively sum up the products of the row entries with the corresponding column entries in the second matrix B. (Note to yourself: come back and visit here again once we have covered dot products in subsection 5.7.1.)

The $1, 1$ entry of C is row number 1 of A lined up with column number 1 of B, we multiply corresponding entries, then add them all up: $au + bw$.

The $1, 2$ entry of C is row number 1 of A lined up with column number 2 of B, we multiply corresponding entries, then add them all up: $av + bx$.

The $2, 1$ entry of C is row number 2 of A lined up with column number 1 of B, we multiply corresponding entries, then add them all up: $cu + dw$.

The 2, 2 entry of C is row number 2 of A lined up with column number 2 of B, we multiply corresponding entries, then add them all up: $cv + dx$.

That was tedious, but you can see how the subscripts on the entries in the product correspond to the rows and columns from the original matrices.

This is informally known as *subscript math*. You can forget why you are doing it, but you follow a procedure to keep the subscripts and the arithmetic straight.

If

$$A = \begin{bmatrix} a_{1,1} & \cdots & a_{1,m} \\ \vdots & \vdots & \vdots \\ a_{n,1} & \cdots & a_{n,m} \end{bmatrix} \quad \text{and} \quad B = \begin{bmatrix} b_{1,1} & \cdots & b_{1,p} \\ \vdots & \vdots & \vdots \\ b_{m,1} & \cdots & b_{m,p} \end{bmatrix}$$

then

$$AB = \begin{bmatrix} a_{1,1}b_{1,1} + \cdots + a_{1,m}b_{m,1} & \cdots & a_{1,1}b_{1,p} + \cdots + a_{1,m}b_{m,p} \\ \vdots & \vdots & \vdots \\ a_{n,1}b_{1,1} + \cdots + a_{n,m}b_{m,1} & \cdots & a_{n,1}b_{1,p} + \cdots + a_{n,m}b_{m,p} \end{bmatrix}.$$

I remember this as "first row by first column, first row by second column, . . ." and so on.

In terms of linear maps, B takes us from a p-dimensional vector space to an m-dimensional one. A takes us from an m-dimensional vector space to an n-dimensional one. Applying B and then A brings us from p dimensions to n by way of m in the middle.

Transpositions operate on products by reversing the order: $(AB)^\mathsf{T} = B^\mathsf{T}A^\mathsf{T}$. Adjoints of complex matrices behave in the same way: $(AB)^\dagger = B^\dagger A^\dagger$. Remember, the adjoint is the conjugate transpose.

5.5.2 Arithmetic of square matrices

Square matrices have additional properties, operations, and structure because of the possibility of computing their inverses.

Inverses

If A is a square matrix and there exists a square matrix B so that $AB = BA = I$, the identity matrix, then B is the inverse of A, denoted A^{-1}.

For a 2 by 2 matrix

$$A = \begin{bmatrix} a & b \\ c & d \end{bmatrix}$$

we can multiply by

$$\begin{bmatrix} d & -b \\ -c & a \end{bmatrix}$$

to get

$$\begin{bmatrix} ad - bc & 0 \\ 0 & ad - bc \end{bmatrix}.$$

If $ad - bc \neq 0$, we can divide the second matrix by $ad - bc$ to get A^{-1}.

$$A^{-1} = \frac{1}{ad - bc} \begin{bmatrix} d & -b \\ -c & a \end{bmatrix} = \begin{bmatrix} \frac{d}{ad-bc} & -\frac{b}{ad-bc} \\ -\frac{c}{ad-bc} & \frac{a}{ad-bc} \end{bmatrix}$$

A^{-1} is the multiplicative inverse of A.

If A is a diagonal matrix and all the entries on the diagonal are non-zero, the inverse of A is diagonal and the entries are the position by position inverses of A's diagonal. This is a mouthful so it is easier to show you an example.

$$\begin{bmatrix} 3 & 0 & 0 \\ 0 & \pi^2 & 0 \\ 0 & 0 & -\frac{1}{4} \end{bmatrix}^{-1} = \begin{bmatrix} \frac{1}{3} & 0 & 0 \\ 0 & \frac{1}{\pi^2} & 0 \\ 0 & 0 & -4 \end{bmatrix}$$

Gaussian elimination

For higher dimensional matrices, the technique of *Gaussian elimination* can be used to compute inverses and solve systems of linear equations. This is covered in most linear algebra texts, but let's look at an example of its use in detail.

Let's find the inverse of the matrix

$$\begin{bmatrix} 1 & 2 & -1 \\ 2 & 0 & 1 \\ 3 & 1 & 0 \end{bmatrix}.$$

We begin by positioning the identity matrix I_3 to the right of the above so we have a 3 by 6 matrix. The vertical line divides the two halves.

$$\left[\begin{array}{ccc|ccc} 1 & 2 & -1 & 1 & 0 & 0 \\ 2 & 0 & 1 & 0 & 1 & 0 \\ 3 & 1 & 0 & 0 & 0 & 1 \end{array}\right]$$

The plan is to perform *elementary row operations* so the submatrix to the left of the vertical line becomes the identity matrix. The inverse will be the submatrix to the right of the line. If we cannot transform the left submatrix to I_3, the matrix is not invertible.

There are three allowable elementary row operations:

- multiply a row by a number that is not 0,
- interchange two different rows, and
- add a multiple of a row to a different row.

We perform these on the complete rows, which in this case have 6 entries each. Mathematical software like MATLAB® and Mathematica® have built-in algorithms to determine which operations to do to which rows in which order. Here we just choose reasonable steps to convert the left-hand side to I_3.

Add -2 times the first row to the second row:

$$\left[\begin{array}{rrr|rrr} 1 & 2 & -1 & 1 & 0 & 0 \\ 2 & 0 & 1 & 0 & 1 & 0 \\ 3 & 1 & 0 & 0 & 0 & 1 \end{array}\right] \mapsto \left[\begin{array}{rrr|rrr} 1 & 2 & -1 & 1 & 0 & 0 \\ 0 & -4 & 3 & -2 & 1 & 0 \\ 3 & 1 & 0 & 0 & 0 & 1 \end{array}\right]$$

Add -3 times the first row to the second row:

$$\left[\begin{array}{rrr|rrr} 1 & 2 & -1 & 1 & 0 & 0 \\ 0 & -4 & 3 & -2 & 1 & 0 \\ 3 & 1 & 0 & 0 & 0 & 1 \end{array}\right] \mapsto \left[\begin{array}{rrr|rrr} 1 & 2 & -1 & 1 & 0 & 0 \\ 0 & -4 & 3 & -2 & 1 & 0 \\ 0 & -5 & 3 & -3 & 0 & 1 \end{array}\right]$$

The first column of the left is now the first column of I_3. Let's work on the second column.

Add -1 times the third row to the second row:

$$\left[\begin{array}{rrr|rrr} 1 & 2 & -1 & 1 & 0 & 0 \\ 0 & -4 & 3 & -2 & 1 & 0 \\ 0 & -5 & 3 & -3 & 0 & 1 \end{array}\right] \mapsto \left[\begin{array}{rrr|rrr} 1 & 2 & -1 & 1 & 0 & 0 \\ 0 & 1 & 0 & 1 & 1 & -1 \\ 0 & -5 & 3 & -3 & 0 & 1 \end{array}\right]$$

Add -2 times the second row to the first:

$$\left[\begin{array}{rrr|rrr} 1 & 2 & -1 & 1 & 0 & 0 \\ 0 & 1 & 0 & 1 & 1 & -1 \\ 0 & -5 & 3 & -3 & 0 & 1 \end{array}\right] \mapsto \left[\begin{array}{rrr|rrr} 1 & 0 & -1 & -1 & -2 & 2 \\ 0 & 1 & 0 & 1 & 1 & -1 \\ 0 & -5 & 3 & -3 & 0 & 1 \end{array}\right]$$

Add 5 times the second row to the third:

$$\left[\begin{array}{rrr|rrr} 1 & 0 & -1 & -1 & -2 & 2 \\ 0 & 1 & 0 & 1 & 1 & -1 \\ 0 & -5 & 3 & -3 & 0 & 1 \end{array}\right] \mapsto \left[\begin{array}{rrr|rrr} 1 & 0 & -1 & -1 & -2 & 2 \\ 0 & 1 & 0 & 1 & 1 & -1 \\ 0 & 0 & 3 & 2 & 5 & -4 \end{array}\right]$$

The second column on the left is complete. Multiply the third row by $\frac{1}{3}$.

$$\left[\begin{array}{ccc|ccc} 1 & 0 & -1 & -1 & -2 & 2 \\ 0 & 1 & 0 & 1 & 1 & -1 \\ 0 & 0 & 3 & 2 & 5 & -4 \end{array}\right] \mapsto \left[\begin{array}{ccc|ccc} 1 & 0 & -1 & -1 & -2 & 2 \\ 0 & 1 & 0 & 1 & 1 & -1 \\ 0 & 0 & 1 & \frac{2}{3} & \frac{5}{3} & -\frac{4}{3} \end{array}\right]$$

We complete the process by adding the third row to the first.

$$\left[\begin{array}{ccc|ccc} 1 & 0 & -1 & -1 & -2 & 2 \\ 0 & 1 & 0 & 1 & 1 & -1 \\ 0 & 0 & 1 & \frac{2}{3} & \frac{5}{3} & -\frac{4}{3} \end{array}\right] \mapsto \left[\begin{array}{ccc|ccc} 1 & 0 & 0 & -\frac{1}{3} & -\frac{1}{3} & \frac{2}{3} \\ 0 & 1 & 0 & 1 & 1 & -1 \\ 0 & 0 & 1 & \frac{2}{3} & \frac{5}{3} & -\frac{4}{3} \end{array}\right]$$

By this process, we have computed

$$\begin{bmatrix} 1 & 2 & -1 \\ 2 & 0 & 1 \\ 3 & 1 & 0 \end{bmatrix}^{-1} = \begin{bmatrix} -\frac{1}{3} & -\frac{1}{3} & \frac{2}{3} \\ 1 & 1 & -1 \\ \frac{2}{3} & \frac{5}{3} & -\frac{4}{3} \end{bmatrix}.$$

Question 5.5.2

(You knew this was coming ...) Verify that the product of the matrix and computed inverse is I_3.

If the matrix had not been invertible, at some point we would have gotten all zeroes in a row or column on the left. You could also have checked if the *determinant* was 0.

To learn more

The computational complexity of inversion of an n by n square matrix by Gaussian elimination is $O(n^3)$. Don Coppersmith and Shmuel Winograd of IBM Research developed an algorithm that brought this down to $O(n^{2.376})$ in 1987. [2] Subsequent optimizations have brought this complexity down only slightly.

Determinants

Ah, the determinant, a function on square matrices that produces values in $\mathbb{F}$. It's so elegant, so useful, tells us so much, and is such an annoying and error prone thing to compute beyond the 2 by 2 case.

Let's look at its properties before we discuss its calculation. Let A and B be n by n square matrices. We denote by $\det(A)$ and $\det(B)$ their determinants.

- $\det(A) \neq 0$ if and only if A is invertible.
- $\det(A^{-1}) = \frac{1}{\det(A)}$ for invertible A.
- $\det(A) = \det(A^\mathsf{T})$.
- $\det(\overline{A}) = \overline{\det(A)}$.
- $\det(A^\dagger) = \overline{\det(A)}$.
- For b a scalar in $\mathbb{F}$, $\det(bA) = b^n \det(A)$.
- If any row or column of A is all zeroes, then $\det(A) = 0$. The determinant being zero does not imply a row or column is zero.
- If all the entries of A above or below the main diagonal are zero, the determinant is the product of the diagonal entries. If one of those diagonal entries is 0, the determinant is thus 0.
- In particular, $\det(I) = 1$ for $A = I$ an identity matrix.
- $\det(AB) = \det(A)\det(B) = \det(BA)$.

Seriously, isn't this a great function?

For a 2 by 2 matrix

$$\det\left(\begin{bmatrix} a & b \\ c & d \end{bmatrix}\right) = ad - bc.$$

Verify the above properties given this formula.

Question 5.5.3

What is the determinant of a rotation matrix in $\mathbb{R}^2$?

After 2 by 2, life gets more complicated. For the 3 by 3 case we must compute some determinants of 2 by 2 submatrices, multiply them by other matrix entries, and remember when to add or subtract them.

$$\det\left(\begin{bmatrix} a_{1,1} & a_{1,2} & a_{1,3} \\ a_{2,1} & a_{2,2} & a_{2,3} \\ a_{3,1} & a_{3,2} & a_{3,3} \end{bmatrix}\right) = a_{1,1}\det\left(\begin{bmatrix} a_{2,2} & a_{2,3} \\ a_{3,2} & a_{3,3} \end{bmatrix}\right) -$$

$$a_{2,1}\det\left(\begin{bmatrix} a_{2,1} & a_{2,3} \\ a_{3,1} & a_{3,3} \end{bmatrix}\right) +$$

$$a_{3,1}\det\left(\begin{bmatrix} a_{2,1} & a_{2,2} \\ a_{3,1} & a_{3,2} \end{bmatrix}\right)$$

If you work this out you get

$$-a_{1,1}\,a_{2,3}\,a_{3,2} + a_{1,1}\,a_{2,2}\,a_{3,3} + a_{1,2}\,a_{2,3}\,a_{3,1}-$$
$$a_{1,2}\,a_{2,1}\,a_{3,3} - a_{1,3}\,a_{2,2}\,a_{3,1} + a_{1,3}\,a_{2,1}\,a_{3,2}$$

There's a formula and a pattern to this and books on linear algebra spell out the general case.

My advice? Use a good calculator or math software to compute determinants.

Consider the matrix

$$A = \begin{bmatrix} 2 & 0 \\ 2 & 3 \end{bmatrix} = \begin{bmatrix} 1 & 0 \\ 1 & 1 \end{bmatrix}\begin{bmatrix} 2 & 0 \\ 0 & 3 \end{bmatrix}.$$

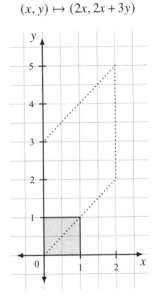

$(x, y) \mapsto (2x, 2x + 3y)$

This is a linear transformation on $\mathbb{R}^2$ that first stretches by 2 in the x direction and 3 in the y direction, followed by a vertical shear. (Remember multiplication corresponds to composition and is done right to left.) It has this effect on the unit square shown on the right.

The area of a parallelogram is its height times its width, which is $3 \times 2 = 6$ in this case.

Using the formula for the 2 by 2 determinant, $\det(A) = 2 \times 3 - 0 \times 2 = 6$. Had we multiplied by a reflection matrix across the x axis we would have gotten the mirror image parallelogram with the same area but the determinant would have been -6.

For a 2 by 2 matrix A operating on $\mathbb{R}^2$, the area of the parallelogram formed by applying A to the unit square is $|\det(A)|$.

Question 5.5.4

Draw the effect on the unit square by multiplying by the product of the matrices

$$\begin{bmatrix} 1 & 0 \\ 0 & -1 \end{bmatrix}\begin{bmatrix} 2 & 0 \\ 2 & 3 \end{bmatrix}$$

and compute its determinant.

In $\mathbb{R}^3$, the unit cube is formed by completing the cube starting with the standard basis $\mathbf{e_1} = (1,0,0)$, $\mathbf{e_2} = (0,1,0)$, and $\mathbf{e_3} = (0,0,1)$ to get the other corners at $(0,0,0)$, $(0,1,1)$, $(1,0,1)$, $(1,1,0)$, and $(1,1,1)$.

> For a 3 by 3 matrix A operating on $\mathbb{R}^3$, the volume of the parallelepiped formed by applying A to the unit cube is $|\det(A)|$.

You get the idea.

In the general case of the n-dimensional real vector space $\mathbb{R}^n$, the points

$$(x_1, x_2, \ldots, x_n)$$

with each coordinate having a value of either 0 or 1 gives us the 2^n corners of the unit *hypercube*. The multidimensional volume of the unit hypercube is 1.

For a n by n matrix A operating on $\mathbb{R}^n$, the multidimensional volume of the *hyperparallelepiped* formed by applying A to the unit hypercube is $|\det(A)|$. I'm not the first to use "hyperparallelepiped," but I wish I had been.

If the area or volume is 0, we have collapsed the image of the linear transformation down to something that is less than n dimensional. Hence it is not invertible. Think of collapsing a cube down to its square base via a projection like $(x, y, z) \mapsto (x, y, 0)$. We cannot recover what the original z value was.

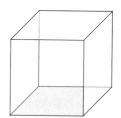

While all matrices with given dimensions over a field are a vector space, subsets of these matrices can have more algebraic structure.

> For a given n in $\mathbb{N}$, the collection of all n by n invertible square matrices with entries in $\mathbb{F}$ form a group under matrix multiplication called the *general linear group* of degree n over $\mathbb{F}$. It is denoted by $\mathbf{GL}(n, \mathbb{F})$.

> The collection of all n by n square matrices with entries in $\mathbb{F}$ and with **determinant 1** form a group under multiplication called the *special linear group* of degree n over $\mathbb{F}$. It is denoted by $\mathbf{SL}(n, \mathbb{F})$ and is a subgroup of the general linear group $\mathbf{GL}(n, \mathbb{F})$.

Traces

Whereas the determinant of a square matrix can be hard to compute, calculating the *trace* for that matrix is easy. Just compute the sum of the diagonal elements. Its value is therefore in $\mathbb{F}$.

If
$$A = \begin{bmatrix} a_{1,1} & \cdots & a_{1,n} \\ \vdots & \vdots & \vdots \\ a_{n,1} & \cdots & a_{n,n} \end{bmatrix}$$

then the trace of A is
$$\text{tr}(A) = a_{1,1} + a_{2,2} + \cdots + a_{n,n}.$$

The trace is a linear map from the vector space of all n by n matrices over $\mathbb{F}$ to $\mathbb{F}$. This follows from the entry-by-entry definitions of addition and scalar multiplication.

For A, B, and C square matrices as d in $\mathbb{F}$,

$$\text{tr}(A + B) = \text{tr}(A) + \text{tr}(B)$$
$$\text{tr}(dA) = d\,\text{tr}(A)$$
$$\text{tr}(-A) = -\text{tr}(A)$$
$$\text{tr}(AB) = \text{tr}(BA)$$
$$\text{tr}(ABC) = \text{tr}(CAB) = \text{tr}(BCA)$$

While it is true that $\text{tr}(AB) = \text{tr}(BA)$ it is **not** generally true that $\text{tr}(AB) = \text{tr}(A)\text{tr}(B)$. Remember this as it is an easy mistake to make. The determinant behaves well under multiplication and the trace behaves well under addition.

The transpose of a square matrix does not change the main diagonal and so

$$\text{tr}(A) = \text{tr}\left(A^{\mathsf{T}}\right).$$

For a complex square matrix,
$$\text{tr}\left(\overline{A}\right) = \text{tr}\left(A^{\dagger}\right) = \overline{\text{tr}(A)}.$$

Question 5.5.5

Why is the trace of a Hermitian matrix real?

The trace does not have a simple geometric interpretation like the determinant, but it enters the picture when we consider changing from one basis of a vector space to another. While I present it here as being for square matrices, the trace is defined for all matrices: just sum the elements on the main diagonal.

5.6 Cartesian products

A Cartesian product of two vectors spaces is a simple construction useful for expressing functions and maps.

If V and W are vector spaces over $\mathbb{F}$ then $V \times W$ is the set of all pairs $(\mathbf{v}, \mathbf{w})$ for $\mathbf{v}$ in V and $\mathbf{w}$ in W.

For example, consider

$$f : V \times W \to U$$

into a third vector space U. When we write $f(\mathbf{v}, \mathbf{w})$ we can either think of this as a function of two variables or a function that maps pairs into U.

If V, W and U are vector spaces, consider $f : V \times W \to U$. If for all scalars a, $\mathbf{v_1}$ and $\mathbf{v_2}$ in V, $\mathbf{w_1}$ and $\mathbf{w_2}$ in W, we have

$$af(\mathbf{v_1}, \mathbf{w_1}) = f(a\mathbf{v_1}, \mathbf{w_1}) = f(\mathbf{v_1}, a\mathbf{w_1})$$
$$f(\mathbf{v_1} + \mathbf{v_2}, \mathbf{w_1}) = f(\mathbf{v_1}, \mathbf{w_1}) + f(\mathbf{v_2}, \mathbf{w_1})$$
$$f(\mathbf{v_1}, \mathbf{w_1} + \mathbf{w_2}) = f(\mathbf{v_1}, \mathbf{w_1}) + f(\mathbf{v_1}, \mathbf{w_2})$$

then f is *bilinear*. Said another way, f is bilinear if it is linear in each coordinate.

Question 5.6.1

Let $h : \mathbb{R}^2 \times \mathbb{R}^2 \to \mathbb{C}^2$ be defined by

$$h((a, b), (c, d)) = (a + bi, c + di).$$

Is h bilinear?

5.7 Length and preserving it

Length is a natural notion in the real world, but it needs to be defined precisely in vector spaces. Using complex numbers complicates things because we need to use conjugation. Length is related to *magnitude*, which is a measure of how big something is. Understanding length and norms is key to the mathematics of quantum algorithms, as we shall see in chapter 10.

5.7.1 Dot products

Let V be a finite dimensional vector space over $\mathbb{R}$ or $\mathbb{C}$ and let $\mathbf{v} = (v_1, v_2, \ldots, v_n)$ and $\mathbf{w} = (w_1, w_2, \ldots, w_n)$ be two vectors in V.

The *dot product* of $\mathbf{v}$ and $\mathbf{w}$ is the sum of the products of the corresponding entries in $\mathbf{v}$ and $\mathbf{w}$.

$$\mathbf{v} \cdot \mathbf{w} = v_1 w_1 + v_2 w_2 + \cdots + v_n w_n$$

If we think of $\mathbf{v}$ and $\mathbf{w}$ as row vectors, and so as 1 by n matrices, then

$$\mathbf{v} \cdot \mathbf{w} = \mathbf{v}\mathbf{w}^{\mathsf{T}}.$$

The dot product of the basis vectors $\mathbf{e}_1 = (1, 0)$ and $\mathbf{e}_2 = (0, 1)$ is 0. When this happens for real vectors we say they are *orthogonal*.

Another such real orthogonal basis pair is $\mathbf{h}_1 = \left(\frac{\sqrt{2}}{2}, \frac{\sqrt{2}}{2}\right)$ and $\mathbf{h}_2 = \left(\frac{\sqrt{2}}{2}, -\frac{\sqrt{2}}{2}\right)$ because

$$\frac{\sqrt{2}}{2}\frac{\sqrt{2}}{2} + \frac{\sqrt{2}}{2}\left(-\frac{\sqrt{2}}{2}\right) = \frac{2}{4} - \frac{2}{4} = 0.$$

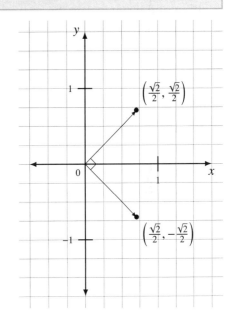

5.7.2 Inner products

Continuing with V over $\mathbb{F} = \mathbb{R}$ or $\mathbb{F} = \mathbb{C}$, we define an *inner product* to be an $\mathbb{F}$-valued function $\langle \mathbf{v}, \mathbf{w} \rangle$ on two vectors such that:

- $\langle \mathbf{v}, \mathbf{v} \rangle$ is real and ≥ 0.

- $\langle \mathbf{v}, \mathbf{v} \rangle = 0$ if and only if $\mathbf{v}$ is zero.
- $\langle \mathbf{v}, \mathbf{w} \rangle = \overline{\langle \mathbf{w}, \mathbf{v} \rangle}$
- $\langle \mathbf{v}, \mathbf{u} + \mathbf{w} \rangle = \langle \mathbf{v}, \mathbf{u} \rangle + \langle \mathbf{v}, \mathbf{w} \rangle$.
- $\langle \mathbf{v}, a\mathbf{w} \rangle = a\langle \mathbf{v}, \mathbf{w} \rangle$ for a in $\mathbb{F}$.

At this point some mathematicians might say "wait, your complex inner product is linear in the second position instead of the first!". I was taught that these should be the fourth and fifth conditions defining an inner product:

- $\langle \mathbf{u} + \mathbf{v}, \mathbf{w} \rangle = \langle \mathbf{u}, \mathbf{w} \rangle + \langle \mathbf{v}, \mathbf{w} \rangle$.
- $\langle a\mathbf{v}, \mathbf{w} \rangle = a\langle \mathbf{v}, \mathbf{w} \rangle$ for a in $\mathbb{F}$.

Alas, physicists use the first forms and it makes conventions like Dirac's bra-ket notation work out more easily. So we adopt the physicists' version of this, though it pains me somehow.

It's useful if you look at this again thinking about a real vector space and hence conjugation being the identity function. The dot product for a vector space over $\mathbb{R}$ is an example of a real inner product. Experiment with the conditions above with respect to the left-hand side of the inner product.

The inner product with the $\mathbf{0}$ vector on either side is 0.

Question 5.7.1

Why is $\langle a\mathbf{v}, \mathbf{w} \rangle = \overline{a}\langle \mathbf{v}, \mathbf{w} \rangle$ for a in $\mathbb{F}$?

Whereas $\langle \mathbf{v}, \mathbf{w} \rangle = \mathbf{v} \cdot \mathbf{w}$ sufficed in the real case, it does not in the complex one. The correct generalization is $\langle \mathbf{v}, \mathbf{w} \rangle = \overline{\mathbf{v}} \cdot \mathbf{w}$. Unless otherwise specified, these are the default inner products we use for real and complex vector spaces.

The properties of the complex inner product imply if A is a square complex matrix then

$$\langle A\mathbf{v}, \mathbf{w} \rangle = \langle \mathbf{v}, A^{\dagger}\mathbf{w} \rangle.$$

That is, you can move the application of a linear transformation from one side of an inner product to the other by taking its conjugate transpose.

If we think of $\mathbf{v}$ and $\mathbf{w}$ as complex row vectors and so 1 by n matrices, then

$$\langle \mathbf{v}, \mathbf{w} \rangle = \overline{\mathbf{v}}\mathbf{w}^{\mathsf{T}}.$$

A vector space with an inner product is called ... wait for it ... an *inner product space*.

Question 5.7.2

If V is a vector space over $\mathbb{F}$ and $\mathbb{F} = \mathbb{R}$, is the inner product

$$\langle \, , \, \rangle : V \times V \to \mathbb{F}$$

a bilinear map? What if $\mathbb{F} = \mathbb{C}$? Don't forget about conjugation.

If A is a Hermitian matrix, then $\langle A\mathbf{v_1}, \mathbf{v_2} \rangle = \langle \mathbf{v_1}, A\mathbf{v_2} \rangle$ for all vectors $\mathbf{v_1}$ and $\mathbf{v_2}$ in V.

5.7.3 Euclidean norm

Let $\mathbf{v} = (v_1, v_2, \ldots, v_n)$ be a vector in V. The *Euclidean norm* of $\mathbf{v}$, also called the *length* of $\mathbf{v}$, is $\|\mathbf{v}\| = \sqrt{\langle \mathbf{v}, \mathbf{v} \rangle}$.

A vector with length 1 is called a *unit vector*. Two unit vectors are *orthonormal* if they are orthogonal, meaning their inner product is 0. A set of unit vectors such as a basis is orthonormal if they are pairwise orthogonal.

The standard basis vectors in any vector space $\mathbb{R}^n$ or $\mathbb{C}^m$ for n and m integers ≥ 1 are orthonormal.

Basis vectors don't need to be orthonormal but it is very helpful in calculations. This is because of the potential cancellation in inner products and the vector lengths not skewing the contributions of the sizes of the coordinates.

For example, if $\mathbf{v}$ is a unit vector then $\|a\mathbf{v}\| = |a| \times \|\mathbf{v}\| = |a|$.

A matrix whose columns are orthonormal with respect to each other, or whose rows are orthonormal to each other, is called an *orthogonal matrix*.

Question 5.7.3

Why is the transpose of an orthogonal matrix A also the inverse of A?

A real or complex finite dimensional vector space with the default inner product and Euclidean norm is an example of a *Hilbert space*, named after the mathematician David Hilbert. In full generality, Hilbert spaces may have infinite dimensions and use other norms. It's very common in quantum mechanics and quantum computing to talk about Hilbert spaces, but you only need to remember the particular examples we discuss here.

We can now give another interpretation of the dot product and Euclidean norm in a vector space V over $\mathbb{R}$. Let **v** and **w** be in V. Then the dot product of these vectors is equal to the product of their lengths times the cosine of the angle θ between them:

$$\mathbf{v} \cdot \mathbf{w} = \|\mathbf{v}\| \, \|\mathbf{w}\| \cos(\theta).$$

David Hilbert in 1912. Photo is in the public domain. ⓟ

As an example, let $\mathbf{v} = (4, 0)$ and $\mathbf{w} = (2, 3)$ in $\mathbb{R}^2$.

$$\mathbf{v} \cdot \mathbf{w} = 4 \times 2 + 0 \times 3 = 8$$
$$\|\mathbf{v}\| = \sqrt{4^2 + 0^2} = 4$$
$$\|\mathbf{w}\| = \sqrt{2^2 + 3^2} = \sqrt{13}$$

and so $8 = 4\sqrt{13}\cos(\theta)$. Solving for $\cos(\theta)$ we find it equals $\frac{2}{\sqrt{13}} \approx 0.5547$. Here "$\approx$" means "approximately equal to."

Applying the *inverse cosine* function arccos via a calculator or software, we find that $\theta \approx 0.9827$.

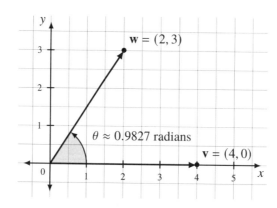

When $\cos(\theta) = 0$ it literally means the two vectors are at a right angle to each other and so are orthogonal. From the formula, this implies the dot product is zero.

5.7.4 Reflections again

In two dimensions, if $\mathbf{v} = (v_1, v_2)$ is a non-zero vector, then $\mathbf{w} = (-v_2, v_1)$ is orthogonal to $\mathbf{v}$ because $\mathbf{v} \cdot \mathbf{w} = 0$. The vectors have the same length. If we set

$$\mathbf{v}' = \left(\frac{v_1}{\|\mathbf{v}\|}, \frac{v_2}{\|\mathbf{v}\|} \right) \quad \text{and} \quad \mathbf{w}' = \left(-\frac{v_2}{\|\mathbf{v}\|}, \frac{v_1}{\|\mathbf{v}\|} \right)$$

then $\mathbf{v}'$ and $\mathbf{w}'$ are orthonormal. because they have length 1 and are orthogonal.

The notation $\mathbf{v}'$ is pronounced "v prime." It's a common way of showing that we have modified a named object like the vector $\mathbf{v}$ in some way.

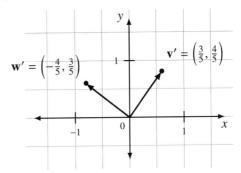

If $\mathbf{v} = (3, 4)$ then $\mathbf{w} = (-4, 3)$. Therefore, $\mathbf{v}' = \left(\frac{3}{5}, \frac{4}{5} \right)$ and $\mathbf{w}' = \left(-\frac{4}{5}, \frac{3}{5} \right)$.

Given a line L through the origin in $\mathbb{R}^2$, let $\mathbf{v}$ be a unit vector on the line. For example, if L is the x axis then we can take $\mathbf{v} = (1, 0)$ or $(-1, 0)$. Let $\mathbf{w}$ be a unit vector orthogonal to $\mathbf{v}$.

The transformation

$$\mathbf{x} \mapsto \mathbf{x} - 2(\mathbf{x} \cdot \mathbf{w})\mathbf{w}$$

is a reflection across the line L.

Let's test this. If L is the x axis then $\mathbf{v} = (1, 0)$ and $\mathbf{w} = (0, 1)$. Then

$$(x, y) \mapsto (x, y) - 2((x, y) \cdot (0, 1))(0, 1) = (x, y) - (0, 2y) = (x, -y)$$

which we know is the right reflection.

Similarly, let L be the line $y = x$. Take $\mathbf{v} = \left(\frac{\sqrt{2}}{2}, \frac{\sqrt{2}}{2}\right)$ and $\mathbf{w} = \left(-\frac{\sqrt{2}}{2}, \frac{\sqrt{2}}{2}\right)$. So

$$
\begin{aligned}
(x, y) \mapsto\ & (x, y) - 2\left((x, y) \cdot \left(-\frac{\sqrt{2}}{2}, \frac{\sqrt{2}}{2}\right)\right)\left(-\frac{\sqrt{2}}{2}, \frac{\sqrt{2}}{2}\right) \\
=\ & (x, y) - 2\left(-x\frac{\sqrt{2}}{2} + y\frac{\sqrt{2}}{2}\right)\left(-\frac{\sqrt{2}}{2}, \frac{\sqrt{2}}{2}\right) \\
=\ & (x, y) - 2\left(\frac{1}{2}x - \frac{1}{2}y, -\frac{1}{2}x + \frac{1}{2}y\right) \\
=\ & (x, y) - (x - y, -x + y) \\
=\ & (y, x)
\end{aligned}
$$

We saw this when we first looked at reflections.

This is a transformation and so it should have a matrix. Let $\mathbf{x} = (x, y)$, $\mathbf{v} = (v_1, v_2)$, and $\mathbf{w} = (-v_2, v_1)$. We transform

$$\mathbf{x} \mapsto \mathbf{x} - 2(\mathbf{x} \cdot \mathbf{w})\mathbf{w}$$

to

$$
\begin{aligned}
(x, y) \mapsto\ & (x, y) - 2((x, y) \cdot (-v_2, v_1))\,(-v_2, v_1) \\
=\ & (x, y) - 2\,(-xv_2 + yv_1)\,(-v_2, v_1) \\
=\ & (x, y) + (2xv_2 - 2yv_1)\,(-v_2, v_1) \\
=\ & (x, y) + \left(-2xv_2^2 + 2yv_1v_2, 2xv_1v_2 - 2yv_1^2\right) \\
=\ & \left(x - 2xv_2^2 + 2yv_1v_2, y + 2xv_1v_2 - 2yv_1^2\right) \\
=\ & \left(x\left(1 - 2v_2^2\right) + 2yv_1v_2, 2xv_1v_2 + y\left(1 - 2v_1^2\right)\right) \\
=\ & \begin{bmatrix} 1 - 2v_2^2 & 2v_1v_2 \\ 2v_1v_2 & 1 - 2v_1^2 \end{bmatrix}\begin{bmatrix} x \\ y \end{bmatrix}
\end{aligned}
$$

Question 5.7.4

Since $\mathbf{v}$ is a unit vector, $v_1^2 + v_2^2 = 1$. What is the determinant of the reflection transformation matrix

$$A = \begin{bmatrix} 1 - 2v_2^2 & 2v_1v_2 \\ 2v_1v_2 & 1 - 2v_1^2 \end{bmatrix}?$$

Show that $A\,A = I_2$, that is, that A is its own inverse. Simplify the expressions as much as possible.

Since $\mathbf{v}$ is a unit vector, there is an angle θ such that $v_1 = \cos(\theta)$ and $v_2 = \sin(\theta)$. θ is the angle that $\mathbf{v}$ makes with the positive x axis. We can rewrite the transformation matrix using the trigonometric forms.

$$A = \begin{bmatrix} 1 - 2v_2^2 & 2v_1v_2 \\ 2v_1v_2 & 1 - 2v_1^2 \end{bmatrix} = \begin{bmatrix} 1 - 2\sin(\theta)^2 & 2\cos(\theta)\sin(\theta) \\ 2\cos(\theta)\sin(\theta) & 1 - 2\cos(\theta)^2 \end{bmatrix}$$

Recalling the trigonometric double angle formulas, and computing

$$1 - 2\cos(\theta)^2 = 1 - 2(1 - \sin(\theta)^2) = -1 + \sin(\theta)^2 = -\cos(2\theta)$$

the last matrix reduces, perhaps surprisingly, to

$$\begin{bmatrix} 1 - 2\sin(\theta)^2 & 2\cos(\theta)\sin(\theta) \\ 2\cos(\theta)\sin(\theta) & 1 - 2\cos(\theta)^2 \end{bmatrix} = \begin{bmatrix} \cos(2\theta) & \sin(2\theta) \\ \sin(2\theta) & -\cos(2\theta) \end{bmatrix}$$

Let L be a line through the origin in $\mathbb{R}^2$. Let θ be the angle the line makes with the positive x axis. Then the transformation matrix for the reflection across the line is

$$\begin{bmatrix} \cos(2\theta) & \sin(2\theta) \\ \sin(2\theta) & -\cos(2\theta) \end{bmatrix}.$$

Let's compose this with another reflection for an angle φ.

$$\begin{bmatrix} \cos(2\theta) & \sin(2\theta) \\ \sin(2\theta) & -\cos(2\theta) \end{bmatrix} \begin{bmatrix} \cos(2\varphi) & \sin(2\varphi) \\ \sin(2\varphi) & -\cos(2\varphi) \end{bmatrix}$$

This equals

$$\begin{bmatrix} \cos(2\theta)\cos(2\varphi) + \sin(2\theta)\sin(2\varphi) & \cos(2\theta)\sin(2\varphi) - \sin(2\theta)\cos(2\varphi) \\ \sin(2\theta)\cos(2\varphi) + \sin(2\theta)\sin(2\varphi) & \sin(2\theta)\sin(2\varphi) + \cos(2\theta)\cos(2\varphi) \end{bmatrix}$$

---

or

$$\begin{bmatrix} \cos(2\theta - 2\varphi) & -\sin(2\theta - 2\varphi) \\ \sin(2\theta - 2\varphi) & \cos(2\theta - 2\varphi) \end{bmatrix}$$

because of the trigonometric sum and difference formulas.

Drum roll please ...

This is a rotation matrix for an angle of $2\theta - 2\varphi$!

> The composition of two reflection transformations in $\mathbb{R}^2$ is a rotation.

5.7.5 Unitary transformations

What are the characteristics of linear transformations that preserve length? That is, if $L : V \to V$ and it is always true that $\|L(\mathbf{v})\| = \|\mathbf{v}\|$, what can we say about the matrix of L?

> If U is a complex square matrix then it is *unitary* if its adjoint $U^\dagger$ is also its inverse U^{-1}. Hence $UU^\dagger = U^\dagger U = I$. This means the columns of U are orthonormal, as are the rows.

Next, $|\det(U)| = 1$. This follows because

$$
\begin{aligned}
1 &= \det(I) & &= \det(UU^\dagger) \\
&= \det(U)\det(U^\dagger) & &= \det(U)\det\left(\overline{U}^\mathsf{T}\right) \\
&= \det(U)\det\left(\overline{U}\right) & &= \det(U)\,\overline{\det(U)} \\
&= |\det(U)|^2
\end{aligned}
$$

To prove unitary matrices preserve length, we need to do more transposition and conjugation math.

$$\|\mathbf{v}\|^2 = \langle U\mathbf{v}, U\mathbf{v}\rangle = \langle \mathbf{v}, U^\dagger U\mathbf{v}\rangle = \langle \mathbf{v}, \mathbf{v}\rangle = \|\mathbf{v}\|^2$$

Since the lengths are non-negative, $\|U\mathbf{v}\| = \|\mathbf{v}\|$. Conversely, if this holds then we must have $U^\dagger U = I$.

Rotations and reflections are unitary. An orthogonal matrix is unitary.

> The matrices corresponding to gates/operations on qubits are all unitary.

Identity matrices are unitary, as are the three Pauli matrices:

$$\sigma_x = \begin{bmatrix} 0 & 1 \\ 1 & 0 \end{bmatrix} \qquad \sigma_y = \begin{bmatrix} 0 & -i \\ i & 0 \end{bmatrix} \qquad \sigma_z = \begin{bmatrix} 1 & 0 \\ 0 & -1 \end{bmatrix}$$

The Greek letter σ is "sigma." The Pauli matrices have great importance in quantum mechanics and satisfy several useful identities.

$$\det(\sigma_x) = \det(\sigma_y) = \det(\sigma_z) = -1$$
$$\mathrm{tr}(\sigma_x) = \mathrm{tr}(\sigma_y) = \mathrm{tr}(\sigma_z) = 0$$
$$\sigma_x^2 = \sigma_y^2 = \sigma_z^2 = -i\sigma_x\sigma_y\sigma_z = I_2$$
$$\sigma_x\sigma_y = i\sigma_z = -\sigma_y\sigma_x$$
$$\sigma_y\sigma_z = i\sigma_x = -\sigma_z\sigma_y$$
$$\sigma_z\sigma_x = i\sigma_y = -\sigma_x\sigma_z$$

This is a Hadamard matrix, also unitary:

$$\begin{bmatrix} \frac{1}{2} & \frac{1}{2} & \frac{1}{2} & \frac{1}{2} \\ \frac{1}{2} & -\frac{1}{2} & \frac{1}{2} & -\frac{1}{2} \\ \frac{1}{2} & \frac{1}{2} & -\frac{1}{2} & -\frac{1}{2} \\ \frac{1}{2} & -\frac{1}{2} & -\frac{1}{2} & \frac{1}{2} \end{bmatrix} = \frac{1}{2}\begin{bmatrix} 1 & 1 & 1 & 1 \\ 1 & -1 & 1 & -1 \\ 1 & 1 & -1 & -1 \\ 1 & -1 & -1 & 1 \end{bmatrix}$$

Computationally, being unitary is a very good property for a matrix to have because computing inverses can be hard but adjoints are straightforward.

The product $U_1 U_2$ of two unitary matrices U_1 and U_2 is unitary.

$$(U_1 U_2)(U_1 U_2)^\dagger = (U_1 U_2)\left(U_2{}^\dagger U_1{}^\dagger\right) = U_1\left(U_2 U_2{}^\dagger\right)U_1{}^\dagger = U_1 U_1{}^\dagger = I$$

For a given integer n in $\mathbb{N}$, the collection of all n by n unitary matrices with entries in $\mathbb{F}$ form a group under multiplication called the *unitary group* of degree n over $\mathbb{F}$. It is denoted by $\mathbf{U}(n, \mathbb{F})$. It is a subgroup of $\mathbf{GL}(n, \mathbb{F})$, the general linear group of degree n over $\mathbb{F}$.

For a given integer n in $\mathbb{N}$, the collection of all n by n unitary matrices with entries in $\mathbb{F}$ and with **determinant 1** form a group under multiplication called the *special unitary group* of degree n over $\mathbb{F}$. It is denoted by $\mathbf{SU}(n, \mathbb{F})$.

$\mathbf{SU}(n, \mathbb{F})$ is a subgroup of $\mathbf{U}(n, \mathbb{F})$. It is also a subgroup of $\mathbf{SL}(n, \mathbb{F})$, the special linear group of degree n over $\mathbb{F}$.

Question 5.7.5

Let P be an n by n square matrix, all of whose entries are either 0 or 1. Further assume that each row and each column of P contains exactly one 1. P is an example of a *permutation matrix*.

- If A is an n by n matrix, how does PA differ from A with respect to the order of its rows?
- How does AP differ from A with respect to the order of its columns?
- Show that P is unitary.

Without question, unitary matrices and transformations are the most important kinds for quantum computing.

To learn more

Though there other concepts in linear algebra yet to cover in this chapter, now is a good time to stop and point you to some other references. Linear algebra can be treated abstractly [4] [6] or it can be approached as a tool for applied mathematics [8] or engineering [7]. Sometimes it is simply presented as manipulations of matrices and vectors. Depending on your background, you may find one text or another more comfortable.

5.7.6 Systems of linear equations

The two equations

$$2x + 3y = 5$$
$$x - 2y = 2$$

together are an example of a system of linear equations. On the left-hand side are linear equations and on the right are constants. In $\mathbb{R}^2$, these represent two lines. The lines may either be the same, be parallel and so never intersect, or intersect in a single point.

If we use subscripted variables, the same relationship is expressed by

$$2x_1 + 3x_2 = 5$$
$$x_1 - 2x_2 = 2.$$

We can further rewrite this in matrix and vector form as

$$\begin{bmatrix} 2 & 3 \\ 1 & -2 \end{bmatrix} \begin{bmatrix} x_1 \\ x_2 \end{bmatrix} = \begin{bmatrix} 5 \\ 2 \end{bmatrix}.$$

If we let

$$A = \begin{bmatrix} 2 & 3 \\ 1 & -2 \end{bmatrix} \qquad \mathbf{x} = \begin{bmatrix} x_1 \\ x_2 \end{bmatrix} \qquad \mathbf{b} = \begin{bmatrix} 5 \\ 2 \end{bmatrix}$$

then our system is simply $A\mathbf{x} = \mathbf{b}$. This is a standard form for writing such systems. This is called a linear equation.

Our goal may be to solve for all of $\mathbf{x}$, learn only some of the x_k, or understand some function f applied to $\mathbf{x}$. If A is invertible, then

$$A\mathbf{x} = \mathbf{b} \quad \Rightarrow \quad A^{-1}A\mathbf{x} = A^{-1}\mathbf{b} \quad \Rightarrow \quad \mathbf{x} = A^{-1}\mathbf{b}.$$

In this case, there is one possible value for $\mathbf{x}$. If A is not invertible, then there might be no solution or a vector space's worth of solutions.

Question 5.7.6

Solve for x_1 and x_2 in

$$2x_1 + 3x_2 = 5$$
$$x_1 - 2x_2 = 2$$

and graph the result.

Question 5.7.7

Find a system of two equations in two variables that has no solution. Find a system whose solution is a line. Find a system whose solution is $\mathbb{R}^2$.

Suppose we have a linear equation $A\mathbf{x} = \mathbf{b}$ with A an invertible square matrix. Assume $\mathbf{b}$ is not quite an exact quantity. This might be because we got it from approximation or reading scientific instruments. If $\mathbf{b_0}$ is the "right answer" then $\mathbf{b} = \mathbf{b_0} + \epsilon$ for some little (we hope) error vector ϵ. So

$$A\mathbf{x} = \mathbf{b} = \mathbf{b_0} + \epsilon$$

and

$$\mathbf{x} = A^{-1}\mathbf{b} = A^{-1}(\mathbf{b_0} + \epsilon) = A^{-1}\mathbf{b_0} + A^{-1}\epsilon.$$

This means that $\mathbf{x}$ differs from the "correct" value $A^{-1}\mathbf{b_0}$ by $A^{-1}\epsilon$.

If $\mathbf{b} \neq \mathbf{0}$, we can look at how large the error ϵ is compared to $\mathbf{b}$ by examining $\frac{\|\epsilon\|}{\|\mathbf{b}\|}$. Similarly, if we look at

$$\frac{\left\|A^{-1}\epsilon\right\|}{\left\|A^{-1}\mathbf{b}\right\|}$$

we can see the relative error in the solution.

How much does the error in $\mathbf{b}$ translate to error in the solution? We take yet another quotient,

$$\frac{\dfrac{\left\|A^{-1}\epsilon\right\|}{\left\|A^{-1}\mathbf{b}\right\|}}{\dfrac{\|\epsilon\|}{\|\mathbf{b}\|}}.$$

Do small errors get magnified or are they kept in check when we look at the solution? We want to know this as a property of A, so we define

$$\kappa(A) = \text{maximum over all non-zero } \epsilon \text{ and non-zero } \mathbf{b} \text{ of } \frac{\dfrac{\left\|A^{-1}\epsilon\right\|}{\left\|A^{-1}\mathbf{b}\right\|}}{\dfrac{\|\epsilon\|}{\|\mathbf{b}\|}}.$$

$\kappa(A)$ is called the *condition number* of A. (κ is the Greek letter "kappa.")

The condition number is ≥ 1. The closer the value is to 1, the better-behaved A is in the presence of errors.

Question 5.7.8

By rearranging the fractions,

$$\frac{\dfrac{\left\|A^{-1}\epsilon\right\|}{\left\|A^{-1}\mathbf{b}\right\|}}{\dfrac{\|\epsilon\|}{\|\mathbf{b}\|}} = \frac{\left\|A^{-1}\epsilon\right\|}{\left\|A^{-1}\mathbf{b}\right\|}\frac{\|\mathbf{b}\|}{\|\epsilon\|} = \frac{\left\|A^{-1}\epsilon\right\|}{\|\epsilon\|}\frac{\|\mathbf{b}\|}{\left\|A^{-1}\mathbf{b}\right\|}.$$

Show that if A is a unitary matrix, the condition number $\kappa(A) = 1$. Hint: if A is unitary, then so is A^{-1}.

When the condition number is very large, we have an "ill-posed" or "ill-conditioned" problem. Small errors in the input can translate to large errors in the answer. Algorithmically, it may translate to very bad performance or wildly inaccurate results. A different algorithm or additional information about the problem may be necessary for more stable computation.

5.8 Change of basis

Given an n-dimensional vector space V and a linear transformation $V \to V$, we can choose different bases for V. Let's call them $X = \{\mathbf{x_1}, \mathbf{x_2}, \ldots, \mathbf{x_n}\}$ and $Y = \{\mathbf{y_1}, \mathbf{y_2}, \ldots, \mathbf{y_n}\}$. With respect to X, the linear transformation has a matrix A_X. With respect to Y it has a different matrix A_Y, but they both implement the same linear transformation. If $\mathbf{v}$ is a vector in V, it has one set of coordinates corresponding to X and another set for Y.

How do we change from one set of coordinates for $\mathbf{v}$ to the other? What is the relationship between A_X and A_Y?

This topic can be confusing because it is easy to get yourself working with the wrong matrix or its inverse. If you start with the basic idea of what you are trying to represent, the rest takes care of itself.

Let's look at an example that demonstrates how choice of basis can make things easier.

Suppose you had city blocks laid out in a nice rectilinear pattern so that we can use the basis vectors $\mathbf{x_1} = (1,0)$ and $\mathbf{x_2} = (0,2)$ to position ourselves. (I've given the coordinates using the standard basis.)

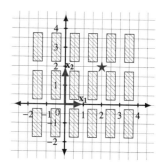

I can give you directions by saying "go north along $\mathbf{x_2}$ for 1 block, turn right and go east along $\mathbf{x_1}$ for 2 blocks." That would position you where the star is in the picture. In terms the X basis, the position is $2\mathbf{x_1} + \mathbf{x_2}$.

$\mathbf{x_1}$ and $\mathbf{x_2}$ are not the same length, on purpose, so the "units" work well with the width and height of the city blocks. Basis vectors need not have length one, of course.

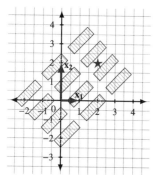

Things don't work out nearly so well if our city grid is laid out at an angle instead of a north-south east-west orientation. Indeed, cities are laid out based on their history, their growth, and their geographic settings. $\mathbf{x_1}$ and $\mathbf{x_2}$ no longer work if we want to walk along one basis vector and then the other. We would have to walk through buildings if we followed those paths.

We need to change to another basis which is more suitable for what we are trying to accomplish.

We want to find one which makes computation and moving around easier to do.

We have selected new basis vectors $\mathbf{y_1}$ and $\mathbf{y_2}$ that line up better with our road layout. Our city and the locations in it now use a new coordinate system.

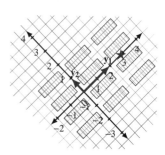

To get to the star we now say "walk $\sqrt{2}$ blocks along $\mathbf{y_1}$." For those of you who don't navigate using square roots, that's about 1.4 blocks.

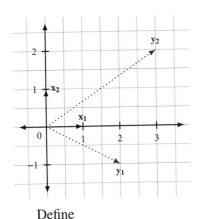

Getting back to bases and moving between them, we choose

$$X = E = \{x_1 = e_1 = (1,0), x_2 = e_2 = (0,1)\}$$

and

$$Y = \{y_1 = (2,-1)_E, y_2 = (3,2)_E\}.$$

Coordinates are with respect to the standard basis in these initial definitions.

Define

$$M_{Y,E} = \begin{bmatrix} 2 & 3 \\ -1 & 2 \end{bmatrix}$$

where the columns are coordinates of each basis vector in Y with respect to E.

If $v_Y = (1,0)_Y$ is the vector y_1 *written with respect to the basis* Y, then

$$M_{Y,E}\, v_Y = \begin{bmatrix} 2 & 3 \\ -1 & 2 \end{bmatrix} \begin{bmatrix} 1 \\ 0 \end{bmatrix}_Y = \begin{bmatrix} 2 & -1 \end{bmatrix}_E = \text{ the coordinates of } y_1 \text{ with respect to } E.$$

We also have for $v_Y = (0,1)_Y$ the vector y_2 *written with respect to the basis* Y,

$$M_{Y,E}\, v_Y = \begin{bmatrix} 2 & 3 \\ -1 & 2 \end{bmatrix} \begin{bmatrix} 0 \\ 1 \end{bmatrix}_Y = \begin{bmatrix} 3 & 2 \end{bmatrix}_E = \text{ the coordinates of } y_2 \text{ with respect to } E.$$

The matrix $M_{Y,E}$ transforms the coordinates of a vector v written with respect to the basis Y to the coordinates for v written with respect to the standard basis E. The columns of $M_{Y,E}$ are the basis vectors X written in coordinates with respect to E.

What if we want to go from coordinates with respect to E to coordinates with respect to X? It turns out this is easy, we use the inverse of $M_{Y,E}$:

$$M_{E,Y} = M_{Y,E}^{-1}.$$

In our example,

$$M_{Y,E} = \begin{bmatrix} 2 & 3 \\ -1 & 2 \end{bmatrix} \quad \text{and} \quad M_{E,Y} = M_{Y,E}^{-1} = \begin{bmatrix} \frac{2}{7} & -\frac{3}{7} \\ \frac{1}{7} & \frac{2}{7} \end{bmatrix}.$$

Question 5.8.1

Verify the product of $M_{E,Y}$ and coordinate vector of $\mathbf{y_1}$ written with respect to E is equal to $(1,0)_X$. What should it be for $\mathbf{y_2}$?

It is not always be the case that we are moving between E and another basis. Even when neither X nor Y is E, we can still make use of the standard basis.

Here is the fundamental thing to remember about changing bases: if you want to go from a representation under basis X to one under Y, go through the standard basis E as an intermediate step.

To go from arbitrary bases X and Y, compute $M_{X,E}$ and $M_{Y,E}$. By composition,

$$M_{X,Y} = M_{E,Y} \, M_{X,E} = M_{Y,E}^{-1} \, M_{X,E}.$$

5.9 Eigenvectors and eigenvalues

Let's pause to review some of the wonderful features of diagonal matrices. Recall that a diagonal matrix has 0s everywhere except maybe on the main diagonal. A simple example for $\mathbb{R}^3$ is

$$A = \begin{bmatrix} 3 & 0 & 0 \\ 0 & 1 & 0 \\ 0 & 0 & -2 \end{bmatrix}.$$

Its effect on the standard basis vectors $\mathbf{e_1}$, $\mathbf{e_2}$, and $\mathbf{e_3}$ is to stretch by a factor of 3 along the first, leave the second alone, and reflect across the xy-plane and then stretch by a factor of 2 along the third.

A general diagonal matrix looks like

$$D = \begin{bmatrix} d_1 & 0 & 0 & \cdots & 0 & 0 \\ 0 & d_2 & 0 & \cdots & 0 & 0 \\ \vdots & \vdots & \vdots & \ddots & \vdots & \vdots \\ 0 & 0 & 0 & \cdots & d_{n-1} & 0 \\ 0 & 0 & 0 & \cdots & 0 & d_n \end{bmatrix}.$$

Of course, we might be dealing with a small matrix and not have quite so many 0s.

For a diagonal matrix D as above,

- $\det(D) = d_1 d_2 \cdots d_n$.
- $\text{tr}(D) = d_1 + d_2 + \cdots + d_n$
- $D^{\mathsf{T}} = D$.
- D is invertible if and only if none of the d_i are 0.
- If D is invertible,

$$D^{-1} = \begin{bmatrix} \frac{1}{d_1} & 0 & 0 & \cdots & 0 & 0 \\ 0 & \frac{1}{d_2} & 0 & \cdots & 0 & 0 \\ \vdots & \vdots & \vdots & \ddots & \vdots & \vdots \\ 0 & 0 & 0 & \cdots & \frac{1}{d_{n-1}} & 0 \\ 0 & 0 & 0 & \cdots & 0 & \frac{1}{d_n} \end{bmatrix}$$

- If $\mathbf{b_1}, ..., \mathbf{b_n}$ is the basis we are using, then $D\mathbf{b_1} = d_1\mathbf{b_1}$, $D\mathbf{b_2} = d_2\mathbf{b_2}$, ..., $D\mathbf{b_n} = d_n\mathbf{b_n}$.

Focusing on this last effect on the basis vectors, given a general and not necessarily diagonal square matrix A, is there a vector $\mathbf{v}$ and a scalar λ such that $A\mathbf{v} = \lambda\mathbf{v}$?

That is, does the linear transformation represented by A have the effect of stretching in the direction of $\mathbf{v}$ by λ?

Two things to note: first, I've tried to limit the use of Greek letters but it is traditional to use lambda = λ in this context. Second, I'm abusing terminology a little here. In case λ is real and negative, we are technically stretching by $|\lambda|$ and doing a reflection.

Let A be a square matrix with entries in $\mathbb{F}$. If there exists a non-zero vector $\mathbf{v}$ and a scalar λ such that $A\mathbf{v} = \lambda\mathbf{v}$, then $\mathbf{v}$ is called an *eigenvector* of A and λ is the corresponding *eigenvalue*. "Eigen" comes from German and means "own" or "inherent."

Eigenvectors are also called *characteristic vectors* and eigenvalues are similarly called *characteristic values*, though the more German versions are in very wide use.

How can we find eigenvectors and their eigenvalues? It goes back to the determinant.

If $A\mathbf{v} = \lambda\mathbf{v}$ then $A\mathbf{v} - \lambda\mathbf{v} = \mathbf{0}$. Here $\mathbf{0}$ is the zero vector of length n. Matrix-wise this is

$$(A - \lambda I_n)\,\mathbf{v} = \mathbf{0}$$

where I_n is the n by n identity matrix. If the matrix $(A - \lambda I_n)$ is invertible then we can multiply both sides of the equation by the inverse and conclude that $\mathbf{v} = \mathbf{0}$, a contradiction. So $(A - \lambda I_n)$ is not invertible and has determinant 0.

$$\det(A - \lambda I_n) = 0$$

This is a polynomial in the single variable λ. Its roots are the eigenvalues of A! An eigenvalue may show up more than once. In that case we say the eigenvalue has *multiplicity* greater than 1.

Question 5.9.1

If A is a 2 by 2 square matrix, show that

$$\det(A - \lambda I_2) = \lambda^2 - \operatorname{tr}(A)\lambda + \det(A).$$

Also show that

$$A^2 - \operatorname{tr}(A)A + \det(A)I_2 = 0.$$

The eigenvalues of a diagonal matrix are the diagonal entries. The corresponding eigenvectors are the basis vectors.

Let's tease this apart in the real and complex 2 by 2 cases. We begin with the case that is only slightly more than trivial,

$$\begin{bmatrix} 1 & 0 \\ 0 & -1 \end{bmatrix}.$$

This is a diagonal matrix. By inspection, 1 is the eigenvalue for eigenvector $\mathbf{e}_1$ and -1 is the eigenvalue for $\mathbf{e}_2$.

Doing the computation via determinants, look at

$$\det\left(\begin{bmatrix} 1 & 0 \\ 0 & -1 \end{bmatrix} - \lambda \begin{bmatrix} 1 & 0 \\ 0 & 1 \end{bmatrix}\right) = \det\left(\begin{bmatrix} 1-\lambda & 0 \\ 0 & -1-\lambda \end{bmatrix}\right)$$
$$= \lambda^2 - 1$$
$$= (\lambda+1)(\lambda-1)$$
$$= 0$$

This confirms the eigenvalues are $\lambda_1 = 1$ and $\lambda_2 = -1$.

Let's consider a more complicated example and also compute the eigenvectors. Let

$$A = \begin{bmatrix} 3 & 3 \\ 2 & 4 \end{bmatrix}.$$

Then

$$\det\left(\begin{bmatrix} 3 & 3 \\ 2 & 4 \end{bmatrix} - \lambda \begin{bmatrix} 1 & 0 \\ 0 & 1 \end{bmatrix}\right) = \det\left(\begin{bmatrix} 3-\lambda & 3 \\ 2 & 4-\lambda \end{bmatrix}\right)$$
$$= \lambda^2 - 7\lambda + 6$$
$$= (\lambda - 6)(\lambda - 1)$$
$$= 0$$

The eigenvalues are $\lambda_1 = 6$ and $\lambda_2 = 1$. To find the eigenvectors for each, we solve

$$\begin{bmatrix} 3-\lambda & 3 \\ 2 & 4-\lambda \end{bmatrix} \begin{bmatrix} x \\ y \end{bmatrix} = \begin{bmatrix} 0 & 0 \end{bmatrix}$$

for $\lambda = \lambda_1 = 6$ and $\lambda = \lambda_2 = 1$ in turn.

For $\lambda = 6$, this is

$$-3x + 3y = 0$$
$$2x - 2y = 0$$

Either case tells us $y = x$ and so we can choose any point along that line to represent the vector. We let $\mathbf{v_1} = (1, 1)$.

For $\lambda = 1$, this is

$$2x + 3y = 0$$
$$2x + 3y = 0$$

Now we have $y = -\frac{2}{3}x$. Choose $\mathbf{v_2} = (1, -\frac{2}{3})$

Let's verify these.

$$A\mathbf{v_1} = \begin{bmatrix} 3 & 3 \\ 2 & 4 \end{bmatrix} \begin{bmatrix} 1 \\ 1 \end{bmatrix} = \begin{bmatrix} 6 & 6 \end{bmatrix} = 6\mathbf{v_1}$$

and

$$A\mathbf{v_2} = \begin{bmatrix} 3 & 3 \\ 2 & 4 \end{bmatrix} \begin{bmatrix} 1 \\ -\frac{2}{3} \end{bmatrix} = \begin{bmatrix} 3 - 3\frac{2}{3} & 2 - 4\frac{2}{3} \end{bmatrix} = \begin{bmatrix} 1 & -\frac{2}{3} \end{bmatrix} = 1\mathbf{v_2}.$$

Neither $\mathbf{v_1}$ nor $\mathbf{v_2}$ are unit vectors. We could have chosen $\mathbf{v_1} = \left(\frac{\sqrt{2}}{2}, \frac{\sqrt{2}}{2}\right)$ on the line $y = x$ to be a unit eigenvector.

Question 5.9.2

What are two choices for $\mathbf{v_2}$ on the line $y = -\frac{2}{3}x$ that would have made $\mathbf{v_2}$ a unit eigenvector?

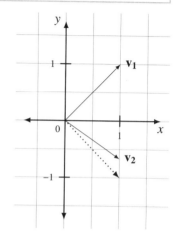

These eigenvectors $\mathbf{v_1}$ and $\mathbf{v_2}$ are not orthogonal as their dot product is $\frac{1}{3}$, which is non-zero.

For comparison I've drawn a dotted line vector which is perpendicular to $\mathbf{v_1}$.

The final example has

$$A = \begin{bmatrix} 1 & 1 \\ -1 & 1 \end{bmatrix}$$

and so

$$\det\left(\begin{bmatrix} 1 & 1 \\ -1 & 1 \end{bmatrix} - \lambda \begin{bmatrix} 1 & 0 \\ 0 & 1 \end{bmatrix}\right) = \det\left(\begin{bmatrix} 1 - \lambda & 1 \\ -1 & 1 - \lambda \end{bmatrix}\right) = \lambda^2 - 2\lambda + 2 = 0.$$

Notice I left out the factorization step. By the quadratic formula, the factorization is

$$(\lambda + (-1 - i))(\lambda + (-1 + i)).$$

The eigenvalues are $\lambda_1 = 1 + i$ and $\lambda_2 = 1 - i$. Whoops.

This simple looking matrix A does not have real eigenvalues, only complex ones. Thus, it does not have eigenvectors in $\mathbb{R}^2$.

If instead of working in $\mathbb{R}^2$ we instead were in $\mathbb{C}^2$, we could proceed and find eigenvectors by the method in the previous example.

Question 5.9.3

What are the eigenvectors for $A = \begin{bmatrix} 1 & 1 \\ -1 & 1 \end{bmatrix}$ on $\mathbb{C}^2$ that correspond to λ_1 and λ_2?

In these 2 by 2 cases it was easy to take determinants and factor the polynomials. It gets much harder as the dimension gets higher. People have developed algorithms and software to compute or estimate eigenvalues and eigenvectors. These are called *eigensolvers*.

Let A be a complex n by n matrix on $\mathbb{C}^n$. The polynomial $\det(A - \lambda I_n)$ in the single variable λ can be completely factored over $\mathbb{C}$. Therefore A has n eigenvalues and n corresponding eigenvectors. Some eigenvalues may appear more than once.

The determinant of A, $\det(A)$, is the product of the n eigenvalues. The trace of A, $\text{tr}(A)$, is the sum of them.

All is not lost for real matrices. The mathematician Augustin-Louis Cauchy proved that if a real square matrix A is symmetric, then it has real eigenvalues.

In fact, we can change to an orthonormal basis where the new matrix of the linear transformation represented by A (in the old basis) is a diagonal matrix. The diagonal entries are none other than the eigenvalues with the orthonormal basis as eigenvalues.

Let's look at some additional facts related to diagonal matrices.

A matrix A is diagonalizable if there exists an invertible matrix V such that $D = V^{-1}AV$ and D is an diagonal matrix. This says there is a change of basis so that the transformation represented by A in the original basis can be represented by D in the new basis.

If we can compute all the eigenvalues of a complex matrix, can we diagonalize it? Not quite.

A complex matrix A is diagonalizable if all its eigenvalues are different. It may be possible to diagonalize a complex matrix with repeated eigenvalues, but this is not true in general.

If A is Hermitian then all its eigenvalues are positive if and only if it is positive definite. For each positive semi-definite matrix A, there is exactly one positive semi-definite matrix B such that $A = B^2$. This is called the square root of A and written $A^{\frac{1}{2}}$.

Let A be a complex n by n Hermitian matrix. All the eigenvalues of A are real, eigenvectors corresponding to different eigenvalues are orthogonal, and we can always find a basis of $\mathbb{C}^n$ consisting of eigenvectors of A. Since we can always normalize eigenvectors, we can make this basis be orthonormal.

Since unitary matrices are so important for quantum computing, I note they have a special property regarding being transformed to a diagonal matrix.

Let U be a complex unitary matrix. There exists another unitary matrix V and a unitary diagonal matrix D so that $U = V^\dagger D V$. Since $V^\dagger = V^{-1}$ because it unitary, this means that $VUV^\dagger = D$.

There is a unitary change of basis for a unitary transformation so that it can be represented by a diagonal matrix.

Question 5.9.4

Are the diagonal elements in D necessarily the eigenvalues of U?

5.10   Direct sums

Our treatment of vector spaces has alternated between the fairly concrete examples of $\mathbb{R}^2$ and $\mathbb{R}^3$ and the more abstract definition presented near the beginning of this chapter. I continue going back and forth with the fundamental idea of the *direct sum* of two vectors spaces over the same field $\mathbb{F}$.

In a vague and neither concrete not abstract sense, a direct sum is when you push two vector spaces together. It's one of the ways you can get a new vector space from existing ones.

Let V and W be two vectors spaces of dimensions n and m over $\mathbb{F}$. If we write $\mathbf{v} = (v_1, v_2, \ldots, v_n)$ and $\mathbf{w} = (w_1, w_2, \ldots, w_m)$ then

$$\mathbf{v} \oplus \mathbf{w} = (v_1, v_2, \ldots, v_n, w_1, w_2, \ldots, w_m)$$

in the direct sum vector space $V \oplus W$. It has dimension $n + m$.

All the necessary requirements regarding addition and scalar multiplication follow directly from this definition because those operations are done coordinate by coordinate. We've simply added more coordinates.

There are four special linear maps for $V \oplus W$. Two are injections and two are projections. $\text{inj}_V : V \to V \oplus W$ maps

$$(v_1, v_2, \ldots, v_n) \to \left(v_1, v_2, \ldots, v_n, \underbrace{0, \ldots, 0}_{m \text{ times}} \right)$$

$\text{inj}_W : W \to V \oplus W$ maps

$$(w_1, w_2, \ldots, w_m) \to \left(\underbrace{0, \ldots, 0}_{n \text{ times}}, w_1, w_2, \ldots, w_m \right)$$

$\text{proj}_V : V \oplus W \to V$ maps

$$(v_1, v_2, \ldots, v_n, w_1, w_2, \ldots, w_m) \to (v_1, v_2, \ldots, v_n)$$

$\text{proj}_W : V \oplus W \to W$ maps

$$(v_1, v_2, \ldots, v_n, w_1, w_2, \ldots, w_m) \to (w_1, w_2, \ldots, w_m)$$

The composition $\text{proj}_V \circ \text{inj}_V$ is the identity transformation on V, as is $\text{proj}_W \circ \text{inj}_W$ for W.

The matrix equation for inj_V is

$$\begin{bmatrix} 1 & 0 & \cdots & 0 \\ 0 & 1 & \cdots & 0 \\ \vdots & \vdots & & \vdots \\ 0 & 0 & \cdots & 1 \\ 0 & 0 & \cdots & 0 \\ \vdots & \vdots & & \vdots \\ 0 & 0 & \cdots & 0 \end{bmatrix} \begin{bmatrix} v_1 \\ v_2 \\ \vdots \\ v_n \end{bmatrix} = \begin{bmatrix} v_1 & v_2 & \cdots & v_n & 0 & \cdots & 0 \end{bmatrix}.$$

Question 5.10.1

What are they for the other three maps?

Starting with $\mathbb{R}^1$ you can create $\mathbb{R}^2 = \mathbb{R}^1 \oplus \mathbb{R}^1$ and

$$\mathbb{R}^3 = \mathbb{R}^1 \oplus \mathbb{R}^1 \oplus \mathbb{R}^1 = \mathbb{R}^2 \oplus \mathbb{R}^1 = \mathbb{R}^1 \oplus \mathbb{R}^2.$$

Question 5.10.2

How many combinations are there that make $\mathbb{R}^4$? $\mathbb{R}^{10}$?

If we have $\mathbb{C}^1$ we can make $\mathbb{C}^2$ in the same way and combine two of them to create $\mathbb{C}^4$. Everything works smoothly because of the way we have concatenated the coordinates.

There's another technique to build a new vector space from two existing ones. It's the *tensor product* and is essential to the mathematics of using multiple qubits for quantum computing. Unfortunately, it's a little messier. We cover it in section 8.1 immediately before we need it to explain entangling qubits.

5.11   Homomorphisms

When functions operate on collections with algebraic structure, we usually require additional properties to be preserved. We can now redefine linear maps and transformations of vector spaces in terms of these functions.

5.11.1   Group homomorphisms

Suppose $(\mathbf{G}, \circ)$ and $(\mathbf{H}, \times)$ are two groups, which we first explored in subsection 3.6.1. The function $f : \mathbf{G} \to \mathbf{H}$ is a *group homomorphism* if for any two elements a and b in $\mathbf{G}$,

$$f(a \circ b) = f(a) \times f(b).$$

This means that f is not just a function, but it preserves the operations on the groups.

We have the following properties for group homomorphisms:

- $f(id_\mathbf{G}) = f(id_\mathbf{G} \circ id_\mathbf{G}) = f(id_\mathbf{G}) \times f(id_\mathbf{G})$, which means $f(id_\mathbf{G}) = id_\mathbf{H}$.
- $id_\mathbf{H} = f(id_\mathbf{G}) = f(a \circ a^{-1}) = f(a) \times f(a^{-1})$, which means $f(a^{-1}) = f(a)^{-1}$.

The set of all elements a in $\mathbf{G}$ such that $f(a) = id_\mathbf{H}$ is called the *kernel of f*. It is a subgroup of $\mathbf{G}$.

If $f(a) = f(b)$ implies that $a = b$ then f is a *monomorphism*. We also then say that f is *one-to-one*. We also say that f is *injective* or that f *embeds* $\mathbf{G}$ in $\mathbf{H}$. f is injective if its kernel contains the single element $id_\mathbf{G}$.

> **Question 5.11.1**
>
> Re-express the monomorphism condition in terms of $f(a \circ b^{-1})$.

The *image of* f is the set of all elements in **H** that look like $f(a)$ for some a in **G**. The image of f is a subgroup of **H**.

If for every h in **H** there is some a in **G** so that $f(a) = h$, then f is *surjective* and is called an *epimorphism*. This means that **H** is the image of f. We also say that f *covers* **H**.

A group homomorphism that is both injective and surjective is an *isomorphism*. This means that **G** and **H** are "the same" subject to the mapping of elements from the first group to the second.

Finally, if **G** = **H** and f is an isomorphism, then we say f is an *automorphism*.

Examples

1. The standard embeddings $\mathbb{Z} \to \mathbb{Q}$, $\mathbb{Q} \to \mathbb{R}$, and $\mathbb{R} \to \mathbb{C}$ are all group monomorphisms.
2. Let n be a positive integer > 1. The set of numbers $\{0, 1, 2, \ldots, n - 1\}$ is a group under addition denoted $\mathbb{Z}/n\mathbb{Z}$. These are the integers mod n. The function

$$g : \mathbb{Z} \to \mathbb{Z}/n\mathbb{Z}$$

which maps an integer j to j mod n is an epimorphism but not a monomorphism.

> **Question 5.11.2**
>
> What is the kernel of g?

3. Let H be the set of elements $\{2^n\}$ in $\mathbb{Q}$ for all integers n, with the multiplication operation "×". Define the function
$$h : (\mathbb{Z}, +) \to (\{2^n\}, \times)$$

by $n \mapsto 2^n$.

> **Question 5.11.3**
>
> Is h a homomorphism? Monomorphism? Epimorphism? Isomorphism?

5.11.2   Ring and field homomorphisms

Suppose $(\mathbf{R}, +, \times)$ and $(\mathbf{S}, +, \times)$ are two rings. The function $f : \mathbf{R} \to \mathbf{S}$ is a *ring homomorphism* if

- f is a group homomorphism for "+" (and so $f(0_{\mathbf{R}}) = 0_{\mathbf{S}}$),

- for any two elements a and b in $\mathbf{R}$, $f(a \times b) = f(a) \times f(b)$, and
- $f(1_{\mathbf{R}}) = 1_{\mathbf{S}}$.

Question 5.11.4

Show that for any three elements a, b, and c in $\mathbf{R}$,

$$f(a \times (b + c)) = f(a) \times (f(b) + f(c)).$$

I now drop the subscripts on terms like $0_{\mathbf{S}}$ because the additive and multiplicative identity elements will be clear from context.

The standard embeddings $\mathbb{Z} \to \mathbb{Q}$, $\mathbb{Q} \to \mathbb{R}$, and $\mathbb{R} \to \mathbb{C}$ are all monomorphisms.

Question 5.11.5

Define monomorphisms, epimorphisms, and isomorphisms for rings.

A field homomorphism is simply a ring homomorphism.

Suppose $(\mathbb{F}_1, +, \times)$ and $(\mathbb{F}_2, +, \times)$ are two fields and f is a homomorphism from the first to the second. Let's examine the kernel of f, the set of elements in $\mathbb{F}_1$ that map to 0 in $\mathbb{F}_2$.

We know that 0 is in the kernel, so suppose we have a non-zero a in $\mathbb{F}_1$ with $f(a) = 0$. Since $\mathbb{F}_1$ is a field, there exists an a^{-1} with $a \times a^{-1} = 1$.

$$1 = f(1) = f(a \times a^{-1}) = f(a) \times f(a^{-1}) = 0 \times f(a^{-1}) = 0.$$

Since $0 \neq 1$, this is a contradiction and no such non-zero a exists.

The kernel of every field homomorphism contains only 0. Every field homomorphism is a monomorphism.

5.11.3   Vector space homomorphisms

A vector space homomorphism is an additive group homomorphism that preserves scalar multiplication by elements in the underlying field. So in addition to rules like

$$f(\mathbf{v} + \mathbf{w}) = f(\mathbf{v}) + f(\mathbf{w})$$

we also insist that $f(a\,\mathbf{v}) = a\,f(\mathbf{v})$.

> Let U and V be vector spaces over the same field $\mathbb{F}$. A linear map $f : U \to V$ as we defined it in section 5.3 is a vector space homomorphism.
>
> A homomorphism from U to itself that preserves scalar multiplication is a *linear transformation*.

The kernel of a linear map f is called the *null space* of f. It is a vector space. The dimension of this vector space is called the *nullity*. If $f : U \to V$, then the *rank* of f is the dimension of V minus the nullity of f.

Question 5.11.6

If U, V, and W are vector spaces over the same field $\mathbb{F}$ and

$$f : U \oplus V \to W$$

is a homomorphism, show that $\mathbf{u} \mapsto f(\mathbf{v}, \mathbf{0})$ is a homomorphism from U to W.

We have the corresponding notions of monomorphism and epimorphism for vector spaces:

- If $f : V \to W$ is a linear map, then f is a monomorphism (is "one-to-one") if $f(\mathbf{v}_1) = f(\mathbf{v}_2)$ implies $\mathbf{v}_1 = \mathbf{v}_2$. One vector in the range is the image of only one vector in the domain.
- If $f : V \to W$ is a linear map then, f is an epimorphism (is "onto") if for any $\mathbf{w}$ in W there is a $\mathbf{v}$ in V such that $f(\mathbf{v}) = \mathbf{w}$. Any vector in W is the image of a vector from V. In this way f "covers" W.

If f is a monomorphism and an epimorphism, then it is an isomorphism. Necessarily, V and W have the same dimension. If f is represented by a matrix A in some basis, then the inverse function f^{-1}, which is a linear map, has matrix A^{-1}.

To learn more

Homomorphism are not just especially nice functions, they are core to mathematics. You can't do group theory without them, and groups are fundamental in the hierarchy of mathematical structures. [3] [5]

5.12 Summary

Linear algebra is the area of mathematics where we gain the language and tools necessary for understanding and working with spaces of arbitrary dimension. General quantum mechanics in physics uses infinite dimensional spaces. Our needs are simpler and we focused on vectors, linear transformations, and matrices in finite dimensions.

The fundamental quantum unit of information is the qubit and its states can be represented by a two-dimensional complex vector space. We now have almost all the tools necessary to jump from the purely mathematical description to one that takes in the evidence from models of the physical world of the very small. One preliminary topic remains, and we tackle that in the next chapter: probability.

References

[1]   Emma Brockes. *Return of the time lord*. 2005. URL: https://www.theguardian.com/science/2005/sep/27/scienceandnature.highereducationprofile.

[2]   D. Coppersmith and S. Winograd. "Matrix Multiplication via Arithmetic Progressions". In: *Proceedings of the Nineteenth Annual ACM Symposium on Theory of Computing*. STOC '87. ACM, 1987, pp. 1–6.

[3]   D. S. Dummit and R. M. Foote. *Abstract Algebra*. 3rd ed. Wiley, 2004.

[4]   Paul R. Halmos. *Finite-Dimensional Vector Spaces*. 1st ed. Undergraduate Texts in Mathematics. Springer Publishing Company, Incorporated, 1993.

[5]   S. Lang. *Algebra*. 3rd ed. Graduate Texts in Mathematics 211. Springer-Verlag, 2002.

[6]   Elizabeth S. Meckes and Mark W. Meckes. *Linear Algebra*. Cambridge Mathematical Textbooks. Cambridge University Press, 2018.

[7]   Ferrante Neri. *Linear Algebra for Computational Sciences and Engineering*. 2nd ed. Springer, 2019.

[8]   Thomas S. Shores. *Applied Linear Algebra and Matrix Analysis*. 2nd ed. Undergraduate Texts in Mathematics. Springer, 2018.

6

What Do You Mean "Probably"?

Any one who considers arithmetical methods of producing random digits is, of course, in a state of sin.

John von Neumann [7]

Here's the key to what we're going to cover in this chapter: in any given situation, the sum of the probabilities of all the different possible things that could happen always add up to 1.

In this short chapter, I cover the basics of practical probability theory to get us started on quantum computing and its applications.

Topics covered in this chapter

6.1   Being discrete

Sometimes it seems like probability is the study of flipping coins or rolling dice, given the number of books that explain it those ways. It's very hard to break away from these convenient examples. An advantage shared by both of them is that they make it is easy to explain discrete events and independence.

For the sake of mixing it up, suppose we have a cookie machine. It's a big box with a button on top. Every time you press the button, a cookie pops out of a slot on the bottom. There are four kinds of cookies: **chocolate**, **sugar**, **oatmeal**, and **coconut**.

Assume for the moment there is no limit to the number of cookies our machine can distribute. If you hit the button a million times, you get a million cookies. Also assume you get a random cookie each time. What does this mean, "random"?

Without a rigorous definition, random here means that the odds of getting any one of the cookies is the same as getting any other. That is, roughly one-fourth of the time I would get **chocolate**, one-fourth of the time I would get **sugar**, one-fourth for **oatmeal**, and one-fourth **coconut**.

The *probability* of getting a **sugar** cookie, say, is 0.25 and the sums of the probabilities for all the cookies is 1.0. Since there are four individual and separate possible outcomes, we say this is a *discrete* situation. We write this as

$$P(\textbf{chocolate}) = P(\textbf{sugar}) = P(\textbf{oatmeal}) = P(\textbf{coconut}) = 0.25$$

and

$$P(\textbf{chocolate}) + P(\textbf{sugar}) + P(\textbf{oatmeal}) + P(\textbf{coconut}) = 1.0$$

We use $P(x)$ to denote the probability of x happening, where $0 \leq P(x) \leq 1$. By definition, if the probability is 1 then it always happens, and if it is 0 then it never happens. If the probability is neither 0 nor 1 then it is neither impossible or certain, respectively.

Question 6.1.1

What is the probability of my *not* getting a **coconut** cookie on one push of the button?

Now let's change this situation. Last time the cookie service-person came, they accidentally loaded the **coconut** slot inside the machine with **chocolate** cookies. This changes the odds:

$$P(\textbf{chocolate}) = 0.5 \qquad P(\textbf{sugar}) = 0.25$$

$$P(\textbf{oatmeal}) = 0.25 \qquad P(\textbf{coconut}) = 0 \, .$$

The sum of the probabilities is still 1, as it must be, but the chance of getting a **chocolate** cookie is now twice as large as it was. The probability of getting a **coconut** cookie is 0, which is another way of saying it is impossible.

Had the service person simply forgotten to fill the **coconut** slot, we would have

$$P(\textbf{chocolate}) = 0.\overline{3} \qquad\qquad P(\textbf{sugar}) = 0.\overline{3}$$

$$P(\textbf{oatmeal}) = 0.\overline{3} \qquad\qquad P(\textbf{coconut}) = 0\,.$$

Since we are only talking about probabilities, it is quite possible for me to get three **chocolate** cookies in a row after I hit the button three times. We do not have enough data points for the observed results to be consistently close to the probabilities. It is only when I perform an action, in this case pressing the button many, many times, that I will start seeing the ratio of the number of times I saw the desired event to the total number of events approach the probability.

If I press the button 100 times I might see the following numbers

Kind of cookie	Number of times seen	Ratio of seen/100
chocolate	22	0.22
sugar	26	0.26
oatmeal	25	0.25
coconut	27	0.27

Assuming everything is as balanced as I think, I'll eventually be approaching the probabilities. If I instead saw

Kind of cookie	Number of times seen	Ratio of seen/100
chocolate	0	0.0
sugar	48	0.48
oatmeal	52	0.52
coconut	0	0.0

I would rightfully suspect something was wrong. I would wonder if there were roughly the same number of **sugar** and **oatmeal** cookies in the machine but no **chocolate** or **coconut** cookies. When experimentation varies significantly from prediction, it makes sense to examine your assumptions, your hardware, and your software.

Now let's understand the probabilities of getting one kind of cookie followed by another. What is the probability of getting an **oatmeal** cookie and then a **sugar** cookie?

This is

$$P(\textbf{oatmeal} \text{ and then } \textbf{sugar}) = P(\textbf{oatmeal}) \times P(\textbf{sugar})$$

$$= 0.25 \times 0.25 = 0.0625 = \frac{1}{16}.$$

Another way to see this is to observe that there are 4 choices for the first cookie and then 4 choices for the second, which means 16 possibilities for both. These look like

chocolate + chocolate	chocolate + sugar	chocolate + oatmeal	chocolate + coconut
sugar + chocolate	sugar + sugar	sugar + oatmeal	sugar + coconut
oatmeal + chocolate	oatmeal + sugar	oatmeal + oatmeal	oatmeal + oatmeal
coconut + chocolate	coconut + sugar	coconut + oatmeal	coconut + coconut

Getting **oatmeal** and then **sugar** is one of the sixteen choices.

What about ending up with one **oatmeal** and one **sugar** cookie? Here it doesn't matter which order the machine gives them to me. Two of the sixteen possibilities yield this combination and so

$$P(\textbf{oatmeal} \text{ and } \textbf{sugar}) = \frac{2}{16} + \frac{1}{8} = 0.125.$$

This is

$$P(\textbf{oatmeal} \text{ and then } \textbf{sugar}) + P(\textbf{sugar} \text{ and then } \textbf{oatmeal}).$$

Question 6.1.2

What is the probability of my getting a **chocolate** cookie on the first push of the button and then my *not* getting a **chocolate** cookie on the second?

6.2 More formally

In the last section, there were initially four different possible outcomes, those being the four kinds of cookies that could pop out of our machine. In this situation, our *sample space* is the collection

$$\{\textbf{chocolate, sugar, oatmeal, coconut}\}.$$

We also say that these four are the values of a *random variable*. Random variables usually have names like X and Y.

The sample space for getting one cookie and then another has size 16. A *probability distribution* assigns a probability to each of the possible outcomes, which are the values of the random variable. The probability distribution for the basic balanced case is

$$
\begin{aligned}
\textbf{chocolate} &\rightarrow 0.25 & \textbf{sugar} &\rightarrow 0.25 \\
\textbf{oatmeal} &\rightarrow 0.25 & \textbf{coconut} &\rightarrow 0.25\,.
\end{aligned}
$$

When the probabilities are all equal, as in this case, we have a *uniform distribution*.

If our sample space is finite or at most *countably infinite*, we say it is *discrete*. A set is countably infinite if it can be put in one-to-one correspondence with $\mathbb{Z}$. One way to think about it is that you are looking at an infinite number of separate things with nothing between them.

The sample space is *continuous* if it can be put in correspondence with some portion of $\mathbb{R}$ or a higher dimensional space. As you boil water, the sample space of its temperatures varies continuously from its starting point to the boiling point. Just because your thermometer only reads out decimals, it does not mean the temperature itself does not change smoothly within its range.

This raises an important distinction. Though the sample space of temperatures is continuous, the sample space of temperatures *as read out on a digital thermometer* is discrete, technically speaking. It is represented by the numbers the thermometer states, to one or two decimal places. The discrete sample space is an approximation to the continuous one.

When we use numeric methods on computers for working in such situations, we are more likely to use continuous techniques. These involve calculus, which I have not made a prerequisite for this book, and so will not cover. We do not need them for our quantum computing discussion.

Our discrete sample spaces will usually be the basis vectors in complex vector spaces of dimension 2^n.

6.3 Wrong again?

Suppose you have a very faulty calculator that does not always compute the correct result.

If the probability of getting the wrong answer is p, the probability of getting the correct answer is $1 - p$. This called the *complementary* probability. Assuming there is no connection between the attempts, the probability of getting the wrong answer two times in a row is p^2 and the probability of getting the correct answer two times in a row is $(1 - p)^2$.

Question 6.3.1

Compute p^2 and $(1-p)^2$ for $p = 0$, $p = 0.5$, and $p = 1.0$.

To make this useful, we want the probability of failure p to be non-zero.

For n independent attempts, the probability of getting the wrong answer is p^n. Let's suppose $p = 0.6$. We get the wrong answer 60% of the time in many attempts. We get the correct answer 40% of the time.

After 10 attempts, the probability of having gotten the wrong answer every time is $0.6^{10} \approx 0.006$.

On the other hand, suppose I want to have a very low probability of having always gotten the wrong answer. It's traditional to use the Greek letter ϵ (pronounced "epsilon") for small numbers such as error rates. For example, I might want the chance of never having gotten the correct answer to be less than $\epsilon = 10^{-6} = 0.000001$.

For this, we solve

$$p^n < \epsilon$$

for n. Having a variable in the exponent is telling you "use a logarithm!". Doing so,

$$n \log(p) < \log(\epsilon).$$

Here is a subtle point: since $0 \leq p < 1$, $\log(p) < 0$. In general, if $a < b$ and $c < 0$, then $ac > bc$. I now divide both sides by $\log(p)$, but I must flip "<" to ">".

$$n > \frac{\log(\epsilon)}{\log(p)}$$

In our example, $p = 0.6$ and $\epsilon = 10^{-6}$. A quick computation yields $n > 27$.

6.4 Probability and error detection

Let's return to our repetition code for error detection from section 2.1. The specific example I want to look at is sending information that is represented in bits. The probability that I will get an error by flipping a 0 to a 1 or a 1 to a 0 is p. The probability that no error occurs is $1 - p$, as above.

This is called a *binary symmetric channel*. We have two representations for information, the bits, and hence "binary." The probability of something bad happening to a 0 or 1 is the same, and that's the symmetry.

Here is the scheme:

- Create a message to be sent to someone.
- Transform that message by encoding it so that it contains extra information. This will allow it to be repaired if it is damaged en route to the other person.
- Send the message. "Noise" in the transmission may introduced errors in the encoded message.
- Decode the message, using the extra information stored in it to try to fix any transmission errors.
- Give the message to the intended recipient.

For this simple example, I send a single bit and I encode it by making three copies. This means I encode 0 as 000 and 1 as 111.

The number of 1s in an encoded message (with or without errors) is called its *weight*. The decoding scheme is

$$\text{decoded received message} = \begin{cases} 0 & \text{if } weight(\text{received message}) \leq 1 \\ 1 & \text{otherwise.} \end{cases}$$

There are eight possible bit triplets I can receive:

$$000 \quad 001 \quad 010 \quad 100 \quad 111 \quad 101 \quad 110 \quad 101$$

If there are more 0s than 1s, I decode, or "fix," the message by making the result 0. Similarly, if there are more 1s than 0s, I decode the message by making the result 1. The result of decoding will be the original if at most one error occurred, in this case. This is why we made an odd number of copies.

The first four would all be correctly decoded as 0 if that's what I sent. They either have zero or only one errors. The last four are the cases where two or three errors occurred.

Question 6.4.1

What's the similar observation if I had sent 1?

Now assume I sent 0. There is one possible received message, 000, where no error popped up and that has probability $(1 - p)^3$.

There are three results with one error and two non-errors: 001, 010, and 100. The total probability of seeing one of these is

$$(1 - p)(1 - p)p + (1 - p)p(1 - p) + p(1 - p)(1 - p) = 3p(1 - p)^2.$$

In total, the probability that I receive the correct message or can fix it to be correct is

$$(1 - p)^3 + 3p(1 - p)^2 = (1 - p)^2(1 - p + 3p) = (1 - p)^2(1 + 2p).$$

If $p = 1.0$, this probability is 0, as you would expect. We can't fix anything. If $p = 0.0$, there are no errors and the probability of ending up with the correct message is 1.0.

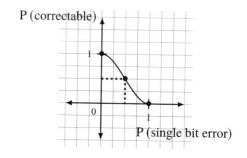

If there is an equal chance of error or no error, $p = 0.5$ and then so too is the probability we can repair the message 0.5. If the chance of error in one in ten, $p = 0.1$, the probability that it was correct or we can fix is 0.972.

Question 6.4.2

What is the maximum value of p I can have so that my chance of getting the right message or being able to correct it is 0.9999?

6.5 Randomness

Many programming languages have functions that return pseudo-random numbers. The prefix "pseudo" is there because they are not truly random numbers but nevertheless do well on statistical measurements of how well-distributed the results are.

Given four possible events E_0, E_1, E_2, and E_3 with associated probabilities p_0, p_1, p_2, and p_3, how might we use random numbers to simulate these events happening?

Suppose

$$p_0 = 0.15 \quad p_1 = 0.37 \quad p_2 = 0.26 \quad p_3 = 0.22 .$$

The probabilities add up to 1.0, as expected.

In Python 3, the `random()` function returns a random real number r such that $0.0 \le r < 1.0$. Based on the value of r we get, we determine that one of the E_0, E_1, E_2, and E_3 events occurred.

If you are not using Python, use whatever similar function is available in your programming language and environment.

The general scheme is to run the following steps in order:

- $r < p_0$ then we have observed E_0 and we **stop**. In the example, $p_0 = 0.15$. If not, ...
- If $r < p_0 + p_1$ then we have observed E_1 and we **stop**. If not, ...
- If $r < p_0 + p_1 + p_2$ then we have observed E_2 and we **stop**. If not, ...
- If $r < p_0 + p_1 + p_2 + p_3$ then we have observed E_3 and we **stop**.

The code in Listing 6.1 simulates this sampling a given number of times. On one run for 100 samples, I saw the events distributed in this way:

```
Results for 100 simulated samples

Event       Actual Probability      Simulated Probability
  0               0.15                      0.11
  1               0.37                      0.36
  2               0.26                      0.29
  3               0.22                      0.24
```

The values are close but not very close. It's probability, after all.

Significantly increasing the number of samples to 1 million produced values much closer to the expected probabilities.

```
Results for 1000000 simulated samples

Event       Actual Probability      Simulated Probability
  0               0.15                      0.1507
  1               0.37                      0.3699
  2               0.26                      0.2592
  3               0.22                      0.2202
```

If you plan to run this frequently, you should ensure you are not always getting the same sequence of random numbers. However, sometimes you do want a repeatable sequence of random numbers so that you can debug your code. In this situation, you want to begin with a specific random "seed" to initialize the production of numbers. In Python this function is `seed()`.

```
#!/usr/bin/env python3

import random

example_probabilities = [0.15, 0.37, 0.26, 0.22]

def sample_n_times(probabilities, n):
    probability_sums = []
    number_of_probabilities = len(probabilities)
    sum = 0.0
    for probability in probabilities:
        sum += probability
        probability_sums.append(sum)
    counts = [0 for probabilities in probabilities]

    for _ in range(n):
        r = random.random()
        for s,j in zip(probability_sums, range(number_of_probabilities)):
            if r < s:
                counts[j] = counts[j] + 1
                break

    print(f"\nResults for {n} simulated samples")
    print("\nEvent     Actual Probability     Simulated Probability")
    for j in range(number_of_probabilities):
        print(f"  {j}{14*' '}{probabilities[j]}{20*' '}" \
            f"{round(float(counts[j])/n, 4)}")

sample_n_times(example_probabilities, 100)
sample_n_times(example_probabilities, 1000000)
```

Listing 6.1: Sample Python 3 code demonstrating sampling

Interestingly enough, one potential use of quantum technology is to generate truly random numbers. It would be odd to use a quantum device to simulate quantum computing in an otherwise classical algorithm!

To learn more

When I talk about Python in this book I always mean version 3. The tutorial and reference material on the official website is excellent. [6] There are many good books that teach how to code in this important, modern programming language for systems, science, and data analysis. [4] [5]

6.6 Expectation

Let's look at the numeric finite discrete distribution of the random variable X with probabilities given in the table on the right.

Value	Probability
2	0.14
5	0.22
9	0.37
13	0.06
14	0.21

If the process that is producing these values continues over time, what value would we "expect" to see?

If they all have the same probability of occurring, the *expected value*, or *expectation*, $E(X)$ is their average:

$$E(X) = \frac{2 + 5 + 9 + 13 + 14}{5} = 8.6$$

Note the answer does not need to be, and often isn't, one of the values in the distribution.

Since each value of X has a given probability, the expected value is instead the weighted average

$$E(X) = 2 \times 0.14 + 5 \times 0.22 + 9 \times 0.37 + 13 \times 0.06 + 14 \times 0.21 = 8.43$$

If you are only given a list of values for a random variable, you can assume a uniform distribution, and the expected value is the usual average, or *mean*. The notation $\mu(X)$ is sometimes used instead of $E(X)$. μ is the lowercase Greek letter "mu."

If X is a random variable with values $\{x_1, x_2, ..., x_n\}$ and corresponding probabilities p_k such that $p_1 + p_2 + \cdots + p_n = 1$, then

$$E(X) = p_1 x_1 + p_2 x_2 + \cdots + p_n x_n.$$

For a uniform distribution, each $p_k = \frac{1}{n}$ and so

$$E(X) = \frac{x_1 + x_2 + \cdots + x_n}{n}.$$

How much do the values in X differ from the expected value $E(X)$? For a uniform distribution, the average of the amounts each x_k varies from $E(X)$ is

$$\frac{|x_1 - E(X)| + |x_2 - E(X)| + \cdots + |x_n - E(X)|}{n}.$$

However, the absolute values can make calculations difficult and unwieldy, particularly those in calculus, so we instead define the variance using squares. This is not that strange since one way of defining $|x|$ is $\sqrt{x^2}$ where we always take the positive real square root.

We now define the variance:

$$Var(X) = \frac{(x_1 - E(X))^2 + (x_2 - E(X))^2 + \cdots + (x_n - E(X))^2}{n}$$
$$= E((X - E(X))^2).$$

The notation $(X - E(X))^2$ refers to the new set of values $(x_k - E(X))^2$ for each x_k in X.

$$(X - E(X))^2 = \left\{ (x_1 - E(X))^2, (x_2 - E(X))^2, \ldots, (x_n - E(X))^2 \right\}$$

For a distribution where each x_k has probability p_k and all the p_k add up to 1, the formula for the variance is

$$Var(X) = p_1 (x_1 - E(X))^2 + p_2 (x_2 - E(X))^2 + \cdots + p_n (x_n - E(X))^2$$
$$= E((X - E(X))^2).$$

The notation $\sigma^2(X)$ is often used instead of $Var(X)$. The standard deviation σ of X is the square root of the variance $\sqrt{Var(X)}$

For a coin flip, $X = \{x_1 = 0 = \text{tails}, x_2 = 1 = \text{heads}\}$ with $p_1 = p_2 = 0.5$. Then

$$E(X) = \frac{0 + 1}{2} = \frac{1}{2}$$

and

$$Var(X) = \sigma^2 = \frac{\left(0 - \frac{1}{2}\right)^2 + \left(1 - \frac{1}{2}\right)^2}{2} = \frac{1}{4}.$$

with the standard deviation

$$\sigma = \sqrt{\frac{1}{4}} = \frac{1}{2}.$$

When I am flipping the coin multiple times, I use the notation X_j to mean "the jth coin flip." So X_1 is the first coin flip and X_{652} is the 652nd. I flip the coin 20 times, the number of heads is $X_1 + \cdots + X_{20}$. More generally, for n flips, the number of heads is $X_1 + X_2 + \cdots + X_n$ and the number of tails is n minus this number. Each of the values taken on by the X_j are identical and they have the same distributions.

Basically, I'm doing the same thing many times, with the same probabilities, and there are no connections between the X_j. The answer I get for any one X_j is independent of what happens for any other.

We return to expectation in the context of measuring qubits in subsection 7.3.4.

6.7 Markov and Chebyshev go to the casino

In this section we work through the math from estimating π in section 1.5. We dropped coins into a square and looked at how many of them had their centers on or inside a circle.

There are two important inequalities involving expected values, variances, and error terms. Let X be a finite random variable with distribution so that

$$E(X) = p_1 x_1 + p_2 x_2 + \cdots + p_n x_n$$

and each $x_k \geq 0$.

> **Markov's Inequality:** For a real number $a > 0$,
>
> $$P(X > a) \leq \frac{E(X)}{a}$$

In Markov's Inequality, the expression $P(X > a)$ means "look at all the x_k in X and for all those where $x_k > a$, add up the p_k to get $P(X > a)$."

> **Question 6.7.1**
>
> Show that Markov's Inequality holds for the distribution at the beginning of this section for the values $a = 3$ and $a = 10$.

Andrey Markov, circa 1875. Photo is in the public domain. Ⓟ

Chebyshev's Inequality: For any finite random variable X with distribution, and real number $\epsilon > 0$,

$$P\left(|X - E(X)| \geq \epsilon\right) \leq \frac{Var(X)}{\epsilon^2} = \frac{\sigma^2}{\epsilon^2}.$$

Equivalently,

$$P\left(|X - E(X)| < \epsilon\right) = 1 - P\left(|X - E(X)| \geq \epsilon\right)$$

$$\geq 1 - \frac{Var(X)}{\epsilon^2}$$

$$= 1 - \frac{\sigma^2}{\epsilon^2}.$$

In Chebyshev's Inequality, the expression $P\left(|X - E(X)|\right) \geq \epsilon$ means "look at all the x_k in X and for those whose distance from the expected value $E(X)$ is greater than or equal to ϵ, add up the p_k to get $P\left(|X - E(X)|\right) \geq \epsilon$." Think of ϵ as an error term related to the probability of being away from the expected value.

From Chebyshev's Inequality we get a result which is useful when we are taking samples X_j, such as coin flips or the Monte Carlo samples described in section 1.5.

Weak Law of Large Numbers: Let the set of X_j for $1 \leq j \leq n$ be independent identical random variables with identical distributions. Let $\mu = E(X_j)$, which is the same for all j. Then

$$P\left(\left|\frac{X_1 + \cdots + X_n}{n} - \mu\right| \geq \epsilon\right) \leq \frac{Var(X)}{n\epsilon^2} = \frac{\sigma^2}{n\epsilon^2}.$$

As n gets larger, the expression on the right gets closer to 0 because ϵ and σ^2 do not change.

What does this tell us about the probability of getting 7 or fewer heads when I do ten coin flips? Remember that in this case $\mu = \frac{1}{2}$ and $\sigma^2 = \frac{1}{4}$.

P(**we get 7 or fewer heads**)

$$= P(X_1 + \cdots + X_{10} \le 7)$$

$$= P\left(\frac{X_1 + \cdots + X_{10}}{10} \le \frac{7}{10}\right)$$

$$= P\left(\frac{X_1 + \cdots + X_{10}}{10} - \frac{5}{10} \le \frac{2}{10}\right)$$

$$= P\left(\left|\frac{X_1 + \cdots + X_{10}}{10} - \frac{1}{2}\right| \le \frac{1}{5}\right)$$

$$= 1 - P\left(\left|\frac{X_1 + \cdots + X_{10}}{10} - \frac{1}{2}\right| \ge \frac{1}{5}\right)$$

$$\ge \frac{\sigma^2}{n\epsilon^2} = \frac{\left(\frac{1}{4}\right)^2}{10\left(\frac{1}{5}\right)^2} = \frac{25}{10 \times 16} = \frac{25}{160}$$

$$= 0.15625$$

Pafnuty Chebyshev, circa 1890.
Image is in the public domain. ⓟ

Here we use $\epsilon = \frac{1}{5}$. The probability of getting 7 or fewer heads in 10 flips is greater than or equal to 0.15625.

Question 6.7.2

Repeat the calculation for 70 or fewer heads in 100 flips, and 700 or fewer heads in 1000 flips.

In section 1.5 we started using a Monte Carlo method to estimate the value of π by randomly placing coins in a 2 by 2 square in which is inscribed a unit circle. It's easier to use the language of points instead of coins, where the point is the center of the coin.

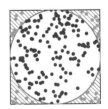

The plot on the right uses 200 random points and yields 3.14 as the approximation.

Let's now analyze this in the same way we did for the coin flips. In this case,

$$X = \{x_1 = 0 = \textbf{point does not land in circle},$$
$$x_2 = 1 = \textbf{point lands on or inside circle}\}$$

So X_1 is dropping the first random point into the square and $X_{1000000}$ is dropping the one millionth. The value is 1 for either of these if the point is in the circle. We have

$$P(X_k = 1) = \frac{\pi}{4} \approx 0.78539816339$$

because the area of the circle is π and the area of the square is 4.

The expected value μ is therefore this probability. For the variance,

$$Var(X) = \sigma^2 = \frac{\left(0 - \frac{\pi}{4}\right)^2 + \left(1 - \frac{\pi}{4}\right)^2}{2} = \frac{\pi^2}{16} - \frac{\pi}{4} + \frac{1}{2} \approx 0.3314521116706366.$$

We put these values into the Weak Law of Large Numbers:

$$P\left(\left|\frac{X_1 + \cdots + X_n}{n} - \frac{\pi}{4}\right| \geq \epsilon\right) \leq \frac{\frac{\pi^2}{16} - \frac{\pi}{4} + \frac{1}{2}}{n\epsilon^2}.$$

Suppose we want the probability of the difference between π and the approximation being greater than $0.01 = \frac{1}{100}$. First,

$$P\left(\left|4\frac{X_1 + \cdots + X_n}{n} - \pi\right| \geq \frac{1}{100}\right) = P\left(\left|\frac{X_1 + \cdots + X_n}{n} - \frac{\pi}{4}\right| \geq \frac{4}{100}\right).$$

So $\epsilon = \frac{4}{100}$ and

$$\frac{\frac{\pi^2}{16} - \frac{\pi}{4} + \frac{1}{2}}{n\epsilon^2} = \frac{\frac{\pi^2}{16} - \frac{\pi}{4} + \frac{1}{2}}{n\left(\frac{4}{100}\right)^2} = \frac{\frac{625}{16}\pi^2 - \frac{625}{4}\pi + \frac{625}{2}}{n} \approx \frac{207.158}{n}$$

If we want the probability of the error being this small to be less than $5\% = 0.05$, we solve for n in

$$\frac{207.158}{n} = 0.05$$

to get $n \approx 4143.1$.

This means if we want the probability of the estimate being off by at most $0.01 = 10^{-2}$ (in the second decimal place) to be less than or equal to 5%, we need the number of points to be at least 4144.

If we repeat the calculation and want the probability of the estimate being off by at most $0.0001 = \frac{1}{10000}$, we need $n \geq 2071576$. Said another way, we need n this large to get the estimate this close with probability 99.9999%,

Finally, if want the probability of the estimate to π being off by at most $0.00001 = 10^{-5}$ (in the fifth decimal place) to be less than $0.0001 = \frac{1}{10000}$, we need $n \geq 82863028$.

As you can see, we need many points to estimate π accurately to a small number of digits, in the last case more than 82 million points.

6.8 Summary

In this chapter we covered the elements of probability necessary for our treatment of quantum computing and its applications. When we work with qubits and circuits in the next two chapters we use discrete sample spaces, though they can get quite large. In these cases, the sizes of the sample spaces will be powers of 2.

Our goal in quantum algorithms is to adjust the probability distribution so that the element in the sample space with the highest probability is the best solution to some problem. Indeed, the manipulation of *probability amplitudes* leads us to find what we hope is the best answer. My treatment of algorithms in chapter 9 and chapter 10 does not go deeply into probability calculations, but does so sufficiently to give you an idea of how probability interacts with interference, complexity, and the number of times we need to run a calculation to be confident we have seen the correct answer.

To learn more

There are hundreds of books about probability and combination texts covering probability and statistics. These are for the applied beginner all the way to the theoretician, with several strong and complete treatments. [1]

Probability has applications in many areas of science and engineering, including AI/machine learning and cryptography. [3] [2, Chapter 5]

An in-depth understanding of continuous methods requires some background in differential and integral calculus.

References

[1]   D.P. Bertsekas and J.N. Tsitsiklis. *Introduction to Probability*. 2nd ed. Athena Scientific optimization and computation series. Athena Scientific, 2008.

[2]   Thomas H. Cormen et al. *Introduction to Algorithms*. 3rd ed. The MIT Press, 2009.

[3] Jeffrey Hoffstein, Jill Pipher, and Joseph H. Silverman. *An Introduction to Mathematical Cryptography*. 2nd ed. Undergraduate Texts in Mathematics 152. Springer Publishing Company, Incorporated, 2014.

[4] Mark Lutz. *Learning Python*. 5th ed. O'Reilly Media, 2013.

[5] Fabrizio Romano. *Learn Python Programming*. 2nd ed. Packt Publishing, 2018.

[6] The Python Software Foundation. *Python.org*. URL: https://www.python.org/.

[7] John von Neumann. "Various techniques used in connection with random digits". In: *Monte Carlo Method*. Ed. by A.S. Householder, G.E. Forsythe, and H.H. Germond. National Bureau of Standards Applied Mathematics Series, 12, 1951, pp. 36–38.

II

Quantum
Computing

7

One Qubit

Anyone who is not shocked by
quantum theory has not understood it.

Niels Bohr [1]

A *quantum bit*, or *qubit*, is the fundamental information unit of quantum computing. In this chapter, I give a mathematical definition of a qubit based on the foundational material in the first part of this book. Together we examine the operations you can perform on a single qubit from both mathematical and computational perspectives.

Despite a single qubit living in a seemingly strange two-dimensional complex Hilbert space, we can visualize it, its superposition, and its behavior by projecting onto the surface of a sphere in $\mathbb{R}^3$.

All vector spaces considered in this chapter are over $\mathbb{C}$, the field of complex numbers introduced in section 3.9. All bases are orthonormal unless otherwise specified.

Topics covered in this chapter

7.1   Introducing quantum bits

If you have seen descriptions of qubits elsewhere, you may have read something like "a qubit implements a two-state quantum mechanical system and is the quantum analog of a classical bit." A bit, as we saw in section 2.1, also has two states, those being 0 and 1.

Those other discussions then usually include one or more of the following: light switches, spinning electrons, polarized light, and rotating coins or donuts. I don't spend my days thinking about electrons but I do like sunglasses and donuts. These approaches all have merit and are the basis for teasing apart the difference between the quantum and classical situations. The electron and polarized light examples do depict quantum systems.

Otherwise, analogies tend to be imperfect and eventually may lead you into corners where their behavior and your understanding are not consistent with the real situation. It's for this reason we developed the essential mathematics and insight to reason accurately about what happens in quantum computing.

Let's begin by thinking about a classical bit and a quantum bit.

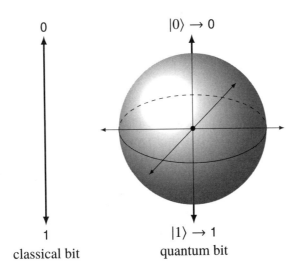

classical bit   quantum bit

On the left, we have the classical situation where a bit can only take on the values 0 and 1. More precisely, a bit can be in one of those *states* and only those. You can look at the bit at any time and, assuming nothing has happened to change the state, it stays in that state.

For the quantum situation on the right, we change the notation slightly. The qubit always becomes the state $|0\rangle$ or $|1\rangle$ when we read information from it by a process called *measurement*. However, it is possible to move it to an infinite number of other states and change from one of them to another while we are computing with the qubit before measurement.

Measurement says "ok, I'm going to peek at the qubit now" and the result is always a 0 or 1 once you do so. We can then read that out as a bit value of 0 or 1, respectively.

Yes, this is weird. This is quantum mechanics and it has amazed, and befuddled, and surprised, and delighted people for close to one hundred years. Quantum computing is based on and takes advantage of this behavior.

Continuing with the right side, we represent all the states the qubit could be in as points on the unit sphere. $|0\rangle$ is at the north pole, and $|1\rangle$ is at the south. Remember: points on the sphere equal quantum states. In the next section I define more precisely what we mean when we write $|0\rangle$ and $|1\rangle$.

I need to state and then clarify something about *superposition*. Mathematically, the qubit is always in superposition because its state can be at the poles $|0\rangle$ or $|1\rangle$ or any point in between. However, by slight abuse of terminology, people typically mean the state is in superposition when it is *not* $|0\rangle$ or $|1\rangle$. If the state is not one of the poles we are speaking about a non-trivial superposition. To "move into superposition" usually means the state is moved to a point on the equator.

A quantum algorithm uses several qubits which together constitute a *quantum register*. From the viewpoint of a single qubit, its activity looks like

1. I, the qubit, am in initial state $|0\rangle$.
2. I get moved into a standard superposition. Don't peek!
3. Steps in the algorithm apply zero or more reversible operations to me.
4. I get measured and always become $|0\rangle$ or $|1\rangle$.
5. When these get read out for classical use, they are converted to 0 or 1.

It's what happens in steps 2 and 3 that make quantum computing interesting. For the unit sphere, which is called the *Bloch sphere* in quantum computing, step 2 means I get moved to a designated point/state on the equator. Step 3 translates to getting moved to other points/states. This movement is a result of what the algorithm is doing and causing to happen.

The final two steps are like telling the qubit "look, you can't stay somewhere in superposition forever. You need to decide whether you will be 0 or 1. This is not optional." Where it is in superposition, in what state on the sphere, determines the relative probabilities of the qubit ultimately yielding a classical bit value of 0 or 1.

If the state lies on the equator, then with perfect randomness it will end up being 0 half of the time and 1 half of the time.

Though the Bloch sphere is a convenient and visual representation, the linear algebraic description of a qubit is more fundamental, in my opinion. We go back and forth between them when one or the other makes it easier to understand some aspect of quantum computing.

Moving from the sphere back to linear algebra, compare these two statements:

1. The variable $\mathbf{q}$ can have either the vector value $\mathbf{v}$ or $\mathbf{w}$, and they are an orthonormal basis.
2. The variable $\mathbf{q}$ can be any linear combination $a\mathbf{v} + b\mathbf{w}$. If a and b are non-zero the linear combination is *non-trivial*.

The second seems much more powerful than the first. When $a = 1$ and $b = 0$ we get $\mathbf{v}$. Reversing these assignments gives us $\mathbf{w}$. The second statement appears to reduce to the first when we make these choices. Alternatively, the first is extended by the second.

We explore and return to this statement throughout the rest of this chapter:

A *qubit*—a *quantum bit*—is the fundamental unit of quantum information. At any given time, it is in a superposition state represented by a linear combination of vectors $|0\rangle$ and $|1\rangle$ in $\mathbb{C}^2$:

$$a|0\rangle + b|1\rangle \quad \text{where} \quad |a|^2 + |b|^2 = 1 .$$

Through *measurement*, a qubit is forced to collapse irreversibly through projection to either $|0\rangle$ or $|1\rangle$. The probability of its doing either is $|a|^2$ and $|b|^2$, respectively. a and b are called *probability amplitudes*.

If necessary, we can convert ("read out") $|0\rangle$ and $|1\rangle$ to classical bit values of 0 and 1.

$|0\rangle$ and $|1\rangle$ are defined in the next section.

The word "collapse" comes from the physical interpretation of a quantum system where measurement causes a superposition to move to one of two choices. We examine what it is that is collapsing when we look at the quantum description of the polarization of photons in section 11.3.

A qubit by itself is not very interesting, despite all the mathematical formalism. Algorithms need to produce results and those results should translate to something meaningful. A qubit really does eventually produce only a single classical 0 or 1 when measured and read, and so we need many more of them to represent useful information.

Quantum algorithms are interesting because of the ways multiple qubits interact when they are in their pre-measured states, when we can use the full power of using linear algebra in $\mathbb{C}^2$. *Entanglement* is the situation where two or more qubits are so tightly correlated that we cannot learn something about one without learning about the other(s).

7.2 Bras and kets

When we previously looked at vector notation in section 5.4.2, we saw several forms like

$$\mathbf{v} = (v_1, v_2, \ldots, v_n)$$

$$= \begin{bmatrix} v_1 & v_2 & \cdots & v_n \end{bmatrix} \text{ as a row vector}$$

$$= \begin{bmatrix} v_1 \\ v_2 \\ \vdots \\ v_n \end{bmatrix} \text{ as a column vector.}$$

We now add two more which were invented by Paul Dirac, an English theoretical physicist, for use in quantum mechanics. They simplify many of the expressions we use in quantum computing.

Given a vector $\mathbf{v} = (v_1, v_2, \ldots, v_n)$, we denote by $\langle v|$, pronounced "bra-v", the row vector

$$\begin{bmatrix} \overline{v_1} & \overline{v_2} & \cdots & \overline{v_n} \end{bmatrix}$$

where we take the complex conjugate of each entry. Similarly, $|w\rangle$, pronounced "ket-w", is the column vector

$$\begin{bmatrix} w_1 \\ w_2 \\ \vdots \\ w_m \end{bmatrix}$$

without the conjugations, where $\mathbf{w} = (w_1, w_2, \ldots, w_m)$.

$|v\rangle\langle w|$ is the *outer product*

$$|v\rangle\langle w| = \begin{bmatrix} v_1\overline{w_1} & v_1\overline{w_2} & \cdots & v_1\overline{w_m} \\ v_2\overline{w_1} & v_2\overline{w_2} & \cdots & v_2\overline{w_m} \\ \vdots & \vdots & \ddots & \vdots \\ v_2\overline{w_1} & v_2\overline{w_2} & \cdots & v_2\overline{w_m} \end{bmatrix}.$$

When $n = m$, the bra-ket $\langle v|w\rangle = \langle v| \, |w\rangle = (\langle v|)(|w\rangle)$ is the usual *inner product*.

$$\langle v|w\rangle = \langle \mathbf{v}, \mathbf{w} \rangle = \overline{v_1}w_1 + \overline{v_2}w_2 + \cdots \overline{v_n}w_n.$$

The length of $\mathbf{v}$ is $\|\mathbf{v}\| = \sqrt{\langle v|v\rangle}$.

To avoid notational overload, I continue to put vector names in bold case like $\mathbf{v}$ when I use them in isolation, but will drop the bold in the bra or ket forms. They already make it clear that a vector is involved.

To learn more

How does mathematical notation get invented? As mathematicians work with concepts, they need a concise way of conveying what they mean. Good notation can make a statement or a proof much clearer and more insightful to the reader. Over time, the symbols and expressions that prove to be most useful win out while the others fade away into the archives. In the case of Dirac's bra-ket notation, it has become ubiquitous across quantum mechanics and now quantum computing. [5, Section 1.6] [6, Section 6.2]

Example

If $\mathbf{v} = (3, -i)$ and $\mathbf{w} = (2 + i, 4)$ then

$$\langle v| = \begin{bmatrix} 3 & i \end{bmatrix} \quad \text{and} \quad |w\rangle = \begin{bmatrix} 2 + i \\ 4 \end{bmatrix}$$

and

$$\langle v|w\rangle = 3(2 + i) + 4i = 6 + 7i$$

and

$$|w\rangle\langle v| = \begin{bmatrix} (2+i)3 & (2+i)i \\ 4 \times 3 & 4i \end{bmatrix} = \begin{bmatrix} 6+3i & 2i-1 \\ 12 & 4i \end{bmatrix}.$$

We use symbols within bras or kets to label them even though we may not list the coordinates. Common examples are $|0\rangle$, $|1\rangle$, $|+\rangle$, $|-\rangle$, $|i\rangle$, $|-i\rangle$, and $|\psi\rangle$.

Just as we early on used n for a generic number in $\mathbb{N}$ or z for a number in $\mathbb{C}$, $|\psi\rangle$ is a general purpose labeled ket. The Greek letter ψ is "psi."

$|0\rangle$ is $(1, 0)$ and $|1\rangle$ is $(0, 1)$ in the standard basis $\mathbf{e_1}$ and $\mathbf{e_2}$. $|0\rangle$ and $|1\rangle$ are an orthonormal basis for $\mathbb{C}^2$. Remember, vectors exist independently of their representation in a given basis, so what we are really saying is:

$|0\rangle$ is the vector in $\mathbb{C}^2$ that has coordinates $(1, 0)$ when we happen to be using the basis $\mathbf{e_1}$ and $\mathbf{e_2}$. $|1\rangle$ has coordinates $(0, 1)$ in this basis.

$|0\rangle$ and $|1\rangle$ are frequently called the *computational basis*.

If we decide to use another orthonormal basis, then $|0\rangle$ and $|1\rangle$ would still be the same vectors but with different coordinates.

In the same way $|+\rangle = \left(\frac{\sqrt{2}}{2}, \frac{\sqrt{2}}{2}\right)$ and $|-\rangle = \left(\frac{\sqrt{2}}{2}, -\frac{\sqrt{2}}{2}\right)$ in the standard basis. Like $|0\rangle$ and $|1\rangle$, these are an orthonormal basis for $\mathbb{C}^2$.

To remember which is $|0\rangle$ and which is $|1\rangle$, look at the second coordinate. For $|+\rangle$ and $|-\rangle$, look at the sign of the second coordinate.

Question 7.2.1

What are the coordinates for $|0\rangle$ in the basis $(\frac{\sqrt{3}}{2}, \frac{1}{2})$ and $(-\frac{1}{2}, \frac{\sqrt{3}}{2})$? What are the coordinates for $|1\rangle$?

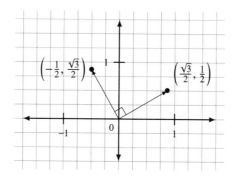

This diagram may help but does it make a difference that it is drawn in $\mathbb{R}^2$ instead of $\mathbb{C}^2$?

Though we use the notation $|0\rangle$ and $|1\rangle$, **they are not equal to the numbers 0 and 1**! I show a connection later when we discuss measuring the state of a qubit, but do keep them separate in your mind. The reason I highlight this is because, as a basis, any vector in $\mathbb{C}^2$ can be written as a linear combination

$$a|0\rangle + b|1\rangle \text{ for } a \text{ and } b \text{ in } \mathbb{C}.$$

You might therefore be tempted to say such a vector is "0 and 1 at the same time!". Don't say that. It's just a linear combination. It's also a cliché that should not be used unless you know what it means.

Just as we saw that a 1 by n matrix or a row vector defines a linear form, so does a bra when combined with a ket. For a vector $\mathbf{a} = (a_1, a_2, \ldots, a_n)$,

$$\langle a| = \begin{bmatrix} \overline{a_1} & \overline{a_2} & \cdots & \overline{a_n} \end{bmatrix}$$

and this corresponds to the linear form

$$\langle a|v \rangle = \langle \mathbf{a}, \mathbf{v} \rangle = \overline{a_1}v_1 + \overline{a_2}v_2 + \cdots + \overline{a_n}v_n$$

for all $\mathbf{v}$ in a complex n-dimensional vector space.

If L is a linear transformation on a complex vector space, then we write $L|w\rangle$ for "L applied to $|w\rangle$." If A is a matrix for L then we use the form $A|w\rangle$ for "A multiplied by $|w\rangle$". Both A and the explicit column representation of $|w\rangle$ use the same basis.

For a bra $\langle v|$, what does $\langle v|L$ mean? If A is again a matrix for L, then $\langle v|L$ is $\langle v|A$ and we do the matrix multiplication between the row vector and the matrix.

These all have the nice property that we can slide L and A from the bra to the ket and back:

$$\langle v|L \times |w\rangle = ((\langle v|L)\,|w\rangle) = \langle v|\,(L|w\rangle) = \langle v| \times L|w\rangle$$
$$\langle v|A \times |w\rangle = ((\langle v|A)\,|w\rangle) = \langle v|\,(A|w\rangle) = \langle v| \times A|w\rangle$$

Let's get rid of unneeded parentheses and "$\times$" symbols: write $\langle v|L|w\rangle$ for the first case and $\langle v|A|w\rangle$ for the matrix version in the second.

Let V be an n-dimensional vector space and let $|e_1\rangle$ to $|e_n\rangle$ be the standard basis $\mathbf{e_1}, \ldots, \mathbf{e_2}$ in ket form. If A is an n by n square matrix then the (i, j)-th entry of A is $A_{i,j} = \langle e_i|A|e_j \rangle$.

$|e_i\rangle\langle e_j|$ is an n by n square matrix with a 1 in the (i, j)-th position and 0 elsewhere.

The sum of the n^2 matrices $|e_i\rangle A_{i,j}\langle e_j|$ is A.

Define matrices

$$P(i) = |e_i\rangle\langle e_i|$$
$$P(i, j) = P(i) + P(j)$$
$$P(i, j, k) = P(i) + P(j) + P(k) \text{ for } i \neq j \neq k \neq i.$$

We can consider P having up to n arguments, no one of argument equal to any other. They are each diagonal matrices and trivially Hermitian. (Though I use P here, this is not for probability as we used it in chapter 6.)

Question 7.2.2

Show that

$$P(i)P(j) = \begin{cases} 1 & \text{if } i = j \\ 0 & \text{otherwise.} \end{cases}$$

A Hermitian matrix A is a *projector* or a *projection matrix* if $A^2 = A$. The rank of the projector is the dimension of its image vector space. $P(i)$ has rank 1, $P(i, j)$ has rank 2, and so on. $P(1, \ldots, n) = I_n$.

Why did I bother introducing this new notation? First, it's what scientists and practitioners in quantum mechanics and quantum computing use, and you have to know their language. Second, while it now only seems like a modest improvement over the vector and linear transformation notation we developed in chapter 5, it dramatically simplifies expressions involving multiple qubits. It provides, I hope, greater clarity and faster understanding for you.

7.3 The complex math and physics of a single qubit

Let's revisit our definition of a qubit from section 7.1. This time we break it into two pieces, a mathematical part and a physical/quantum mechanical part.

Mathematics

A *qubit*—a *quantum bit*—is the fundamental unit of quantum information. At any given time, it is in a superposition state represented by a linear combination of vectors $|0\rangle$ and $|1\rangle$ in $\mathbb{C}^2$:

$$a|0\rangle + b|1\rangle \quad \text{where} \quad |a|^2 + |b|^2 = 1 .$$

Physics

Through *measurement*, a qubit is forced to collapse irreversibly to either $|0\rangle$ or $|1\rangle$. The probability of its doing either is $|a|^2$ and $|b|^2$, respectively. a and b are called *probability amplitudes*.

The mathematical portion is linear algebra of a two-dimensional complex vector space. As a vector, the qubit state has length 1. Linear transformations must preserve this length and are isometries. Their matrices are unitary. Being unitary, they are invertible: moving a qubit from one state to another is always reversible.

When we built circuits that manipulated bits in section 2.4, only one of the core gates, **not**, was reversible. When we build quantum circuits using qubits in chapter 9, all the gates are reversible with the exception of measurement. Mathematically, we can think of a quantum circuit as the application of unitary matrices to vectors representing qubit states.

Mathematics is elegant and beautiful in its own right but it is often only a tool used in other fields. We frequently use it to build models that allow us to reason about how things may work in the so-called "real world." These models themselves are not the real world but are our best effort to formalize the relationships that seem to make things behave the way they do.

Linear algebra by itself does not say the probability amplitudes affect whether we get $|0\rangle$ or $|1\rangle$ when a qubit is measured. Linear algebra, complex numbers, and probability are used to represent much of quantum computing. We must use our understanding of the physical system to interpret what the elements of the mathematics mean. Mathematical formalism can only get us so far in the development of our structure: physics must give us additional relationships that bring us further.

In particular, measurement is a physical action. It causes us to drop from a superposition state involving $|0\rangle$ and $|1\rangle$ to one or the other. Mathematics describes the probability of either being produced but it does not force one value or another.

7.3.1 Quantum state representation

If $|\psi\rangle = a|0\rangle + b|1\rangle$ is a quantum state, we know $|\psi|^2 = |a|^2 + |b|^2 = 1$. This is the same as saying $\langle\psi|\psi\rangle = 1$.

The values $|a|^2$ and $|b|^2$ are non-negative numbers in $\mathbb{R}$ that are the probabilities that $|\psi\rangle$ will go to $|0\rangle$ or $|1\rangle$, respectively, when measured.

Consider the following example in which, for illustrative purposes, I have made some simplifications and taken some labeling liberties. Suppose a and b can only be real.

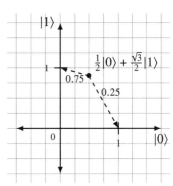

Let $a = \frac{1}{2}$ and $b = \frac{\sqrt{3}}{2}$. Then $|a|^2 = \left|\frac{1}{2}\right|^2 = 0.25$ and $|b|^2 = \left|\frac{\sqrt{3}}{2}\right|^2 = 0.75$. This means that the state of the qubit has a 25% chance of collapsing to $|0\rangle$ when measured and a 75% chance of collapsing to $|1\rangle$.

If c is in $\mathbb{C}$ with $|c| = 1$, then

$$c|\psi\rangle = |c\psi\rangle = ca|0\rangle + cb|1\rangle$$

then

$$|ca|^2 = |c|^2|a|^2 = |a|^2 \quad \text{and} \quad |cb|^2 = |c|^2|b|^2 = |b|^2 .$$

This means the probability of obtaining $|0\rangle$ or $|1\rangle$ upon measurement is not changed by multiplying a quantum state by a complex number of absolute value 1. We say there is no *observable* difference in doing this multiplication. As we know from section 4.5, all complex numbers with absolute value 1 look like $e^{\varphi i} = \cos(\varphi) + i\sin(\varphi)$.

We identify two quantum states if they only differ by a multiple of a complex unit. Recall that a unit has absolute value equal to 1. Such numbers look like $e^{i\varphi}$ for $0 \le \varphi < 2\pi$.

This has the consequence that when we say "$|\psi\rangle$ is not changed," we really mean that after doing whatever we are doing, "it is fine if we have $e^{i\varphi}|\psi\rangle$." This point is easy to miss because you expect $|\psi\rangle$ to be literally unchanged.

Let's re-express a and b in polar form:

$$|\psi\rangle = a|0\rangle + b|1\rangle = r_1 e^{\varphi_1 i}|0\rangle + r_2 e^{\varphi_2 i}|1\rangle$$

If either a or b is 0, then the other is 1 and we have the basis kets. Assume $a \ne 0$ and so $r_1 \ne 0$. Then

$$r_1 e^{\varphi_1 i}|0\rangle + r_2 e^{\varphi_2 i}|1\rangle = \left(e^{\varphi_1 i}\right)\left(r_1|0\rangle + r_2 e^{(\varphi_2 - \varphi_1)i}|1\rangle\right)$$

with $e^{\varphi_1 i}$ in $\mathbb{C}$ with absolute value equal to 1.

There's a whole lot of subscripting and superscripting going on there! The net is this: from the perspective of measurement and observable results,

$$a|0\rangle + b|1\rangle = r_1 e^{\varphi_1 i}|0\rangle + r_2 e^{\varphi_2 i}|1\rangle$$

is effectively the same as

$$r_1|0\rangle + r_2 e^{(\varphi_2 - \varphi_1)i}|1\rangle$$

and *the coefficient a of $|0\rangle$ is real.*

We are left with two degrees of freedom for the state of a qubit:

1. one related to the magnitudes since r_1 and r_2 are dependent via $|r_1|^2 + |r_2|^2 = 1$, and
2. the *relative phase* $\varphi_2 - \varphi_1$.

The relative phase is significant and is important when we see *interference* in section 9.6, a technique used in quantum algorithms.

The state of a single qubit $|\psi\rangle$ can be represented by

$$|\psi\rangle = r_1|0\rangle + r_2 e^{\varphi i}|1\rangle$$

with r_1 and r_2 in $\mathbb{R}$, $r_1^2 + r_2^2 = 1$, and $0 \leq \varphi < 2\pi$. Moreover, we can find $0 \leq \theta \leq \pi$ with

$$r_1 = \cos\left(\tfrac{\theta}{2}\right) \quad \text{and} \quad r_2 = \sin\left(\tfrac{\theta}{2}\right)$$

so that

$$|\psi\rangle = \cos\left(\frac{\theta}{2}\right)|0\rangle + \sin\left(\frac{\theta}{2}\right)e^{\varphi i}|1\rangle.$$

When two quantum states $|\psi\rangle_1$ and $|\psi\rangle_2$ differ only by a complex unit multiple u,

$$|\psi\rangle_2 = u|\psi\rangle_1,$$

we cannot tell the difference between them when we measure. That is, u is not observable. u is called a *global phase*.

7.3.2 Unitary matrices mapping to standard form

If $|\psi\rangle = \cos\left(\tfrac{\theta}{2}\right)|0\rangle + \sin\left(\tfrac{\theta}{2}\right)e^{\varphi i}|1\rangle$ is an arbitrary quantum state where we have made the first probability amplitude real, then what does a 2 by 2 unitary matrix U look like that maps $|0\rangle$ to $|\psi\rangle$?

$$U|0\rangle = U\begin{bmatrix}1\\0\end{bmatrix} = |\psi\rangle = \cos\left(\frac{\theta}{2}\right)|0\rangle + \sin\left(\frac{\theta}{2}\right)e^{\varphi i}|1\rangle$$

By inspection,

$$U = \begin{bmatrix} \cos\left(\frac{\theta}{2}\right) & a \\ \sin\left(\frac{\theta}{2}\right)e^{\varphi i} & b \end{bmatrix}$$

for some complex numbers a and b. Since U is unitary, $I_2 = UU^{-1} = UU^\dagger$ where

$$U^\dagger = \begin{bmatrix} \overline{\cos\left(\frac{\theta}{2}\right)} & \overline{\sin\left(\frac{\theta}{2}\right)e^{\varphi i}} \\ \overline{a} & \overline{b} \end{bmatrix} = \begin{bmatrix} \cos\left(\frac{\theta}{2}\right) & \sin\left(\frac{\theta}{2}\right)e^{-\varphi i} \\ \overline{a} & \overline{b} \end{bmatrix}.$$

Using the explicit form of the inverse of a 2 by 2 matrix,

$$U^{-1} = \frac{1}{\det(U)}\begin{bmatrix} b & -a \\ -\sin\left(\frac{\theta}{2}\right)e^{\varphi i} & \cos\left(\frac{\theta}{2}\right) \end{bmatrix}.$$

Since U is unitary, $|\det(U)| = 1$. Choose δ in $\mathbb{R}$ so that $\det(U) = e^{\delta i}$. Then $U^{-1} = U^\dagger$ means

$$e^{-i\delta} \begin{bmatrix} b & -a \\ -\sin\left(\frac{\theta}{2}\right) e^{\varphi i} & \cos\left(\frac{\theta}{2}\right) \end{bmatrix} = \begin{bmatrix} \cos\left(\frac{\theta}{2}\right) & \sin\left(\frac{\theta}{2}\right) e^{-\varphi i} \\ \overline{a} & \overline{b} \end{bmatrix}.$$

Thus

$$e^{-i\delta} b = \cos\left(\frac{\theta}{2}\right) \quad \text{and} \quad -e^{-i\delta} a = \sin\left(\frac{\theta}{2}\right) e^{-\varphi i}$$

and

$$b = \cos\left(\frac{\theta}{2}\right) e^{\delta i} \quad \text{and} \quad a = -\sin\left(\frac{\theta}{2}\right) e^{-\varphi i} e^{\delta i} \ .$$

This makes

$$U = \begin{bmatrix} \cos\left(\frac{\theta}{2}\right) & -\sin\left(\frac{\theta}{2}\right) e^{i\delta - \varphi i} \\ \sin\left(\frac{\theta}{2}\right) e^{\varphi i} & \cos\left(\frac{\theta}{2}\right) e^{\delta i} \end{bmatrix}.$$

An alternative form defines $\lambda = \delta - \varphi$. We can then parametrize U by three numbers θ, φ, and λ with

$$U(\theta, \varphi, \lambda) = \begin{bmatrix} \cos\left(\frac{\theta}{2}\right) & -\sin\left(\frac{\theta}{2}\right) e^{i\lambda} \\ \sin\left(\frac{\theta}{2}\right) e^{\varphi i} & \cos\left(\frac{\theta}{2}\right) e^{i\lambda + i\varphi} \end{bmatrix}.$$

7.3.3 The density matrix

For $|\psi\rangle = a|0\rangle + b|1\rangle$, we define the *density matrix* of $|\psi\rangle$ to be

$$\rho = |\psi\rangle\langle\psi| = \begin{bmatrix} a \\ b \end{bmatrix} \otimes \begin{bmatrix} \overline{a} & \overline{b} \end{bmatrix} = \begin{bmatrix} a\overline{a} & a\overline{b} \\ b\overline{a} & b\overline{b} \end{bmatrix} = \begin{bmatrix} |a|^2 & a\overline{b} \\ b\overline{a} & |b|^2 \end{bmatrix}$$

We commonly use the Greek letter ρ ("rho") as the symbol representing the density matrix. Note that $\mathrm{tr}(\rho) = \mathrm{tr}(|\psi\rangle\langle\psi|) = 1$ and that ρ is Hermitian.

Question 7.3.1

What is $\det(|\psi\rangle\langle\psi|)$?

For $|\psi\rangle$ expressed by $r_1|0\rangle + r_2 e^{\varphi i}|1\rangle$, with r_1 and r_2 in $\mathbb{R}$ and non-negative, the density matrix is

$$\rho = |\psi\rangle\langle\psi| = \begin{bmatrix} r_1 \\ r_2 e^{\varphi i} \end{bmatrix} \otimes \begin{bmatrix} \overline{r_1} & \overline{r_2 e^{\varphi i}} \end{bmatrix}$$

$$= \begin{bmatrix} r_1 \\ r_2 e^{\varphi i} \end{bmatrix} \otimes \begin{bmatrix} r_1 & r_2 e^{-\varphi i} \end{bmatrix}$$

$$= \begin{bmatrix} r_1^2 & r_1 r_2 e^{-\varphi i} \\ r_1 r_2 e^{\varphi i} & r_2^2 \end{bmatrix}.$$

From this we can compute $r_1 = \sqrt{\rho_{1,1}}$, $r_2 = \sqrt{\rho_{2,2}}$, and $e^{\varphi i} = \frac{\rho_{2,1}}{r_1 r_2}$. Up to a global phase, we have not lost anything from going from the ket to the density matrix.

Question 7.3.2

What can you tell about the original ket if given its density matrix

$$\rho = \begin{bmatrix} d_{1,1} & d_{1,2} \\ d_{2,1} & d_{2,2} \end{bmatrix}?$$

For $|\psi\rangle$ expressed generally by $r_1 e^{\varphi_1 i}|1\rangle|0\rangle + r_2 e^{\varphi_2 i}|1\rangle$, with r_1 and r_2 in $\mathbb{R}$ and non-negative, the density matrix is

$$\rho = |\psi\rangle\langle\psi| = \begin{bmatrix} r_1 e^{\varphi_1 i} \\ r_2 e^{\varphi_2 i} \end{bmatrix} \otimes \begin{bmatrix} \overline{r_1 e^{\varphi_1 i}} & \overline{r_2 e^{\varphi_2 i}} \end{bmatrix}$$

$$= \begin{bmatrix} r_1 e^{\varphi_1 i} \\ r_2 e^{\varphi_2 i} \end{bmatrix} \otimes \begin{bmatrix} r_1 e^{-\varphi_1 i} & r_2 e^{-\varphi_2 i} \end{bmatrix}$$

$$= \begin{bmatrix} r_1^2 & r_1 r_2 e^{i(\varphi_1 - \varphi_2)} \\ r_1 r_2 e^{-i(\varphi_1 - \varphi_2)} & r_2^2 \end{bmatrix}.$$

If you let $\varphi = \varphi_1 - \varphi_2$, this reduces to the previous case where we had already made the probability amplitude of $|0\rangle$ real.

The density matrix of $|\psi\rangle$ is independent of any global phase.

7.3.4 Observables and expectation

The matrices $M_0 = |0\rangle\langle 0|$ and $M_1 = |1\rangle\langle 1|$ are projectors and hence Hermitian matrices. If $|\psi\rangle = a|0\rangle + b|1\rangle$,

$$\langle\psi|M_0|\psi\rangle = \left\langle\psi\left|\left(|0\rangle\langle 0|\right)\right|\psi\right\rangle$$
$$= \langle\psi|0\rangle\langle 0|\psi\rangle$$
$$= \begin{bmatrix}\overline{a} & \overline{b}\end{bmatrix}\begin{bmatrix}1 \\ 0\end{bmatrix}\begin{bmatrix}1 & 0\end{bmatrix}\begin{bmatrix}a \\ b\end{bmatrix}$$
$$= \begin{bmatrix}\overline{a} & \overline{b}\end{bmatrix}\begin{bmatrix}1 & 0 \\ 0 & 0\end{bmatrix}\begin{bmatrix}a \\ b\end{bmatrix}$$
$$= \begin{bmatrix}\overline{a} & \overline{b}\end{bmatrix}\begin{bmatrix}a \\ 0\end{bmatrix} = |a|^2$$

Similarly, $\langle\psi|M_1|\psi\rangle = |b|^2$.

To be explicit, $\langle\psi|M_0|\psi\rangle = |a|^2$ is the probability of measuring $|0\rangle$ and $\langle\psi|M_1|\psi\rangle = |b|^2$ is the probability of measuring $|1\rangle$.

The eigenvalues of M_0 are 0 and 1, corresponding to eigenvectors $|1\rangle$ and $|0\rangle$, respectively. M_1 has the same eigenvalues, with the eigenvectors reversed. M_0 and M_1 are both examples of *observables*: Hermitian matrices whose eigenvectors form a basis for the quantum state space.

Question 7.3.3

What is the relationship between the coefficients of the basis elements formed by the eigenvectors and the probabilities of seeing those basis elements when we measure?

Now let's turn this around and suppose A is an observable. A is a Hermitian matrix with eigenvectors $|e_1\rangle$ and $|e_2\rangle$ corresponding to eigenvalues e_1 and e_2. $|e_1\rangle$ and e_1 are different objects! We use $|e_1\rangle$ as the label for the eigenvector corresponding to e_1. Also remember that an expression like $\langle e_1|\psi\rangle$ means $\langle e_1||\psi\rangle$. It is not any sort of multiplication by e_1.

By definition,

$$A|e_1\rangle = e_1|e_1\rangle \quad \text{and} \quad A|e_2\rangle = e_2|e_2\rangle .$$

As stated at the end of section 5.9, we may assume that the eigenvectors $|e_1\rangle$ and $|e_2\rangle$ form an orthonormal basis of $\mathbb{C}^2$.

If $|\psi\rangle = a|e_1\rangle + b|e_2\rangle$ with $\langle\psi|\psi\rangle = 1$, by the properties of the inner product,

$$a = \langle e_1|\psi\rangle \quad \text{and} \quad b = \langle e_2|\psi\rangle$$

and thus
$$|\psi\rangle = \langle e_1|\psi\rangle|e_1\rangle + \langle e_2|\psi\rangle|e_1\rangle.$$

This basis is related to A. So when we *measure* the observable A, the probability of getting e_1 is $|\langle e_1|\psi\rangle|^2$ and the probability of getting e_2 is $|\langle e_2|\psi\rangle|^2$.

Now that we have a set of values and corresponding probabilities for when we might get them, it is reasonable to talk about expectation as we first discussed in section 6.6.

The expected value, or expectation, $\langle A\rangle$ of A given the state $|\psi\rangle$ is

$$\langle A\rangle = |\langle e_1|\psi\rangle|^2 e_1 + |\langle e_2|\psi\rangle|^2 e_2.$$

Question 7.3.4

Why does $|\langle e_1|\psi\rangle|^2 = \langle\psi|e_1\rangle\langle e_1|\psi\rangle$?

We can simplify this:

$$\begin{aligned}
\langle A\rangle &= |\langle e_1|\psi\rangle|^2 e_1 + |\langle e_2|\psi\rangle|^2 e_2 \\
&= \langle\psi|e_1\rangle\langle e_1|\psi\rangle e_1 + \langle\psi|e_2\rangle\langle e_2|\psi\rangle e_2 \quad \text{because } |\langle e_1|\psi\rangle|^2 = \langle\psi|e_1\rangle\langle e_1|\psi\rangle \\
&= \langle\psi|A|e_1\rangle\langle e_1|\psi\rangle + \langle\psi|A|e_2\rangle\langle e_2|\psi\rangle \quad \text{because } A|e_1\rangle = e_1|e_1\rangle \\
&= \left\langle\psi\left| A\Big(|e_1\rangle\langle e_1| + |e_2\rangle\langle e_2|\Big)\right|\psi\right\rangle \\
&= \langle\psi|A|\psi\rangle.
\end{aligned}$$

I think this calculation shows the elegance of the bra-ket notation.

Question 7.3.5

What are $\langle M_0\rangle$ and $\langle M_1\rangle$ for $|\psi\rangle = a|e_1\rangle + b|e_2\rangle$?

7.4 A non-linear projection

In chapter 5 we saw examples of linear projections, such as mapping any point in the real plane to the line $y = x$. Now we look at a special kind of projection that is non-linear. We map almost every point on the unit circle onto a line.

Here is a unit circle and the line $y = -1$ that sits right below it.

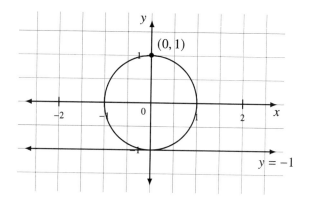

We can map every point on the circle except $(0, 1)$, the north pole, to a point on the line $y = -1$. We simply draw a line from $(0, 1)$ through the point on the circle. The result is where that line intersects $y = -1$.

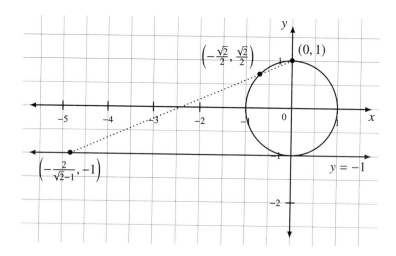

The point where it intersects the line is $\left(-\frac{2}{\sqrt{2}-1}, -1\right)$. We compute this using the slope-intercept form.

- We know two points on the line: the north pole $(0, 1)$ and the point on the circle $\left(-\frac{\sqrt{2}}{2}, \frac{\sqrt{2}}{2}\right)$.
- The slope m is the difference in y values divided by the difference in x values.

$$m = \frac{1 - \frac{\sqrt{2}}{2}}{0 - -\frac{\sqrt{2}}{2}} = \frac{1 - \frac{\sqrt{2}}{2}}{\frac{\sqrt{2}}{2}} = \frac{2 - \sqrt{2}}{\sqrt{2}} = \frac{2\sqrt{2} - 2}{2} = \sqrt{2} - 1$$

- When $x = 0$, $y = 1$. The equation of the line is

$$y = \left(\sqrt{2} - 1\right) x + 1$$

- To see where it intersects the line $y = -1$, we just set $y = -1$ and solve.

$$-1 = \left(\sqrt{2} - 1\right) x + 1$$

and so

$$-2 = \left(\sqrt{2} - 1\right) x$$

and so

$$-\frac{2}{\sqrt{2} - 1} = x$$

By construction, the y coordinate is always -1.

Let's generalize this to a point (x_0, y_0) on the circle.

- The slope is

$$m = \frac{y_0 - 1}{x_0 - 0} = \frac{y_0 - 1}{x_0}$$

- The line must intersect the y axis at $(0, 1)$ and so the equation of the line is

$$y = \frac{y_0 - 1}{x_0} x + 1$$

- Setting $y = -1$ and solving for x yields

$$-\frac{2}{\frac{y_0 - 1}{x_0}} = -\frac{2x_0}{y_0 - 1} = x$$

- The image of the projection is $\left(-\frac{2x_0}{y_0 - 1}, -1\right)$.

Question 7.4.1

Does this generalized method give us the same answer we calculated earlier for $(x_0, y_0) = \left(-\frac{\sqrt{2}}{2}, \frac{\sqrt{2}}{2}\right)$?

This general formula works for every point on the circle except for $y_0 = 1$, the north pole.

Question 7.4.2

What point on the circle maps to $(0, -1)$?

Is this projection f invertible?

$$f : \text{unit circle} - \{(0, 1)\} \to \mathbb{R}$$

where the image is the x coordinate on the line $y = -1$. What is the inverse function f^{-1} so $f^{-1} \circ f = $ the identity function? This means we can map the real line to the unit circle with the point $(0, 1)$ removed.

To construct f^{-1}, we begin with an a in $\mathbb{R}$. We want $f^{-1}(a)$ to be a point on the unit circle (x_0, y_0). What's the most defining aspect of a point on the unit circle? It's the relationship

$$x_0^2 + y_0^2 = 1.$$

- Given a, consider the point $(a, -1)$ on the line $y = -1$.
- Draw the line from this point to the north pole $(0, 1)$ and ask: where does this line intersect the unit circle?
- The slope m is $-\frac{2}{a}$ (verify this and see the note below) and the full equation of the line is

$$y = -\frac{2}{a}x + 1$$

- Squaring both sides:

$$y^2 = \frac{4}{a^2}x^2 - \frac{4}{a}x + 1$$

- Substituting in x_0 and y_0:

$$y_0^2 = \frac{4}{a^2}x_0^2 - \frac{4}{a}x_0 + 1$$

- But $x_0^2 + y_0^2 = 1$!

$$1 - x_0^2 = \frac{4}{a^2}x_0^2 - \frac{4}{a}x_0 + 1$$

- Negating and moving the terms on the left side to the right and then simplifying:

$$0 = \left(\frac{4}{a^2} + 1\right)x_0^2 - \frac{4}{a}x_0$$

$$= x_0\left(\left(\frac{4}{a^2} + 1\right)x_0 - \frac{4}{a}\right)$$

- We can rule out $x_0 = 0$ because this would give us the north pole. Therefore

$$0 = \left(\frac{4}{a^2} + 1\right)x_0 - \frac{4}{a} = \frac{4 + a^2}{a^2}x_0 - \frac{4}{a}$$

or

$$0 = \left(4 + a^2\right)x_0 - 4a$$

or

$$\frac{4a}{4 + a^2} = x_0$$

If $a = 2$ then $x_0 = 1$. Trivially, $y_0 = 0$. So $f^{-1}(2) = (1, 0)$.

What about if $a = -4$? Then $x_0 = \frac{-16}{4+16} = -\frac{4}{5}$. Given the equation of the unit circle from subsection 4.2.3,

$$\left(-\frac{4}{5}\right)^2 + y_0^2 = 1$$

and so

$$y_0^2 = 1 - \frac{16}{25} = \frac{9}{25}.$$

And so we inconclusively conclude $y_0 = \pm\frac{3}{5}$. Which is it?

By inspection of the graph,

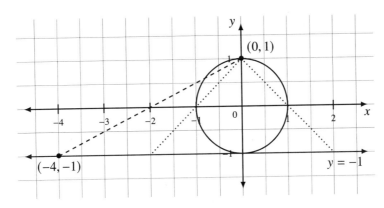

When $|a| \geq 1$ then $y_0 \geq 0$. Similarly, $|a| < 1$ implies $y_0 < 0$. We conclusively conclude $y_0 = \frac{3}{5}$.

There were two problems with the above analysis. First, it's not really acceptable to say "I looked at the graph and decided this was the situation." Therefore, I leave it to you:

Question 7.4.3

Show algebraically that $|a| \geq 1$ means $y_0 \geq 0$, and $|a| < 1$ means $y_0 < 0$.

The second problem is much more serious. Though I said the "slope m of the line is $-\frac{2}{a}$," **I never said anything about a being non-zero.**

The case $a = 0$ is special because it corresponds to the point directly under the north and south poles on $y = -1$. The line through the north pole is vertical and so is ∞. Here $f^{-1}(0) = (0, -1)$.

We can now fully describe f and f^{-1}.

If f is the projection of a point (x_0, y_0), $y_0 \neq 1$, on the unit circle onto the line $y = -1$, then its image is $(a, -1)$ where

$$a = -\frac{2x_0}{y_0 - 1}.$$

The inverse function $f^{-1}(a)$ for an a in $\mathbb{R}$ is defined by

$$f^{-1}(a) = \begin{cases} (0, -1) & \text{when } a = 0 \\[2ex] \left(\dfrac{4a}{4 + a^2} + \sqrt{1 - \left(\dfrac{4a}{4 + a^2} \right)^2} \right) & \text{when } |a| \geq 1 \\[3ex] \left(\dfrac{4a}{4 + a^2} - \sqrt{1 - \left(\dfrac{4a}{4 + a^2} \right)^2} \right) & \text{otherwise} \end{cases}$$

Our intermediate computations disallowed the case $a = 0$ though it is covered by the third case.

Though we did use trigonometry in the above analysis, we did not use angles. To conclude this section, let's quickly recast what we worked out but this time give the point on the unit circle by its angle $0 \leq \theta < 2\pi$ rather than Cartesian coordinates.

We need to exclude $\theta = \frac{\pi}{4}$ because this is the north pole. To keep our functions straight, we call this one g:

$$g : \left(0 \leq \theta < \frac{\pi}{4} \right) \text{ or } \left(\frac{\pi}{4} < \theta < 2\pi \right) \to \mathbb{R}$$

where the image is the x coordinate on the line $y = -1$. The definition is very simple:

$$g(\theta) = f(\cos(\theta), \sin(\theta)).$$

If g is the projection of a point given by an angle θ such that $0 \le \theta < 2\pi$, $\theta \ne \frac{\pi}{4}$, on the unit circle onto the line $y = -1$, then its image is $(a, -1)$ where

$$a = -\frac{2\cos(\theta)}{\sin(\theta) - 1}.$$

What about g^{-1}? In this example, we want to go from -4 to the value of θ.

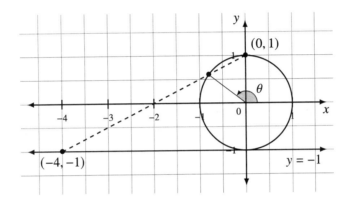

Let $\theta = \arccos(x_0) = \arccos\left(\frac{4a}{4+a^2}\right)$. By the definition of the inverse cosine, this has a value between 0 and π, inclusive. This alone does not tell us if the point is on the upper or lower half of the unit circle.

Let a be in $\mathbb{R}$ and $\theta = \arccos\left(\frac{4a}{4+a^2}\right)$. The inverse function $g^{-1}(a)$ is defined by

$$g^{-1}(a) = \begin{cases} \theta & \text{when } |a| \ge 1 \\ 2\pi - \theta & \text{otherwise} \end{cases}$$

In this projection, we took an object that lived in two dimensions but was determined by one variable θ and mapped it to the one-dimensional line $\mathbb{R}^1$. Here dimensions map to the idea of "degrees of freedom" as they would be expressed through the linear independence of a basis. Though the unit circle lives in $\mathbb{R}^2$, the relationship

$$x_0^2 + y_0^2 = 1$$

means

$$x_0 = \pm\sqrt{1 - y_0^2}$$

and so x_0 is not independent of y_0. Under what circumstances can we take an object in a higher dimensional space and map it to an object in a lower dimensional one?

In three real dimensions, the equivalent process of mapping a sphere onto a plane is called *stereographic projection*.

7.5 The Bloch sphere

We describe the state of a qubit by a vector

$$a|0\rangle + b|1\rangle = r_1 e^{\varphi_1 i}|0\rangle + r_2 e^{\varphi_2 i}|1\rangle$$

in $\mathbb{C}^2$ with r_1 and r_2 non-negative numbers in $\mathbb{R}$.

The magnitudes r_1 and r_2 are related by $r_1^2 + r_2^2 = 1$. This is a mathematical condition. We saw in section 7.3 that it's the relative phase of $\varphi_2 - \varphi_1$ that is significant and not the individual phases φ_1 and φ_2. This is a physical condition and it also means we can take a to be real.

We also saw that we could represent a quantum state as

$$|\psi\rangle = \cos\left(\frac{\theta}{2}\right)|0\rangle + \sin\left(\frac{\theta}{2}\right)e^{\varphi i}|1\rangle.$$

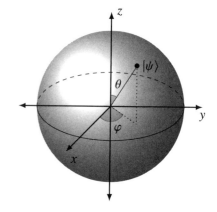

We do this via a non-linear projection and a change of coordinates, and get a point on the surface of the *Bloch sphere*.

The two angles have the ranges $0 \le \theta \le \pi$ and $0 \le \varphi < 2\pi$. θ is measured from the positive z axis and φ from the positive x axis in the xy-plane.

The non-linear projection is from the three-dimensional surface of the hypersphere in $\mathbb{C}^2$, thought of as $\mathbb{R}^4$, to the two-dimensional surface of the Bloch sphere in $\mathbb{R}^2$. They key property that allows us to do this is that we can ignore global phase.

We start by asking "what are all the vectors in $\mathbb{C}^2$ of length 1?" to get the hypersphere. We then state "we're going to say that any two points on the hypersphere are the 'same' if they are different only by a complex number multiple of magnitude 1." With these properties and equivalences, along with some algebra and geometry, we get the points on the Bloch sphere as the possible quantum states of a qubit.

The eponymous sphere is named after Felix Bloch, a scientist who won the 1962 Nobel Prize in physics for his work in nuclear magnetic resonance (NMR). [2]

It's easy to get confused here when we see $|\psi\rangle$. Is this on the Bloch sphere or is it in $\mathbb{C}^2$? We do not distinguish these cases by notation and so I will make it clear by context.

Given a point/state on the Bloch sphere given by θ and φ, we can go back to $|\psi\rangle = a|0\rangle + b|1\rangle$ in $\mathbb{C}^2$ by

$$a = \cos\left(\tfrac{\theta}{2}\right) \quad \text{and} \quad b = e^{\varphi i} \sin\left(\tfrac{\theta}{2}\right) .$$

Note a is real.

On the other hand, if

Felix Bloch (1905-1983) in 1961. Photo subject to use via the Creative Commons Attribution 3.0 Unported license ⓒⓘ. [3]

$$a|0\rangle + b|1\rangle = r_1 e^{\varphi_1 i}|0\rangle + r_2 e^{\varphi_2 i}|1\rangle$$

what are θ and φ?

If $a = 0$ then $b = 1$ and we have $|1\rangle$. We map this to $\theta = \pi$ and $\varphi = 0$. If $a = 1$ then $b = 0$ and we have $|0\rangle$, which goes to $\theta = 0$ and $\varphi = 0$.

Assume neither a nor b is 0 and so the magnitudes r_1 and r_2 are positive real numbers. We rewrite

$$r_1 e^{\varphi_1 i}|0\rangle + r_2 e^{\varphi_2 i}|1\rangle = \left(e^{\varphi_1 i}\right)\left(r_1|0\rangle + r_2 e^{i(\varphi_2 - \varphi_2)}|1\rangle\right)$$

and can discard the $e^{\varphi_1 i}$ as it is not physically observable. We are left with

$$r_1|0\rangle + r_2 e^{i(\varphi_2 - \varphi_1)}|1\rangle.$$

We let $\varphi = \varphi_2 - \varphi_1$ adjusted to be an equivalent angle between 0 and 2π by adding 2π if the difference is negative.ß

Since $r_1^2 + r_2^2 = 1$, the point (r_1, r_2) is on the unit circle *in the first quadrant*. We can identify an angle $\theta_0 = \arccos(r_1)$ with $0 < \theta_0 < \frac{\pi}{2}$ so that

$$r_1 = \cos(\theta_0) \quad \text{and} \quad r_2 = \sin(\theta_0) .$$

Let $\theta = 2\theta_0$.

Let's go back to $|0\rangle$ and $|1\rangle$ and examine the choices we made above for θ and φ.

$$|0\rangle = r_1 e^{\varphi_1 i}|0\rangle + r_2 e^{\varphi_2 i} = 1 e^{\varphi_1 i}|0\rangle + 0 e^{\varphi_2 i}|1\rangle$$

By pulling out and discarding $e^{\varphi_1 i}$ as we have done several times, we are considering

$$1|0\rangle + 0 e^{(\varphi_2 - \varphi_1) i}|1\rangle.$$

Our choice of θ_0 is $\arccos(1) = 0$ and thus θ also is 0. This means we are at the north pole of the sphere. The rotation around the z axis, φ is meaningless and so we choose it to be 0.

We cannot choose φ_2 uniquely because of the multiplication by 0 in front of $e^{\varphi_2 i}$. We might as well choose $\varphi_2 = \varphi_1$ and $\varphi = 0$.

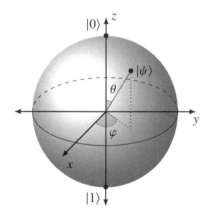

Question 7.5.1

Are our choices $\theta = \pi$ and $\varphi = 0$ for $|1\rangle$ also reasonable?

From all this, we have worked out the mappings to and from the Bloch sphere for qubit states in $\mathbb{C}^2$. $|0\rangle$ maps to the north pole and $|1\rangle$ maps to the south pole.

What about $|+\rangle$ and $|-\rangle$?

Question 7.5.2

Show $|+\rangle$ maps to $\theta = \frac{\pi}{2}$ and $\varphi = 0$ on the Bloch sphere. Then show $|-\rangle$ maps to $\theta = \frac{\pi}{2}$ and $\varphi = \pi$. What are these points using Cartesian coordinates in $\mathbb{R}^3$?

The kets $|0\rangle$ and $|1\rangle$ lie on the z axis and $|+\rangle$ and $|-\rangle$ lie on the x axis.

What is the orthonormal basis that would lie on the y axis? Let's call them $|\mathbf{A}\rangle$ and $|\mathbf{B}\rangle$ for now.

The two points in which we are interested have Cartesian coordinates $(0, 1, 0)$ and $(0, -1, 0)$.

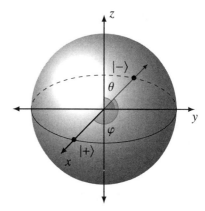

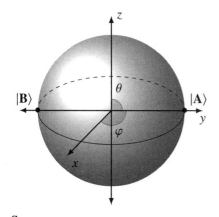

For each of them we can take $\theta = \frac{\pi}{2}$. For $|\mathbf{A}\rangle$, we use $\varphi = \frac{\pi}{2}$, and for $|\mathbf{B}\rangle$, $\varphi = \frac{3\pi}{2}$ gets us to $(0, -1, 0)$.

Now we use these to go back to $|\psi\rangle = a|0\rangle + b|1\rangle$ in $\mathbb{C}^2$. We set

$$ a = \cos\left(\tfrac{\theta}{2}\right) \quad \text{and} \quad b = e^{\varphi i} \sin\left(\tfrac{\theta}{2}\right) $$

to get

$$ a = \tfrac{\sqrt{2}}{2} \quad \text{and} \quad b = e^{\frac{\pi}{2}i}\tfrac{\sqrt{2}}{2} = \tfrac{\sqrt{2}}{2}i \; . $$

So

$$ |\mathbf{A}\rangle = \tfrac{\sqrt{2}}{2}\left(|0\rangle + i|1\rangle\right) \quad \text{and} \quad |\mathbf{B}\rangle = \tfrac{\sqrt{2}}{2}\left(|0\rangle - i|1\rangle\right) \; . $$

With this we officially rename these kets as

$$ |i\rangle = \tfrac{\sqrt{2}}{2}\left(|0\rangle + i|1\rangle\right) \quad \text{and} \quad |-i\rangle = \tfrac{\sqrt{2}}{2}\left(|0\rangle - i|1\rangle\right) \; . $$

Question 7.5.3

Verify $|i\rangle$ and $|-i\rangle$ are orthonormal.

We can now fully label the Bloch sphere.

From the evidence we've seen in three cases so far, an orthonormal basis in $\mathbb{C}^2$ maps to two opposite points on the Bloch sphere. Start thinking about why this is the case.

Every quantum state on the equator has an equal probability of producing $|0\rangle$ or $|1\rangle$ when measured. That is, if a state is on the equator then in terms of its coordinates a and b in $a|0\rangle + b|1\rangle$ in $\mathbb{C}^2$, $|a|^2 = |b|^2 = 0.5$.

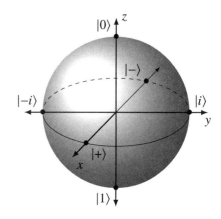

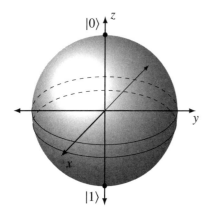

Naturally, the state $|0\rangle$ has probability 1.0 of yielding itself when measured and probability 0.0 of producing $|1\rangle$. These probabilities are reversed for $|1\rangle$.

For a latitude on the Bloch sphere that is not the equator, the probabilities of collapsing to $|0\rangle$ or $|1\rangle$ are different but still add up to 1.0. The probability of yielding $|0\rangle$ is the same for every point on that latitude. Ditto for $|1\rangle$.

We have now seen three pairs of basis elements and they go by different names:

Basis	Common name	Bloch sphere axis name
$\{|0\rangle, |1\rangle\}$	Computational	Z
$\{|+\rangle, |-\rangle\}$	Hadamard	X
$\{|i\rangle, |-i\rangle\}$	Circular	Y

The Y basis is the same as the circular basis. In this case we also have the alternative notation $|i\rangle = |\circlearrowleft\rangle$ and $|-i\rangle = |\circlearrowright\rangle$.

7.6   Professor Hadamard, meet Professor Pauli

Other than mapping the state of a qubit to a Bloch sphere and so looking at it in a different way, what can you do with a qubit? In this section we look at the operations, also called gates, you can apply to a single qubit. Later we expand our exploration to gates that have multiple qubits as inputs and outputs. In chapter 9 we build circuits with these gates to implement algorithms.

Here, for example, is a circuit with one qubit initialized to $|0\rangle$ that performs one operation, **X**, and then measures the qubit.

$$q_0: |0\rangle - \boxed{\textbf{X}} - \boxed{\nearrow} - |m_0\rangle \quad = |1\rangle$$

Gates are always reversible, but some other operations are not. This comes from quantum mechanics and gates corresponding to unitary transformations. Measurement is irreversible, and so is the $|0\rangle$ reset operation described in subsection 7.6.13.

When I include the measurement operation in a circuit in the next chapter and beyond, it looks like the graphic on the right.

Measurement returns a $|0\rangle$ or $|1\rangle$. Mathematically, we do not worry about how it happens other than the probability. In chapter 11 I discuss how measurement is done in the different physical qubit technologies.

Since a qubit state is a two-dimensional complex ket, or vector, all quantum gates have 2 by 2 matrices with complex entries according to some basis. This makes them small and very easy to manipulate. They are unitary matrices.

> If A is a complex square matrix then it is *unitary* if its adjoint $A^\dagger$ is also its inverse A^{-1}. Hence $AA^\dagger = A^\dagger A = I$.
>
> This means the columns of A are orthonormal, as are the rows.
>
> $|\det(A)| = 1$.

Note the last statement says the **absolute value of the determinant** is 1, not that the determinant is 1. Since it is in $\mathbb{C}$, the determinant is of the form $e^{\varphi i}$ for $0 \le \varphi < 2\pi$.

The remainder of this section is a catalog of some of the most useful and commonly used 1-qubit quantum gates. As you will see, there is often more than one way to accomplish the same qubit state change. Why you would want to do so is the topic of chapter 9.

When we looked at classical circuits in section 2.4, only the **not** operation was reversible. That's also the only operation which carries over to quantum. We start there.

7.6.1 The quantum X gate

The **X** gate has the matrix

$$\sigma_x = \begin{bmatrix} 0 & 1 \\ 1 & 0 \end{bmatrix}$$

and this is the Pauli **X** matrix, named after Wolfgang Pauli. By "abuse of notation" I often use the same name (in this case **X**) for both the gate and its unitary matrix in the standard basis kets.

It has the property that

$$\sigma_x|0\rangle = |1\rangle \quad \text{and} \quad \sigma_x|1\rangle = |0\rangle.$$

It "flips" between $|0\rangle$ and $|1\rangle$. The classic **not** gate is

The **not** gate is a "bit flip" and by analogy we also say **X** is a bit flip.

For $|\psi\rangle = a|0\rangle + b|1\rangle$ in $\mathbb{C}^2$,

$$\mathbf{X}|\psi\rangle = b|0\rangle + a|1\rangle.$$

This reverses the probabilities of measuring $|0\rangle$ and $|1\rangle$. In $\mathbb{C}^2$, σ_x has eigenvalues $+1$ and -1 for eigenvectors $|+\rangle$ and $|-\rangle$, respectively.

In terms of the Bloch sphere, the **X** gate rotates by π around the x axis. So not only are the poles flipped but points in the lower hemisphere move to the upper and vice versa.

> Since $\mathbf{X}\,\mathbf{X} = I_2$, the **X** gate is its own inverse. This is reasonable because in the classical case it is also true that **not** $\circ$ **not** is the identity operation.

When we considered rotations we saw that the matrix for $\mathbb{R}^3$ that does a rotation around the x axis by θ radians works like

$$\begin{bmatrix} 1 & 0 & 0 \\ 0 & \cos(\theta) & -\sin(\theta) \\ 0 & \sin(\theta) & \cos(\theta) \end{bmatrix} \begin{bmatrix} x \\ y \\ z \end{bmatrix} = \begin{bmatrix} x \\ y\cos(\theta) - z\sin(\theta) \\ y\sin(\theta) + z\cos(\theta) \end{bmatrix}$$

Plugging in $\theta = \pi$, the rotation matrix is

$$\begin{bmatrix} 1 & 0 & 0 \\ 0 & -1 & 0 \\ 0 & 0 & -1 \end{bmatrix}$$

In $\mathbb{R}^3$ standard coordinates, $|0\rangle = (0,0,1)$ and $|1\rangle = (0,0,-1)$. Applying the above matrix:

The physicist Wolfgang Pauli, 1900-1958. Pauli won the Nobel Prize in Physics in 1945. Photo is in the public domain.

$$\begin{bmatrix} 1 & 0 & 0 \\ 0 & -1 & 0 \\ 0 & 0 & -1 \end{bmatrix} \begin{bmatrix} 0 \\ 0 \\ 1 \end{bmatrix} = \begin{bmatrix} 0 & 0 & -1 \end{bmatrix} \quad \text{and} \quad \begin{bmatrix} 1 & 0 & 0 \\ 0 & -1 & 0 \\ 0 & 0 & -1 \end{bmatrix} \begin{bmatrix} 0 \\ 0 \\ -1 \end{bmatrix} = \begin{bmatrix} 0 & 0 & 1 \end{bmatrix} .$$

You can see this flipped $|0\rangle$ and $|1\rangle$. What about $|+\rangle$ and $|-\rangle$?

$$\begin{bmatrix} 1 & 0 & 0 \\ 0 & -1 & 0 \\ 0 & 0 & -1 \end{bmatrix} \begin{bmatrix} 1 \\ 0 \\ 0 \end{bmatrix} = \begin{bmatrix} 1 & 0 & 0 \end{bmatrix} \quad \text{and} \quad \begin{bmatrix} 1 & 0 & 0 \\ 0 & -1 & 0 \\ 0 & 0 & -1 \end{bmatrix} \begin{bmatrix} -1 \\ 0 \\ 0 \end{bmatrix} = \begin{bmatrix} -1 & 0 & 0 \end{bmatrix} .$$

As you would expect from looking at the geometry of the Bloch sphere, it leaves these alone.

Question 7.6.1

What does the **X** gate do to $|i\rangle$ and $|-i\rangle$?

Question 7.6.2

What are the eigenvectors and eigenvalues of σ_x?

When I include the **X** gate in a circuit it looks like the graphic on the right. The horizontal line, called a *wire*, represents the qubit and its state.

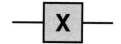

The input state enters on the left, the **X** unitary transformation is applied, and the new quantum state result comes out the right side.

7.6.2 The quantum Z gate

The **Z** gate has the matrix

$$\sigma_z = \begin{bmatrix} 1 & 0 \\ 0 & -1 \end{bmatrix}$$

and this is the Pauli **Z** matrix. It rotates qubit states by π around the z axis on the Bloch sphere.

The **Z** gate swaps $|+\rangle$ and $|-\rangle$ as well as $|i\rangle$ and $|-i\rangle$. It leaves $|0\rangle$ and $|1\rangle$ alone on the Bloch sphere.

Question 7.6.3

Show by calculation that **X Z** = −**Z X**.

Since **Z Z** = I_2, the **Z** gate is its own inverse. If you rotate by π and then rotate by π again, you end up back where you started.

For $|\psi\rangle = a|0\rangle + b|1\rangle$ in $\mathbb{C}^2$,

$$\mathbf{Z}|\psi\rangle = a|0\rangle - b|1\rangle.$$

The probabilities of measuring $|0\rangle$ and $|1\rangle$ do not change after applying **Z**.

In $\mathbb{C}^2$, σ_z has eigenvalues +1 and −1 for eigenvectors $|0\rangle$ and $|1\rangle$, respectively. Per subsection 7.3.4, **Z** or σ_z is the observable for the standard computational basis $|0\rangle$ and $|1\rangle$.

Remember that $|1\rangle$ and $-|1\rangle = e^{\pi i}|1\rangle$ in $\mathbb{C}^2$ map to the same point we are also calling $|1\rangle$ on the Bloch sphere. If we express

$$|\psi\rangle = r_1 e^{\theta_1 i}|0\rangle + r_2 e^{\theta_2 i}|1\rangle$$

then

$$\sigma_z|\psi\rangle = r_1 e^{\theta_1 i}|0\rangle - r_2 e^{\theta_2 i}|1\rangle$$
$$= r_1 e^{\theta_1 i}|0\rangle + e^{\pi i} r_2 e^{\theta_2 i}|1\rangle$$
$$= r_1 e^{\theta_1 i}|0\rangle + r_2 e^{(\pi+\theta_2)i}|1\rangle$$
$$= e^{\theta_1 i}\left(r_1|0\rangle + r_2 e^{(\pi+\theta_2-\theta_1)i}|1\rangle\right)$$

The relative phase of $|\psi\rangle$ is changed to π plus that relative phase, adjusted to be between 0 and 2π. This is a *phase flip* and **Z** is called a phase flip gate. Since it reverses the sign of the second amplitude, it is also called a *sign flip* gate.

When I include the **Z** gate in a circuit it looks like the graphic on the right.

7.6.3 The quantum Y gate

The **Y** gate has the matrix

$$\sigma_y = \begin{bmatrix} 0 & -i \\ i & 0 \end{bmatrix} = i\begin{bmatrix} 0 & -1 \\ 1 & 0 \end{bmatrix}$$

and this is the Pauli **Y** matrix. It rotates qubit states by π around the y axis on the Bloch sphere.

It swaps $|0\rangle$ and $|1\rangle$ and so is a bit flip. It also swaps $|+\rangle$ and $|-\rangle$ but leaves $|i\rangle$ and $|-i\rangle$ alone.

Question 7.6.4

Show by calculation that **X Y** = $-$**Y X** and **Z Y** = $-$**Y Z**.

Since **Y Y** = I_2, the **Y** gate is its own inverse.

For $|\psi\rangle = a|0\rangle + b|1\rangle$ in $\mathbb{C}^2$,

$$\mathbf{Y}|\psi\rangle = -bi|0\rangle + ai|1\rangle = e^{\frac{3\pi}{2}i}\left(b|0\rangle - a|1\rangle\right).$$

From this we can see directly that **Y** does a bit flip and a phase flip at the same time. In $\mathbb{C}^2$, σ_y has eigenvalues $+1$ and -1 for eigenvectors $|i\rangle$ and $|-i\rangle$, respectively.

If $|\psi\rangle = a|0\rangle + b|1\rangle$ in $\mathbb{C}^2$, then a bit flip interchanges the coefficients of $|0\rangle$ and $|1\rangle$. A phase flip changes the sign of the coefficient of $|1\rangle$. A simultaneous bit and phase flip does both:

$$a|0\rangle + b|1\rangle \mapsto b|0\rangle - a|1\rangle$$

Question 7.6.5

What is the 3 by 3 rotation matrix for the **Y** gate?

When I include the **Y** gate in a circuit it looks like the graphic on the right.

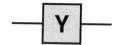

7.6.4 The quantum ID gate

The **ID** gate does nothing and is typically used when constructing or drawing circuits where we want to show something happening to every qubit at every step. If we did construct an implementation, it would be multiplication by I_2, the 2 by 2 identity matrix.

When I include the **ID** gate in a circuit it looks like the graphic on the right.

The **ID** gate is also used in circuits to indicate a place to pause or delay. This allows, for example, researchers to calculate measurements of the decoherence of a qubit.

7.6.5 The quantum H gate

The **H** gate, or $\mathbf{H}^{\otimes 1}$ or Hadamard gate, has the matrix

$$\mathbf{H} = \begin{bmatrix} \frac{\sqrt{2}}{2} & \frac{\sqrt{2}}{2} \\ \frac{\sqrt{2}}{2} & -\frac{\sqrt{2}}{2} \end{bmatrix} = \frac{\sqrt{2}}{2} \begin{bmatrix} 1 & 1 \\ 1 & -1 \end{bmatrix}$$

operating on $\mathbb{C}^2$.

By matrix multiplication,

$$\mathbf{H}|0\rangle = \tfrac{\sqrt{2}}{2}(|0\rangle + |1\rangle) = |+\rangle \quad \text{and} \quad \mathbf{H}|1\rangle = \tfrac{\sqrt{2}}{2}(|0\rangle - |1\rangle) = |-\rangle.$$

By linearity,

$$\mathbf{H}|+\rangle = \mathbf{H}\left(\frac{\sqrt{2}}{2}(|0\rangle + |1\rangle)\right) = \frac{\sqrt{2}}{2}(\mathbf{H}|0\rangle + \mathbf{H}|1\rangle)$$

$$= \frac{\sqrt{2}}{2}\left(\frac{\sqrt{2}}{2}(|0\rangle + |1\rangle) + \frac{\sqrt{2}}{2}(|0\rangle - |1\rangle)\right) = \frac{1}{2}(|0\rangle + |1\rangle + |0\rangle - |1\rangle) = |0\rangle$$

and $\mathbf{H}|-\rangle = |1\rangle$.

> The Hadamard gate is one of the most frequently used gates in quantum computing. **H** is often the first gate applied in a circuit. When you read "put the qubit in superposition" it usually means "take the qubit initialized in the $|0\rangle$ state and apply **H** to it."

> The Hadamard matrix is the change of basis matrix from $\{|0\rangle, |1\rangle\}$ to $\{|+\rangle, |-\rangle\}$.
>
> Since $\mathbf{H}\mathbf{H} = I_2$, the **H** gate is its own inverse.

Question 7.6.6

What is the 3 by 3 matrix for the **H** gate on the Bloch sphere? It is the product of two 3 by 3 rotation matrices. What are they?

The mathematician Jacques Hadamard, 1865-1963. Photo is in the public domain. ℗

When I include the **H** gate in a circuit it looks like the graphic on the right.

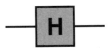

In quantum computing there are a lot of zeroes and ones floating around. We use these in interesting ways. For example, suppose we have $|x\rangle$ where x is either 0 or 1. Thinking of x as being in $\mathbb{Z}$, look at expressions like $(-1)^x$, which is 1 or -1 as x is 0 or 1.

For our **H** gate, we can stare at

$$\mathsf{H}|0\rangle = \tfrac{\sqrt{2}}{2}\left(|0\rangle + |1\rangle\right) \quad \text{and} \quad \mathsf{H}|1\rangle = \tfrac{\sqrt{2}}{2}\left(|0\rangle - |1\rangle\right)$$

and notice that

$$\mathsf{H}|u\rangle = \frac{\sqrt{2}}{2}\left(|0\rangle + (-1)^u|1\rangle\right)$$

when u is one of $\{0, 1\}$. When $u = 0$ we have $|0\rangle$ going to $\tfrac{\sqrt{2}}{2}\left(|0\rangle + |1\rangle\right)$, as expected. For $u = 1$ we end up with $\tfrac{\sqrt{2}}{2}\left(|0\rangle - |1\rangle\right)$.

Question 7.6.7

Show using matrix calculations that **X = H Z H**.

A change of basis from the computational basis $\{|0\rangle, |1\rangle\}$ to the Hadamard basis $\{|+\rangle, |-\rangle\}$ changes a bit flip **X** to a phase flip **Z**.

Question 7.6.8

What is **H X H**?

7.6.6 The quantum $\mathsf{R}^{\mathsf{z}}_{\varphi}$ gates

We can generalize the phase changing behavior of the **Z** gate by noting

$$\sigma_z = \begin{bmatrix} 1 & 0 \\ 0 & -1 \end{bmatrix} = \begin{bmatrix} 1 & 0 \\ 0 & e^{\pi i} \end{bmatrix} = \begin{bmatrix} 1 & 0 \\ 0 & e^{\varphi i} \end{bmatrix} \quad \text{where } \varphi = \pi.$$

This last form is the template for the collection of gates given the name $\mathsf{R}^{\mathsf{z}}_{\varphi}$:

$$\mathsf{R}^{\mathsf{z}}_{\varphi} = \begin{bmatrix} 1 & 0 \\ 0 & e^{\varphi i} \end{bmatrix}$$

This collection is infinite as φ can take on any radian value greater than or equal to 0 and less than 2π. These gates change the phase of a qubit state by φ.

Question 7.6.9

What is the 3 by 3 rotation matrix for $\mathbf{R}^z_\varphi$? For $\mathbf{Z}$?

The inverse of $\mathbf{R}^z_\varphi$ is $\mathbf{R}^z_{2\pi-\varphi}$.

$\mathbf{R}^z_0 = \mathbf{ID}$.

$\mathbf{R}^z_\pi = \mathbf{Z}$.

When I include the $\mathbf{R}^z_\varphi$ gate in a circuit it looks like the graphic on the right for a particular value of φ.

An alternative form of the matrix for $\mathbf{R}^z_\varphi$ is

$$\mathbf{R}^z_\varphi = \begin{bmatrix} e^{-\frac{\varphi i}{2}} & 0 \\ 0 & e^{\frac{\varphi i}{2}} \end{bmatrix} = \cos\left(\frac{\varphi}{2}\right) I_2 - \cos\left(\frac{\varphi}{2}\right) i\sigma_z$$

where I_2 is the 2 by 2 identity matrix and σ_z is the Pauli $\mathbf{Z}$ matrix.

This is the same as the first matrix multiplied by $e^{-\frac{\varphi i}{2}}$. We can do this because $e^{-\frac{\varphi i}{2}}$ is a complex unit and multiplication by it is not observable when we measure.

7.6.7 The quantum S gate

The **S** gate is a shorthand for $\mathbf{R}^z_{\frac{\pi}{2}}$. After applying, the phase is adjusted to be greater than or equal to 0 and less than 2π.

$$\mathbf{S} = \mathbf{R}^z_{\frac{\pi}{2}} = \begin{bmatrix} 1 & 0 \\ 0 & e^{\frac{\pi i}{2}} \end{bmatrix} = \begin{bmatrix} 1 & 0 \\ 0 & i \end{bmatrix}$$

Question 7.6.10

What is the 3 by 3 rotation matrix for the **S** gate?

When I include the **S** gate in a circuit it looks like the graphic on the right.

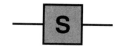

Traditionally and confusingly, the **S** gate is also known as the $\frac{\pi}{4}$ gate. This is because we can express the matrix in this way:

$$\begin{bmatrix} 1 & 0 \\ 0 & e^{\frac{\pi i}{2}} \end{bmatrix} = e^{\frac{\pi}{4}} \begin{bmatrix} e^{-\frac{\pi i}{4}} & 0 \\ 0 & e^{\frac{\pi i}{4}} \end{bmatrix}$$

The unit factor $e^{\frac{\pi}{4}}$ in front does not have an observable effect on the quantum state of the result of applying **S**. Some authors call **S** "the phase gate," but I won't.

7.6.8 The quantum $S^\dagger$ gate

The $S^\dagger$ gate is a shorthand for $R^z_{\frac{3\pi}{2}} = R^z_{-\frac{\pi}{2}}$. After applying, the phase is adjusted to be greater than or equal to 0 and less than 2π.

$$S^\dagger = R^z_{\frac{3\pi}{2}} = \begin{bmatrix} 1 & 0 \\ 0 & e^{\frac{3\pi i}{2}} \end{bmatrix} = \begin{bmatrix} 1 & 0 \\ 0 & -i \end{bmatrix}$$

It gets its name because the matrix for $S^\dagger$ is the adjoint of the **S** matrix.

$$\begin{bmatrix} 1 & 0 \\ 0 & i \end{bmatrix}^\dagger = \begin{bmatrix} 1 & 0 \\ 0 & -i \end{bmatrix}$$

Question 7.6.11

What is the 3 by 3 rotation matrix for the $S^\dagger$ gate?

When I include the $S^\dagger$ gate in a circuit it looks like the graphic on the right.

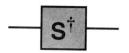

7.6.9 The quantum T gate

The **T** gate is a shorthand for $R^z_{\frac{\pi}{4}}$. After applying, the phase is adjusted to be greater than or equal to 0 and less than 2π.

$$T = R^z_{\frac{\pi}{4}} = \begin{bmatrix} 1 & 0 \\ 0 & e^{\frac{\pi i}{4}} \end{bmatrix} = \begin{bmatrix} 1 & 0 \\ 0 & \cos\left(\frac{\pi}{4}\right) + \sin\left(\frac{\pi}{4}\right)i \end{bmatrix} = \begin{bmatrix} 1 & 0 \\ 0 & \frac{\sqrt{2}}{2} + \frac{\sqrt{2}}{2}i \end{bmatrix}$$

Question 7.6.12

What is the 3 by 3 rotation matrix for the **T** gate?

We can get the **S** by applying the **T** twice: **S** = **T** ∘ **T**.

When I include the **T** gate in a circuit it looks like the graphic on the right.

The **T** gate is also known as the $\frac{\pi}{8}$ gate. We can write its matrix as

$$\begin{bmatrix} 1 & 0 \\ 0 & e^{\frac{\pi i}{4}} \end{bmatrix} = e^{\frac{\pi}{8}} \begin{bmatrix} e^{-\frac{\pi i}{8}} & 0 \\ 0 & e^{\frac{\pi i}{8}} \end{bmatrix}$$

The unit factor $e^{\frac{\pi}{8}}$ in front does not have an observable effect on the quantum state of the result of applying **T**.

7.6.10   The quantum T† gate

The **T**† gate is a shorthand for $R^z_{\frac{7\pi}{4}} = R^z_{-\frac{\pi}{4}}$. After applying, the phase is adjusted to be greater than or equal to 0 and less than 2π.

$$T^\dagger = R^z_{\frac{7\pi}{4}} = \begin{bmatrix} 1 & 0 \\ 0 & e^{\frac{7\pi i}{4}} \end{bmatrix} = \begin{bmatrix} 1 & 0 \\ 0 & \cos\left(\frac{7\pi}{4}\right) + \sin\left(\frac{7\pi}{4}\right)i \end{bmatrix} = \begin{bmatrix} 1 & 0 \\ 0 & \frac{\sqrt{2}}{2} - \frac{\sqrt{2}}{2}i \end{bmatrix}$$

It gets its name because the matrix for **T**† is the adjoint of the **T** matrix.

$$\begin{bmatrix} 1 & 0 \\ 0 & \frac{\sqrt{2}}{2} + \frac{\sqrt{2}}{2}i \end{bmatrix}^\dagger = \begin{bmatrix} 1 & 0 \\ 0 & \frac{\sqrt{2}}{2} - \frac{\sqrt{2}}{2}i \end{bmatrix}$$

Question 7.6.13

What is the 3 by 3 rotation matrix for the $\mathbf{T}^\dagger$ gate?

We can get the $\mathbf{S}^\dagger$ by applying the $\mathbf{T}^\dagger$ twice: $\mathbf{S}^\dagger = \mathbf{T}^\dagger \circ \mathbf{T}^\dagger$.

When I include the $\mathbf{T}^\dagger$ gate in a circuit it looks like the graphic on the right.

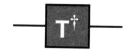

7.6.11 The quantum $\mathbf{R}_\varphi^x$ and $\mathbf{R}_\varphi^y$ gates

Just as $\mathbf{R}_\varphi^z$ is an arbitrary rotation around the z axis, we can define gates that rotate around the x and y axes.

$$\mathbf{R}_\varphi^x = \begin{bmatrix} \cos\left(\frac{\varphi}{2}\right) & -\sin\left(\frac{\varphi}{2}\right)i \\ -\sin\left(\frac{\varphi}{2}\right)i & \cos\left(\frac{\varphi}{2}\right) \end{bmatrix}$$

$$= \cos\left(\frac{\varphi}{2}\right)I_2 - \cos\left(\frac{\varphi}{2}\right)i\sigma_x$$

and

$$\mathbf{R}_\varphi^y = \begin{bmatrix} \cos\left(\frac{\varphi}{2}\right) & -\sin\left(\frac{\varphi}{2}\right) \\ \sin\left(\frac{\varphi}{2}\right) & \cos\left(\frac{\varphi}{2}\right) \end{bmatrix}$$

$$= \cos\left(\frac{\varphi}{2}\right)I_2 - \cos\left(\frac{\varphi}{2}\right)i\sigma_y$$

where I_2 is the 2 by 2 identity matrix, σ_x is the Pauli **X** matrix, and σ_y is the Pauli **Y** matrix.

7.6.12 The quantum $\sqrt{\text{NOT}}$ gate

Another gate used in some of the quantum computing literature is the "square root of **NOT**" gate. It has the matrix

$$\frac{1}{2}\begin{bmatrix} 1+i & 1-i \\ 1-i & 1+i \end{bmatrix} = \begin{bmatrix} \frac{1}{2}+\frac{1}{2}i & \frac{1}{2}-\frac{1}{2}i \\ \frac{1}{2}-\frac{1}{2}i & \frac{1}{2}+\frac{1}{2}i \end{bmatrix}.$$

Squaring this we get

$$\frac{1}{2}\begin{bmatrix} 1+i & 1-i \\ 1-i & 1+i \end{bmatrix} \times \frac{1}{2}\begin{bmatrix} 1+i & 1-i \\ 1-i & 1+i \end{bmatrix} = \begin{bmatrix} 0 & 1 \\ 1 & 0 \end{bmatrix} = \mathbf{X}.$$

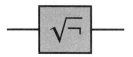

The **X** gate is the quantum version of **not**, and that's how this gate gets its name.

Question 7.6.14

Show that $\sqrt{\mathbf{NOT}}$ is unitary. What is its determinant? What does it do to $|0\rangle$ and $|1\rangle$?

The adjoint of this gate has matrix

$$\frac{1}{2}\begin{bmatrix} 1-i & 1+i \\ 1+i & 1-i \end{bmatrix} = \begin{bmatrix} \frac{1}{2} - \frac{1}{2}i & \frac{1}{2} + \frac{1}{2}i \\ \frac{1}{2} + \frac{1}{2}i & \frac{1}{2} - \frac{1}{2}i \end{bmatrix}.$$

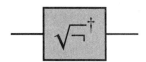

What is its square?

7.6.13 The quantum $|0\rangle$ RESET operation

Though it is not a reversible unitary operation, and hence not a gate, some quantum computing software environments allow you to reset a qubit in the middle of a circuit to $|0\rangle$, Use of this operation may make your code non-portable.

7.7 Gates and unitary matrices

The collection of all 2 by 2 unitary matrices (subsection 5.7.5) with entries in $\mathbb{C}$ form a group under multiplication called the *unitary group* of degree 2. It is denoted by $\mathbf{U}(2, \mathbb{C})$. It is a subgroup of $\mathbf{GL}(2, \mathbb{C})$, the general linear group of degree 2 over $\mathbb{C}$.

Every 1-qubit gate corresponds to such a unitary matrix. We can create all 2 by 2 unitary matrices from the identity and Pauli matrices.

Any U in $\mathbf{U}(2, \mathbb{C})$ can be written as a product of a complex unit times a linear combination of unitary matrices

$$U = e^{\theta i}\left(c_{I_2}I_2 + c_{\sigma_x}\sigma_x + c_{\sigma_y}\sigma_y + c_{\sigma_z}\sigma_z \right)$$

where we have the following definitions, properties, and identities:

- I_2 is the 2 by 2 identity matrix
- c_{σ_x} c_{σ_y}, and c_{σ_z} are Pauli matrices
- $0 \leq \theta < 2\pi$
- c_{I_2} is in $\mathbb{R}$
- c_{σ_x}, c_{σ_y}, and c_{σ_z} are in $\mathbb{C}$

- $\left|c_{I_2}\right|^2 + \left|c_{\sigma_x}\right|^2 + \left|c_{\sigma_y}\right|^2 + \left|c_{\sigma_z}\right|^2 = 1$
- $\mathrm{Re}\left(c_{I_2}\right)\overline{c_{\sigma_x}} + \mathrm{Im}\left(c_{\sigma_y}\right)\overline{c_{\sigma_z}} = 0$
- $\mathrm{Re}\left(c_{I_2}\right)\overline{c_{\sigma_y}} + \mathrm{Im}\left(c_{\sigma_x}\right)\overline{c_{\sigma_z}} = 0$
- $\mathrm{Re}\left(c_{I_2}\right)\overline{c_{\sigma_z}} + \mathrm{Im}\left(c_{\sigma_x}\right)\overline{c_{\sigma_y}} = 0$

The complex unit only affects the global phase of the qubit state and so is not observable. This means that we do not see its effect when we measure because it does not affect the probabilities of seeing one thing or another. [4]

Question 7.7.1

What does this mean in terms of the **ID**, **X**, **Y**, and **Z** gates?

7.8 Summary

The quantum states of a qubit are the unit vectors in $\mathbb{C}^2$, where we identify two states as being equivalent if they differ only by a multiple of a complex unit. To better visualize actions on a qubit, we introduced the Bloch sphere in $\mathbb{R}^3$ and showed where special orthonormal bases map onto the sphere.

Any new idea seems to deserve its own notation and we did not disappoint when we introduced Dirac's bra-ket representation of vectors. This significantly simplifies calculation when working with multiple qubits.

Given the ket form of qubit states, we introduced the standard 1-qubit gate operations. In the classical case in section 2.4, there was only one operation we could perform on a single bit, **not**. In the quantum case, there are many, in fact an infinite number, of single qubit operations.

We next look at how to work with two or more qubits and the quantum gates that operate on them. We also introduce entanglement, an essential notion from quantum mechanics.

References

[1]   Karen Barad. *Meeting the Universe Halfway. Quantum Physics and the Entanglement of Matter and Meaning.* 2nd ed. Duke University Press Books, 2007.

[2]   F. Bloch. "Nuclear Induction". In: *Physical Review* 70 (7-8 Oct. 1946), pp. 460–474.

[3]   Creative Commons. *Attribution 3.0 Unported (CC BY 3.0)* ⓒⓘ. URL: https://creativecommons.org/licenses/by/3.0/legalcode.

[4]   Simon J. Devitt, William J. Munro, and Kae Nemoto. "Quantum error correction for beginners". In: *Reports on Progress in Physics* 76.7, 076001 (July 2013), p. 076001.

[5]   P. A. M. Dirac. *The Principles of Quantum Mechanics*. Clarendon Press, 1930.

[6]   A. Yu. Kitaev, A. H. Shen, and M. N. Vyalyi. *Classical and Quantum Computation*. American Mathematical Society, 2002.

8

Two Qubits, Three

Not only is the Universe stranger than we think,
it is stranger than we can think.

Werner Heisenberg [2]

In the previous chapter we defined qubits and saw what we could do with just one of them. Things now start to get exponential with every additional qubit added, because entanglement allows the size of the working state space to double.

This chapter is about seeing how multiple qubits can behave together and then building a collection of tools to manipulate those qubits. These include the concept of entanglement, which is a requirement to do quantum computing. We also examine important 2-qubit gates like **CNOT** and **SWAP**. This will lead us into chapter 9 and chapter 10 where we look at algorithms and build circuits that use this machinery.

All vector spaces considered in this chapter are over $\mathbb{C}$, the field of complex numbers introduced in section 3.9. All bases are orthonormal unless otherwise specified.

Topics covered in this chapter

8.1   Tensor products

In this section I introduce the linear algebra construction of a *tensor product*. If the direct sum seems to concatenate two vector spaces, then the tensor product interleaves them. In the first case, if we start with dimensions n and m, we end up with a new vector space of $n + m$ dimensions. For the tensor product, we get nm dimensions.

We can quickly get vector spaces with high dimensions through this multiplicative effect. This means we need to use our algebraic intuition and tools more than our geometric ones.

The initial construction is straight linear algebra but we specialize it to quantum computing and working with multiple qubits in the next section.

Let V and W be two finite dimensional vector spaces over $\mathbb{F}$. Define a new vector space $V \otimes W$, pronounced "V tensor W" or "the tensor product of V and W," to be the vector space generated by addition and scalar multiplication of all objects $\mathbf{v} \otimes \mathbf{w}$ for each $\mathbf{v}$ in V and $\mathbf{w}$ in W.

> Note I said "generated by." Not all vectors in $V \otimes W$ look like $\mathbf{v} \otimes \mathbf{w}$ for some $\mathbf{v}$ in V and $\mathbf{w}$ in W. A vector might look like $2\mathbf{v}_1 \otimes \mathbf{w}_3 + 9\mathbf{v}_4 \otimes \mathbf{w}_7$, for example.

Tensor products have the following properties:

- If a is a scalar in $\mathbb{F}$ then

$$a \left(\mathbf{v} \otimes \mathbf{w} \right) = \left(a\mathbf{v} \right) \otimes \mathbf{w} = \mathbf{v} \otimes \left(a\mathbf{w} \right).$$

- For $\mathbf{v}_1$ in V and $\mathbf{w}_1$ and $\mathbf{w}_2$ in W,

$$\mathbf{v}_1 \otimes (\mathbf{w}_1 + \mathbf{w}_2) = \mathbf{v}_1 \otimes \mathbf{w}_1 + \mathbf{v}_1 \otimes \mathbf{w}_2.$$

- For $\mathbf{v}_1$ and $\mathbf{v}_2$ in V and $\mathbf{w}_1$ in W,

$$(\mathbf{v}_1 + \mathbf{v}_2) \otimes \mathbf{w}_1 = \mathbf{v}_1 \otimes \mathbf{w}_1 + \mathbf{v}_2 \otimes \mathbf{w}_1.$$

- If $f : V \times W \to U$ is a bilinear map, then $f^{\otimes} : V \otimes W \to U$ is a linear map defined by

$$f^{\otimes}(\mathbf{v} \otimes \mathbf{w}) = f(\mathbf{v}, \mathbf{w}).$$

- If $\{\mathbf{v}_1, \mathbf{v}_2, \ldots, \mathbf{v}_n\}$ and $\{\mathbf{w}_1, \mathbf{w}_2, \ldots, \mathbf{w}_m\}$ are bases for V and W, then

$$\mathbf{v}_1 \otimes \mathbf{w}_1, \mathbf{v}_1 \otimes \mathbf{w}_2, \ldots, \mathbf{v}_1 \otimes \mathbf{w}_m,$$

$$\vdots$$

$$\mathbf{v}_n \otimes \mathbf{w}_1, \mathbf{v}_n \otimes \mathbf{w}_2, \ldots, \mathbf{v}_n \otimes \mathbf{w}_m$$

are basis vectors of $V \otimes W$. There are nm of them and this is the dimension of $V \otimes W$.

For two additional vectors spaces X and Y over $\mathbb{F}$, if

$$f : V \to X \quad \text{and} \quad g : W \to Y$$

are linear maps, then so is

$$f \otimes g : V \otimes W \to X \otimes Y$$

where we define

$$(f \otimes g)(\mathbf{v} \otimes \mathbf{w}) = f(\mathbf{v}) \otimes g(\mathbf{w}).$$

Question 8.1.1

Confirm $f \otimes g$ is a linear map.

If f and g are both monomorphisms, so is $f \otimes g$. If they are both epimorphisms, so is $f \otimes g$.

If we have

$$A = \begin{bmatrix} a_{1,1} & a_{1,2} \\ a_{2,1} & a_{2,2} \end{bmatrix} \quad \text{and} \quad B = \begin{bmatrix} b_{1,1} & b_{1,2} \\ b_{2,1} & b_{2,2} \end{bmatrix}$$

then the regular matrix product is

$$AB = \begin{bmatrix} a_{1,1}b_{1,1} + a_{1,2}b_{2,1} & a_{1,1}b_{1,2} + a_{1,2}b_{2,2} \\ a_{2,1}b_{1,1} + a_{2,2}b_{2,1} & a_{2,1}b_{1,2} + a_{2,2}b_{2,2} \end{bmatrix}$$

The matrix tensor product for 2 by 2 matrices is

$$A \otimes B = \begin{bmatrix} a_{1,1}B & a_{1,2}B \\ a_{2,1}B & a_{2,2}B \end{bmatrix}$$

$$= \begin{bmatrix} a_{1,1}\begin{bmatrix} b_{1,1} & b_{1,2} \\ b_{2,1} & b_{2,2} \end{bmatrix} & a_{1,2}\begin{bmatrix} b_{1,1} & b_{1,2} \\ b_{2,1} & b_{2,2} \end{bmatrix} \\ a_{2,1}\begin{bmatrix} b_{1,1} & b_{1,2} \\ b_{2,1} & b_{2,2} \end{bmatrix} & a_{2,2}\begin{bmatrix} b_{1,1} & b_{1,2} \\ b_{2,1} & b_{2,2} \end{bmatrix} \end{bmatrix}$$

$$= \begin{bmatrix} a_{1,1}b_{1,1} & a_{1,1}b_{1,2} & a_{1,2}b_{1,1} & a_{1,2}b_{1,2} \\ a_{1,1}b_{2,1} & a_{1,1}b_{2,2} & a_{1,2}b_{2,1} & a_{1,2}b_{2,2} \\ a_{2,1}b_{1,1} & a_{2,1}b_{1,2} & a_{2,2}b_{1,1} & a_{2,2}b_{1,2} \\ a_{2,1}b_{2,1} & a_{2,1}b_{2,2} & a_{2,2}b_{2,1} & a_{2,2}b_{2,2} \end{bmatrix}$$

and this is respect to the $\mathbf{e_1} \otimes \mathbf{e_1}$ $\mathbf{e_1} \otimes \mathbf{e_2}$, $\mathbf{e_2} \otimes \mathbf{e_1}$, and $\mathbf{e_2} \otimes \mathbf{e_1}$ basis. I'm abusing the notation a bit in the second line to show we get a block of elements in the new matrix by multiplying an entry of the first by every entry in the second. The same kind of construction rules applies for larger matrices.

We do something similar with vectors. If $\mathbf{v} = (v_1, v_2, v_3)$ in V and $\mathbf{w} = (w_1, w_2)$ in W, then

$$\mathbf{v} \otimes \mathbf{w} = (v_1\mathbf{w}, v_2\mathbf{w}) = (v_1 w_1, v_1 w_2, \ v_2 w_1, v_2 w_2, \ v_3 w_1, v_3 w_2).$$

I've again taken liberties in the second term to show we get three entries by multiplying the second vector by an entry in the first. If V and W have Euclidean norms, then

$$\begin{aligned} \|\mathbf{v} \otimes \mathbf{w}\|^2 &= \|(v_1 w_1, v_1 w_2, \ v_2 w_1, v_2 w_2, \ v_3 w_1, v_3 w_2)\|^2 \\ &= |v_1 w_1|^2 + |v_1 w_2|^2 + |v_2 w_1|^2 + |v_2 w_2|^2 + |v_3 w_1|^2 + |v_3 w_2|^2 \\ &= |v_1|^2 |w_1|^2 + |v_1|^2 |w_2|^2 + |v_2|^2 |w_1|^2 + |v_2|^2 |w_2|^2 + |v_3|^2 |w_1|^2 + |v_3|^2 |w_2|^2 \\ &= |v_1|^2 \left(|w_1|^2 + |w_2|^2\right) + |v_2|^2 \left(|w_1|^2 + |w_2|^2\right) + |v_3|^2 \left(|w_1|^2 + |w_2|^2\right) \\ &= \left(|v_1|^2 + |v_2|^2 + |v_3|^2\right) \left(|w_1|^2 + |w_2|^2\right) \\ &= \|\mathbf{v}\|^2 \|\mathbf{w}\|^2 \end{aligned}$$

This is true in general for finite dimensional vector spaces V and W with Euclidean norms.

For $\mathbb{R}^2 \otimes \mathbb{R}^2$ or $\mathbb{C}^2 \otimes \mathbb{C}^2$,

$$\mathbf{e}_1 \otimes \mathbf{e}_1 = (1, 0, 0, 0)$$
$$\mathbf{e}_1 \otimes \mathbf{e}_2 = (0, 1, 0, 0)$$
$$\mathbf{e}_2 \otimes \mathbf{e}_1 = (0, 0, 1, 0)$$
$$\mathbf{e}_2 \otimes \mathbf{e}_2 = (0, 0, 0, 1)$$

Tensor products combine nicely with traditional products. For two additional matrices,

$$(A \otimes B)(C \otimes D) = AC \otimes BD.$$

We now take a brief culinary diversion to again compare direct sums and tensor products. Let V be the vector space with basis **chocolate ice cream**, **vanilla ice cream**, and **mint chocolate chip ice cream**. For the second vector space W, the basis is **chocolate fudge sauce**, **caramel sauce**, **mango sauce**, and **raspberry sauce**.

Vectors in $V \oplus W$ are linear combinations of the 7 $(= 3 + 4)$ foods

chocolate ice cream
vanilla ice cream
mint chocolate chip ice cream
chocolate fudge sauce
caramel sauce
mango sauce
raspberry sauce

Vectors in $V \otimes W$ are linear combinations of the 12 $(= 3 \times 4)$ food combinations:

chocolate ice cream $\otimes$ **chocolate fudge sauce**
chocolate ice cream $\otimes$ **caramel sauce**
chocolate ice cream $\otimes$ **mango sauce**
chocolate ice cream $\otimes$ **raspberry sauce**

vanilla ice cream $\otimes$ **chocolate fudge sauce**
vanilla ice cream $\otimes$ **caramel sauce**
vanilla ice cream $\otimes$ **mango sauce**

vanilla ice cream ⊗ raspberry sauce

mint chocolate chip ice cream ⊗ chocolate fudge sauce
mint chocolate chip ice cream ⊗ caramel sauce
mint chocolate chip ice cream ⊗ mango sauce
mint chocolate chip ice cream ⊗ raspberry sauce

We get every possible pairing with the tensor product. I think the last three are the most questionable flavor mixtures. Now back to the math.

If A and B are unitary matrices then so is $A \otimes B$.

Question 8.1.2

For the two unitary matrices

$$H = \begin{bmatrix} \frac{\sqrt{2}}{2} & \frac{\sqrt{2}}{2} \\ \frac{\sqrt{2}}{2} & -\frac{\sqrt{2}}{2} \end{bmatrix} \quad \text{and} \quad \sigma_y = \begin{bmatrix} 0 & -i \\ i & 0 \end{bmatrix}$$

which are the Hadamard and Pauli σ_y matrices, respectively, show

$$H \otimes \sigma_y = \begin{bmatrix} 0 & -\frac{\sqrt{2}}{2}i & 0 & -\frac{\sqrt{2}}{2}i \\ \frac{\sqrt{2}}{2}i & 0 & \frac{\sqrt{2}}{2}i & 0 \\ 0 & -\frac{\sqrt{2}}{2}i & 0 & \frac{\sqrt{2}}{2}i \\ \frac{\sqrt{2}}{2}i & 0 & -\frac{\sqrt{2}}{2}i & 0 \end{bmatrix}$$

and it is unitary.

Consider $\mathbb{C}^2$. Both $\mathbb{C}^2 \oplus \mathbb{C}^2$ and $\mathbb{C}^2 \otimes \mathbb{C}^2$ have four dimensions. These are *isomorphic*: we have an invertible linear map from all of the first vector space to all of the second. What is it?

Let $\mathbf{e_1} = (1, 0)$ and $\mathbf{e_2} = (0, 1)$ be the standard basis on $\mathbb{C}^2$. Since $\mathbb{C}^2 \oplus \mathbb{C}^2 = \mathbb{C}^4$, it has the standard basis $\mathbf{f_1} = (1, 0, 0, 0)$, $\mathbf{f_2} = (0, 1, 0, 0)$, $\mathbf{f_3} = (0, 0, 1, 0)$, and $\mathbf{f_4} = (0, 0, 0, 1)$. (I'm using "$\mathbf{f}$" instead of "$\mathbf{e}$" in this second case to avoid confusion.)

Define $S : \mathbb{C}^2 \oplus \mathbb{C}^2 = \mathbb{C}^4 \to \mathbb{C}^2 \otimes \mathbb{C}^2$ by

$$\mathbf{f_1} \mapsto \mathbf{e_1} \otimes \mathbf{e_1} \qquad \mathbf{f_3} \mapsto \mathbf{e_2} \otimes \mathbf{e_1}$$
$$\mathbf{f_2} \mapsto \mathbf{e_1} \otimes \mathbf{e_2} \qquad \mathbf{f_4} \mapsto \mathbf{e_2} \otimes \mathbf{e_2}$$

This is not the only isomorphism but it is natural given our calculation of the coordinates of the vectors like $\mathbf{e_1} \otimes \mathbf{e_2}$ above.

Another interesting basis for $\mathbb{C}^2 \otimes \mathbb{C}^2$ is

$$\frac{\sqrt{2}}{2} (\mathbf{e_1} \otimes \mathbf{e_1} + \mathbf{e_2} \otimes \mathbf{e_2}) \qquad \frac{\sqrt{2}}{2} (\mathbf{e_1} \otimes \mathbf{e_1} - \mathbf{e_2} \otimes \mathbf{e_2})$$

$$\frac{\sqrt{2}}{2} (\mathbf{e_1} \otimes \mathbf{e_2} + \mathbf{e_2} \otimes \mathbf{e_1}) \qquad \frac{\sqrt{2}}{2} (\mathbf{e_1} \otimes \mathbf{e_2} - \mathbf{e_2} \otimes \mathbf{e_1})$$

Continuing on, $\mathbb{C}^2 \oplus \mathbb{C}^2 \oplus \mathbb{C}^2 = \mathbb{C}^6$ but $\mathbb{C}^2 \otimes \mathbb{C}^2 \otimes \mathbb{C}^2$ has 8 dimensions. If we tensor ten copies of $\mathbb{C}^2$ together we get $2^{10} = 1024$ dimensions.

> The process of tensoring with more copies of $\mathbb{C}^2$ is exponential in the number of dimensions.

To learn more

> Tensor products are not always included in introductory texts on linear algebra, and so you may not have seen them before even if you have studied the subject. More complete mathematical treatments cover tensor products of vectors and matrices, as I have, but may also generalize them using category theory. [1] [3] [4, Monoids]

In the next section, we use the tensor product as the underpinning of qubit entanglement and show what they look like and how they behave with bra-ket notation.

8.2 Entanglement

We've now seen many gate operations that you can apply to a single qubit to change its state. In section 2.5 we worked through how to apply classical logic gates to build a circuit for addition.

While we can apply **not** to a single bit, all the other operations require at least two bits for input. In the same way, we need to work with multiple qubits to produce interesting and useful results.

8.2.1 Moving from one to two qubits

As discussed above, the states of a single qubit are represented by vectors of length 1 in $\mathbb{C}^2$ and all such states that differ only by multiplication by a complex unit are considered equivalent. Each qubit starts by having its own associated copy of $\mathbb{C}^2$.

When we have a quantum system with two qubits, we do not consider their collective states in a single $\mathbb{C}^2$ instance. Instead, we use the tensor product of the two copies of $\mathbb{C}^2$ and the tensor products of the quantum state vectors. This gives us a four-dimensional complex vector space where this "4" is 2×2 rather than the arithmetically equal $2 + 2$.

The tensor product is the machinery that allows up to build quantum systems from two or more other systems. The notation for working with these tensor products starts out as fairly bulky but there are significant simplifications that demonstrate the advantages of bras and kets.

> Let q_1 and q_2 be two qubits and let $\{|0\rangle_1, |1\rangle_1\}$ $\{|0\rangle_2, |1\rangle_2\}$ be the standard orthonormal basis kets for each of their $\mathbb{C}^2$ state spaces. Let
>
> $$|\psi\rangle_1 = a_1|0\rangle_1 + b_1|1\rangle_1 \text{ with } |a_1|^2 + |b_1|^2 = 1 \text{ and}$$
> $$|\psi\rangle_2 = a_2|0\rangle_2 + b_2|1\rangle_2 \text{ with } |a_2|^2 + |b_2|^2 = 1.$$
>
> The four kets
>
> $$|0\rangle_1 \otimes |0\rangle_2, |0\rangle_1 \otimes |1\rangle_2, |1\rangle_1 \otimes |0\rangle_2, \text{ and } |1\rangle_1 \otimes |1\rangle_2,$$
>
> are a basis for the combined state space $\mathbb{C}^2 \otimes \mathbb{C}^2$ for q_1 and q_2.

From the standard properties of tensor products,

$$|\psi\rangle_1 \otimes |\psi\rangle_2 = a_1 a_2 |0\rangle_1 \otimes |0\rangle_2 + a_1 b_2 |0\rangle_1 \otimes |1\rangle_2 +$$
$$b_1 a_2 |1\rangle_1 \otimes |0\rangle_2 + b_1 b_2 |1\rangle_1 \otimes |1\rangle_2.$$

First simplification: we can assume there is a tensor product between basis kets coming from the original but different qubit state spaces. We omit the "$\otimes$" symbols on the right-hand side.

$$|\psi\rangle_1 \otimes |\psi\rangle_2 = a_1 a_2 |0\rangle_1 |0\rangle_2 + a_1 b_2 |0\rangle_1 |1\rangle_2 +$$
$$b_1 a_2 |1\rangle_1 |0\rangle_2 + b_1 b_2 |1\rangle_1 |1\rangle_2.$$

Second simplification: we do not multiply kets in the same state space and so we can drop the subscripts on the basis kets. We use the order in which they are listed to determine where they came from.

$$|\psi\rangle_1 \otimes |\psi\rangle_2 = a_1 a_2 |0\rangle |0\rangle + a_1 b_2 |0\rangle |1\rangle +$$
$$b_1 a_2 |1\rangle |0\rangle + b_1 b_2 |1\rangle |1\rangle.$$

Third simplification: we can merge adjacent basis kets inside a single ket. This is new notation for us but shows the conciseness of what Dirac conceived.

$$|\psi\rangle_1 \otimes |\psi\rangle_2 = a_1 a_2 |00\rangle + a_1 b_2 |01\rangle + b_1 a_2 |10\rangle + b_1 b_2 |11\rangle.$$

When we use general coordinates, it looks like

$$a_{00} |00\rangle + a_{01} |01\rangle + a_{10} |10\rangle + a_{11} |11\rangle.$$

Since we're going to be looking at and applying matrices, it's handy to determine the column vector forms for the 2-qubit basis kets in $\mathbb{C}^2 \otimes \mathbb{C}^2$:

$$|00\rangle = \begin{bmatrix} 1 \\ 0 \\ 0 \\ 0 \end{bmatrix} \quad |01\rangle = \begin{bmatrix} 0 \\ 1 \\ 0 \\ 0 \end{bmatrix} \quad |10\rangle = \begin{bmatrix} 0 \\ 0 \\ 1 \\ 0 \end{bmatrix} \quad |11\rangle = \begin{bmatrix} 0 \\ 0 \\ 0 \\ 1 \end{bmatrix}.$$

We compute these from

$$|0\rangle = \begin{bmatrix} 1 \\ 0 \end{bmatrix} \quad |1\rangle = \begin{bmatrix} 0 \\ 1 \end{bmatrix}$$

and observe that, for example,

$$|01\rangle = |0\rangle \otimes |1\rangle = \begin{bmatrix} 1 \\ 0 \end{bmatrix} \otimes \begin{bmatrix} 0 \\ 1 \end{bmatrix} = \begin{bmatrix} 1 \begin{bmatrix} 0 \\ 1 \end{bmatrix} \\ 0 \begin{bmatrix} 0 \\ 1 \end{bmatrix} \end{bmatrix} = \begin{bmatrix} 1 \times 0 \\ 1 \times 1 \\ 0 \times 0 \\ 0 \times 1 \end{bmatrix} = \begin{bmatrix} 0 \\ 1 \\ 0 \\ 0 \end{bmatrix}.$$

Question 8.2.1

Similarly verify that the column vector forms of $|00\rangle$, $|10\rangle$, and $|11\rangle$ are correct.

Note that $\langle 01|01\rangle = 1$ but $\langle 01|11\rangle = 0$. This is generally true: when both sides are equal for these four vectors, their $\langle|\rangle$ is 1. When the are unequal, they are 0. This is a restatement of saying they are an orthonormal basis.

There is a fourth form I will show you when we look at the general case.

Look at the coefficients in $a_1a_2|00\rangle + a_1b_2|01\rangle + b_1a_2|10\rangle + b_1b_2|11\rangle$. Is it still true that the sum of the squares of the absolute values of the coefficients equals 1? Why would we expect this to be the case?

When we measure the 2-qubit system, each of their states will drop to $|0\rangle$ or $|1\rangle$. There are four possible outcomes: $|00\rangle$, $|01\rangle$, $|10\rangle$, and $|11\rangle$. The sum of the probabilities of each case occurring must add up to 1.0. By extension from the 1-qubit case, we would expect that in

$$|\psi\rangle_1 \otimes |\psi\rangle_2 = a_1a_2|00\rangle + a_1b_2|01\rangle + b_1a_2|10\rangle + b_1b_2|11\rangle$$

we would have the coefficients be probability amplitudes. This means the probability of getting $|01\rangle$, for example, is $|a_1b_2|^2$. The sum is therefore

$$|a_1a_2|^2 + |a_1b_2|^2 + |b_1a_2|^2 + |b_1b_2|^2 = 1.$$

Does the math support this? Lo and behold, it does:

$$
\begin{aligned}
|a_1a_2|^2 + |a_1b_2|^2 + |b_1a_2|^2 + |b_1b_2|^2 &= |a_1|^2|a_2|^2 + |a_1|^2|b_2|^2 + |b_1|^2|a_2|^2 + |b_1|^2|b_2|^2 \\
&= |a_1|^2 \left(|a_2|^2 + |b_2|^2\right) + |b_1|^2 \left(|a_2|^2 + |b_2|^2\right) \\
&= |a_1|^2 \, (1) + |b_1|^2 \, (1) \\
&= |a_1|^2 + |b_1|^2 \\
&= 1
\end{aligned}
$$

This is pretty spectacular in my view. The mathematical model we have been building up is aligning with the physical interpretation. While it is not surprising from the perspective of why I am talking about it in the first place, the pieces fall together nicely.

Measurement causes the state of each qubit to become, or collapse to, $|0\rangle$ or $|1\rangle$. Do they operate independently or can there be combined qubit states that express a greater linkage than might seem obvious?

At any given time, two qubits are in superposition states represented by a linear combination of vectors $|00\rangle$, $|01\rangle$, $|10\rangle$, and $|11\rangle$ in $\mathbb{C}^2 \otimes \mathbb{C}^2$:

$$a_{00}|00\rangle + a_{01}|01\rangle + a_{01}|10\rangle + a_{11}|11\rangle$$

where

$$|a_{00}|^2 + |a_{01}|^2 + |a_{10}|^2 + |a_1|^2 = 1.$$

Through *measurement*, the qubits are forced to collapse irreversibly through projection to $|00\rangle$, $|01\rangle$, $|10\rangle$, or $|11\rangle$. The probability of their doing so is $|a_{00}|^2$, $|a_{01}|^2$, $|a_{10}|^2$, or $|a_{11}|^2$, respectively. The a_{00}, a_{01}, a_{10}, and a_{11} are called *probability amplitudes*.

If necessary, we can convert ("read out") $|00\rangle$, $|01\rangle$, $|10\rangle$, and $|11\rangle$ to classical bit string values of 00, 01, 10, and 11.

If our qubits q_1 and q_2 are in the combined state

$$0|00\rangle + \frac{\sqrt{2}}{2}|01\rangle + \frac{\sqrt{2}}{2}|10\rangle + 0|11\rangle = \frac{\sqrt{2}}{2}|01\rangle + \frac{\sqrt{2}}{2}|10\rangle$$

Then when we measure we expect to get $|10\rangle$ half the time and $|01\rangle$ the rest of the time, given a large number of measurements. We never get $|00\rangle$ or $|11\rangle$.

I now give you q_1 and I keep q_2. I'm very excited to have my qubit and so I immediately measure it. I get a $|1\rangle$! What do you get?

There are two possible states that could be measured to get a $|1\rangle$ for q_2: $|01\rangle$ and $|11\rangle$. But the probability of getting the second is 0! So when measured, you must get $|0\rangle$.

Before measurement, the qubits were in a state of *entanglement*. They were so tightly correlated so that when the measurement value of one is known, it uniquely determines the second. You cannot do this with bits. With superposition, entanglement is one of the key differentiators between quantum and classical computing.

The entangled state we just used is known as a *Bell state* and there are four of them:

$$|\Phi^+\rangle = \tfrac{\sqrt{2}}{2}|00\rangle + \tfrac{\sqrt{2}}{2}|11\rangle \qquad |\Psi^+\rangle = \tfrac{\sqrt{2}}{2}|01\rangle + \tfrac{\sqrt{2}}{2}|10\rangle$$

$$|\Phi^-\rangle = \tfrac{\sqrt{2}}{2}|00\rangle - \tfrac{\sqrt{2}}{2}|11\rangle \qquad |\Psi^-\rangle = \tfrac{\sqrt{2}}{2}|01\rangle - \tfrac{\sqrt{2}}{2}|10\rangle$$

We used $|\Psi^+\rangle$ in the example above.

Φ is the capital Greek letter "phi" and Ψ is the capital Greek letter "psi". φ and ψ are their lowercase counterparts.

Question 8.2.2

Show that

$$|00\rangle = \tfrac{\sqrt{2}}{2}\left(|\Phi^+\rangle + |\Phi^-\rangle\right) \qquad\qquad |11\rangle = \tfrac{\sqrt{2}}{2}\left(|\Phi^+\rangle - |\Phi^-\rangle\right)$$

$$|01\rangle = \tfrac{\sqrt{2}}{2}\left(|\Psi^+\rangle + |\Psi^-\rangle\right) \qquad\qquad |10\rangle = \tfrac{\sqrt{2}}{2}\left(|\Psi^+\rangle - |\Psi^-\rangle\right)$$

Together, the four states $|\Phi^+\rangle$, $|\Phi^-\rangle$, $|\Psi^+\rangle$, and $|\Psi^-\rangle$ are an orthonormal basis for $\mathbb{C}^2 \otimes \mathbb{C}^2$. They are named after John Stewart Bell, a physicist from Northern Ireland.

Let $|\Psi\rangle$ be a 2-qubit quantum state in $\mathbb{C}^2 \otimes \mathbb{C}^2$. $|\Psi\rangle$ is *entangled* if and only if it **cannot** be written as the tensor products of two 1-qubit kets:

$$|\psi\rangle_1 \otimes |\psi\rangle_2 = (a_1|0\rangle_1 + b_1|1\rangle_1) \otimes (a_1|0\rangle_1 + b_2|1\rangle_1)$$

where

$$|\psi\rangle_1 = a_1|0\rangle_1 + b_1|1\rangle_1 \quad \text{and} \quad |\psi\rangle_2 = a_2|0\rangle_2 + b_2|1\rangle_2.$$

Suppose $|\Psi^+\rangle$ is not entangled. Then there exist a_1, b_1, a_2, and b_2 in $\mathbb{C}$ as above with

$$|\Psi^+\rangle = 0|00\rangle + \frac{\sqrt{2}}{2}|01\rangle + \frac{\sqrt{2}}{2}|10\rangle + 0|11\rangle$$

$$= a_1 a_2|00\rangle + a_1 b_2|01\rangle + b_1 a_2|10\rangle + b_1 b_2|11\rangle.$$

This gives us four relationships

$$a_1 a_2 = 0 \qquad a_1 b_2 = \tfrac{\sqrt{2}}{2} \qquad b_1 a_2 = \tfrac{\sqrt{2}}{2} \qquad b_1 b_2 = 0$$

From the first, either a_1 or a_2 is 0. Assume $a_1 = 0$. But then $0 = a_1 b_2 = \tfrac{\sqrt{2}}{2}$. This is a contradiction. So a_2 must be 0. Again, though, $0 = b_1 a_2 = \tfrac{\sqrt{2}}{2}$ and we have another impossibility. This means we cannot write $|\Psi^+\rangle$ as the tensor product of two 1-qubit kets and it is an entangled state.

If a 2-qubit quantum state is not entangled then we can separate it into the tensor product of two 1-qubit states. For this reason, if a quantum state is not entangled then it is *separable*.

Question 8.2.3

Is $\tfrac{\sqrt{2}}{2}|00\rangle + \tfrac{\sqrt{2}}{2}|01\rangle$ an entangled state?

There is something I want to highlight that may be confusing at first. If we start with two 1-qubit states like $a_1|0\rangle_1 + b_1|1\rangle_1$ and $a_2|0\rangle_2 + b_2|1\rangle_2$ then we use all such tensor products $(a_1|0\rangle_1 + b_1|1\rangle_1) \otimes (a_2|0\rangle_2 + b_2|1\rangle_2)$ for all possible complex number coefficients to *generate* the state vector space $\mathbb{C}^2 \otimes \mathbb{C}^2$.

This means we construct all possible sums of such tensor product forms. All the forms together *do not* constitute all of $\mathbb{C}^2 \otimes \mathbb{C}^2$ but together with their sums they do.

Question 8.2.4

There is an infinite number of entangled states and an infinite number of separable states in $\mathbb{C}^2 \otimes \mathbb{C}^2$. Given that, in what sense are there more entangled states than separable ones?

8.2.2 The general case

Every time we add a qubit to a quantum system to create a new one, the state space doubles in dimension. This is because we multiply the dimension of the original system's state space by 2 when we do the tensor product. A 3-qubit quantum system has a state space of dimension 8. An n-qubit system's state space has 2^n dimensions.

Let n in $\mathbb{N}$ be greater than 1 and let Q be an n-qubit quantum system. The state space associated with Q has 2^n dimensions. We write it as $\left(\mathbb{C}^2\right)^{\otimes n}$, which means $\mathbb{C}^2$ tensored with itself n times.

It is generated by the elementary tensor products of the 1-qubit states for each of the n qubits:
$$(a_1|0\rangle_1 + b_1|1\rangle_1) \otimes \cdots \otimes (a_n|0\rangle_n + b_n|1\rangle_n)$$
A quantum state in $\mathbb{C}^2 \otimes^n \mathbb{C}^2$ is *separable* if it can be written as such an elementary state, and is *entangled* otherwise.

It's much easier to use ket notation like $|11010111\rangle$ than

$$|1\rangle \otimes |1\rangle \otimes |0\rangle \otimes |1\rangle \otimes |0\rangle \otimes |1\rangle \otimes |1\rangle \otimes |1\rangle$$

Knowing we have 8 qubits, we could also write this as $|215\rangle_8$ where the number inside the ket is a whole number *in base 10*. The subscript indicates how many qubits there are. In the 2-qubit case,

$$|00\rangle = |0\rangle_2 \qquad |01\rangle = |1\rangle_2 \qquad |10\rangle = |2\rangle_2 \qquad |11\rangle = |3\rangle_2$$

The sets of n-qubit kets for $\left(\mathbb{C}^2\right)^{\otimes n}$ that are composed of only 0s and 1s are called *computational bases*. For example,

$$|000\rangle \quad |001\rangle \quad |010\rangle \quad |011\rangle$$
$$|100\rangle \quad |101\rangle \quad |110\rangle \quad |111\rangle$$

is the computational basis for $\mathbb{C}^2 \otimes \mathbb{C}^2 \otimes \mathbb{C}^2$.

These are the same as

$$|0\rangle_3 \quad |1\rangle_3 \quad |2\rangle_3 \quad |3\rangle_3$$
$$|4\rangle_3 \quad |5\rangle_3 \quad |6\rangle_3 \quad |7\rangle_3.$$

When we use the digital form of ket, we number the probability amplitudes/coefficients using the number inside the ket symbol. For a general quantum state with n qubits,

$$|\psi\rangle = a_0|0\rangle_n + a_1|1\rangle_n + a_2|2\rangle_n + \cdots + a_{2^n-1}|2^n - 1\rangle_n$$

and we have

$$|a_0|^2 + |a_1|^2 + |a_2|^2 + \cdots + |2^n - 1|^2 = 1.$$

However we write the quantum state, the sum of the squares of the absolute values of the probability amplitudes add up to 1.

If we want to write a ket like

$$| \underbrace{00000000000 \ldots 00000000000}_{n} >,$$

we shorthand it to $|0\rangle^{\otimes n}$.

For a given n, let $|\varphi\rangle$ and $|\psi\rangle$ be two computational basis kets. Then

$$\langle\varphi|\psi\rangle = \begin{cases} 1 & \text{if } |\varphi\rangle = |\psi\rangle \\ 0 & \text{otherwise.} \end{cases}$$

If $|\varphi\rangle$ has a 1 in the jth position and 0 elsewhere in its full vector expansion, and $|\psi\rangle$ has a 1 in the kth position and 0 elsewhere, then $|\varphi\rangle\langle\psi|$ is the n by n square matrix which has 0s everywhere except for the (j, k) position, where it is 1. For example,

$$|0\rangle\langle0| = \begin{bmatrix} 1 & 0 \\ 0 & 0 \end{bmatrix} \quad \text{and} \quad |1\rangle\langle0| = \begin{bmatrix} 0 & 1 \\ 0 & 0 \end{bmatrix}.$$

Note that $|0\rangle\langle0| + |1\rangle\langle1| = I_2$, the 2 by 2 identity matrix.

8.2.3   The density matrix again

If $|\psi\rangle$ is a multi-qubit quantum state, its *density matrix* is defined in the same way as the 1-qubit case:

$$\rho = |\psi\rangle\langle\psi|.$$

That is, if

$$|\psi\rangle = a_0|0\rangle_n + a_1|1\rangle_n + a_2|2\rangle_n + \cdots + a_{2^n-1}|2^n-1\rangle_n$$

then

$$\rho = |\psi\rangle\langle\psi| = \begin{bmatrix} a_0 & a_1 & a_2 & \cdots & a_{2^2-1} \end{bmatrix} \otimes \begin{bmatrix} \overline{a_0} \\ \overline{a_1} \\ \overline{a_2} \\ \vdots \\ \overline{a_{2^n-1}} \end{bmatrix}$$

$$= \begin{bmatrix} a_0\overline{a_0} & a_1\overline{a_0} & a_2\overline{a_0} & \cdots & a_{2^n-1}\overline{a_0} \\ a_0\overline{a_1} & a_1\overline{a_1} & a_2\overline{a_1} & \cdots & a_{2^n-1}\overline{a_1} \\ a_0\overline{a_2} & a_1\overline{a_2} & a_2\overline{a_2} & \cdots & a_{2^n-1}\overline{a_2} \\ \vdots & \vdots & \vdots & \ddots & \vdots \\ a_0\overline{a_{2^n-1}} & a_1\overline{a_{2^n-1}} & a_2\overline{a_{2^n-1}} & \cdots & a_{2^n-1}\overline{a_{2^n-1}} \end{bmatrix}$$

$$= \begin{bmatrix} |a_0|^2 & a_1\overline{a_0} & a_2\overline{a_0} & \cdots & a_{2^n-1}\overline{a_0} \\ a_0\overline{a_1} & |a_1|^2 & a_2\overline{a_1} & \cdots & a_{2^n-1}\overline{a_1} \\ a_0\overline{a_2} & a_1\overline{a_2} & |a_2|^2 & \cdots & a_{2^n-1}\overline{a_2} \\ \vdots & \vdots & \vdots & \ddots & \vdots \\ a_0\overline{a_{2^n-1}} & a_1\overline{a_{2^n-1}} & a_2\overline{a_{2^n-1}} & \cdots & |a_{2^n-1}|^2 \end{bmatrix}$$

The diagonal elements are real, $\mathrm{tr}(\rho) = 1$, and ρ is Hermitian and positive semi-definite. This means that ρ has a unique positive semi-definite square root matrix $\rho^{\frac{1}{2}}$.

8.3   Multi-qubit gates

A quantum gate operation that operates on one qubit has a 2 by 2 unitary square matrix in a given basis. For two qubits, the matrix is 4 by 4. For ten, it is 2^{10} by 2^{10}, which is 1024 by 1024. I show by example how to work with common lower dimensional gates and allow you to extrapolate to larger ones.

8.3.1 The quantum $H^{\otimes n}$ gate

We start by looking at what it means to apply a Hadamard **H** to each qubit in a 2-qubit system. The **H** gate, or Hadamard gate, has the matrix

$$\mathbf{H} = \begin{bmatrix} \frac{\sqrt{2}}{2} & \frac{\sqrt{2}}{2} \\ \frac{\sqrt{2}}{2} & -\frac{\sqrt{2}}{2} \end{bmatrix} = \frac{\sqrt{2}}{2} \begin{bmatrix} 1 & 1 \\ 1 & -1 \end{bmatrix}$$

operating on $\mathbb{C}^2$. Starting with the two qubit states

$$|\psi\rangle_1 = a_1|0\rangle_1 + b_1|1\rangle_1 \quad \text{and} \quad |\psi\rangle_2 = a_2|0\rangle_2 + b_2|1\rangle_2 \ .$$

Applying **H** to each qubit means to compute

$$(\mathbf{H}|\psi\rangle_1) \otimes (\mathbf{H}|\psi\rangle_2)$$

which is the same as

$$(\mathbf{H} \otimes \mathbf{H})\,(|\psi\rangle_1 \otimes |\psi\rangle_2) = \mathbf{H}^{\otimes 2}\,(a_1 a_2|00\rangle + a_1 b_2|01\rangle + b_1 a_2|10\rangle + b_1 b_2|11\rangle)$$

for some 4 by 4 unitary matrix $\mathbf{H}^{\otimes 2}$. Given the definition of **H** and the technique of creating a matrix tensor product in section 8.1, we can compute

$$\mathbf{H}^{\otimes 2} = \begin{bmatrix} \frac{1}{2} & \frac{1}{2} & \frac{1}{2} & \frac{1}{2} \\ \frac{1}{2} & -\frac{1}{2} & \frac{1}{2} & -\frac{1}{2} \\ \frac{1}{2} & \frac{1}{2} & -\frac{1}{2} & -\frac{1}{2} \\ \frac{1}{2} & -\frac{1}{2} & -\frac{1}{2} & \frac{1}{2} \end{bmatrix} = \frac{1}{2} \begin{bmatrix} 1 & 1 & 1 & 1 \\ 1 & -1 & 1 & -1 \\ 1 & 1 & -1 & -1 \\ 1 & -1 & -1 & 1 \end{bmatrix} = \frac{\sqrt{2}}{2} \begin{bmatrix} \mathbf{H} & \mathbf{H} \\ \mathbf{H} & -\mathbf{H} \end{bmatrix} .$$

This matrix puts both qubits in a 2-qubit system that are each initially initialized to $|0\rangle$ into superposition.

Note the recursive definition of $\mathbf{H}^{\otimes 2}$ in terms of matrix blocks of **H** matrices.

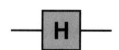

Though we could draw $\mathbf{H}^{\otimes 2}$ as having two inputs and two outputs, we instead show it by applying **H** to each qubit in a circuit.

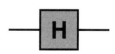

For a 3-qubit system, the corresponding $\mathbf{H}^{\otimes 3}$ matrix is

$$\frac{\sqrt{2}}{4}\begin{bmatrix} 1 & 1 & 1 & 1 & 1 & 1 & 1 & 1 \\ 1 & -1 & 1 & -1 & 1 & -1 & 1 & -1 \\ 1 & 1 & -1 & -1 & 1 & 1 & -1 & -1 \\ 1 & -1 & -1 & 1 & 1 & -1 & -1 & 1 \\ 1 & 1 & 1 & 1 & -1 & -1 & -1 & -1 \\ 1 & -1 & 1 & -1 & -1 & 1 & -1 & 1 \\ 1 & 1 & -1 & -1 & -1 & -1 & 1 & 1 \\ 1 & -1 & -1 & 1 & -1 & 1 & 1 & -1 \end{bmatrix} = \frac{\sqrt{2}}{2}\begin{bmatrix} \mathbf{H}^{\otimes 2} & \mathbf{H}^{\otimes 2} \\ \mathbf{H}^{\otimes 2} & -\mathbf{H}^{\otimes 2} \end{bmatrix}.$$

Earlier I asked you to show that $\mathbf{H}\left|0\right\rangle = \left|+\right\rangle$. That is,

$$\mathbf{H} = \frac{\sqrt{2}}{2}\left(\left|0\right\rangle + \left|1\right\rangle\right)$$

From this it follows that

$$\mathbf{H}^{\otimes 2}\left|00\right\rangle = \left(\mathbf{H}\otimes\mathbf{H}\right)\left(\left|0\right\rangle\otimes\left|0\right\rangle\right)$$

$$= \left(\frac{\sqrt{2}}{2}\left(\left|0\right\rangle + \left|1\right\rangle\right)\right)\otimes\left(\frac{\sqrt{2}}{2}\left(\left|0\right\rangle + \left|1\right\rangle\right)\right)$$

$$= \frac{1}{2}\left(\left|00\right\rangle + \left|01\right\rangle + \left|10\right\rangle + \left|11\right\rangle\right)$$

Also,

$$\mathbf{H}^{\otimes 3}\left|000\right\rangle = \frac{\sqrt{2}}{4}\left(\left|000\right\rangle + \left|001\right\rangle + \left|010\right\rangle + \left|011\right\rangle + \left|100\right\rangle + \left|101\right\rangle + \left|110\right\rangle + \left|111\right\rangle\right)$$

The patterns continues. This shows that applying the Hadamard gates to each qubit initialized to $\left|0\right\rangle$ creates a balanced superposition involving all the ket basis vectors. The number out front, $\frac{\sqrt{2}}{4}$ in the last case, is there to ensure that the square of the absolute value of the ket is 1. It is the *normalization constant*.

Question 8.3.1

Show that the normalization constant is $\frac{1}{\sqrt{2^n}}$ where n is the number of qubits.

If you have three classical bits, you can represent all the following but *only one at a time:*

$$000 \quad 001 \quad 010 \quad 011 \quad 100 \quad 101 \quad 110 \quad 111$$

In contrast, the 3-qubit state $H^{\otimes 3}|0\rangle_3$ contains each of the corresponding ket basis forms *at the same time*.

This is a situation where the decimal expression for a basis ket is concise. We can rewrite the last equality as

$$H^{\otimes 3}|0\rangle_3 = \frac{\sqrt{2}}{4} \left(|0\rangle_3 + |1\rangle_3 + |2\rangle_3 + |3\rangle_3 + |4\rangle_3 + |5\rangle_3 + |6\rangle_3 + |7\rangle_3\right)$$

The Hadamard gate matrices $H^{\otimes n}$ can be defined recursively by

$$H^{\otimes n} = \frac{\sqrt{2}}{2} \begin{bmatrix} H^{\otimes n-1} & H^{\otimes n-1} \\ H^{\otimes n-1} & -H^{\otimes n-1} \end{bmatrix}$$

where $H^{\otimes 1} = H$.

Now is a good time to introduce summation notation. The capital Greek letter sigma is used to express a sum based on a formula.

$$\sum_{j=1}^{4} j = 1 + 2 + 3 + 4 = 10$$

This means we start j at 1 and in turn consider $j = 2$, $j = 3$, and $j = 4$. We let the initial value of the sum be 0 and then add in each evaluation of the formula to the right of Σ with the given value of j. 1 is the lower bound for j and 4 is the upper bound.

Here is another example:

$$\sum_{j=0}^{2} \cos(j\pi) = \cos(0\pi) + \cos(1\pi) + \cos(2\pi) = 1$$

We don't only need a constant value for the upper bound and we don't have to solely use the variable j.

$$\sum_{m=1}^{n} 2^m = 2^1 + 2^2 + \cdots + 2^{n-1} + 2^n$$

The lower bound need not be constant either.

The general form for an n-qubit register state using decimal ket notation is

$$|\psi\rangle = \sum_{j=0}^{n-1} a_j |j\rangle_n$$

where

$$1 = \sum_{j=0}^{n-1} |a_j|^2.$$

With this we can express of the formula for the balanced superposition of n qubits.

$$\mathbf{H}^{\otimes n} |0\rangle_n = \frac{1}{\sqrt{2^n}} \sum_{j=0}^{n-1} |j\rangle_n$$

We can drop the subscript n on the kets if we know we are working with a specific number of qubits.

$$\mathbf{H}^{\otimes n} |0\rangle = \frac{1}{\sqrt{2^n}} \sum_{j=0}^{n-1} |j\rangle$$

Question 8.3.2

Fully write out

$$\mathbf{H}^{\otimes 4} |0\rangle = \frac{1}{\sqrt{2^4}} \sum_{j=0}^{3} |j\rangle$$

Expand the kets using binary notation.

Incidentally, we have similar notation for products.

$$\prod_{j=1}^{4} j = 1 \times 2 \times 3 \times 4$$

For example, if we factor a positive N in $\mathbb{Z}$ into a set of primes $\{p_1, p_2, \ldots, p_n\}$ and each prime p_j occurs e_i times, then

$$N = \prod_{j=1}^{n} p_j^{e_j}.$$

Now let's turn to 2-qubit gates that are not tensor products of 1-qubit ones. As when we considered those small gates, we examine some of the most commonly used 2-qubit operations.

8.3.2   The quantum SWAP gate

In subsection 7.6.1 we showed the **X** gate is a bit flip: given $|\psi\rangle = a|0\rangle + b|1\rangle$, $\mathbf{X}|\psi\rangle = b|0\rangle + a|1\rangle$. Now that we are considering two qubits, is there a gate that switches the qubits? What does this even mean?

As we have seen before, given two qubits

$$|\psi\rangle_1 = a_1|0\rangle_1 + b_1|1\rangle_1 \quad \text{and} \quad |\psi\rangle_2 = a_2|0\rangle_2 + b_2|1\rangle_2,$$

their tensor product is

$$|\psi\rangle_1 \otimes |\psi\rangle_2 = a_1a_2|00\rangle + a_1b_2|01\rangle + b_1a_2|10\rangle + b_1b_2|11\rangle.$$

If we tensor them in the reverse order, we get

$$|\psi\rangle_2 \otimes |\psi\rangle_1 = a_2a_1|00\rangle + a_2b_1|01\rangle + b_2a_1|10\rangle + b_2b_1|11\rangle.$$

The first and the fourth coefficients are the same but the second and third are switched.

The matrix

$$M = \begin{bmatrix} 1 & 0 & 0 & 0 \\ 0 & 0 & 1 & 0 \\ 0 & 1 & 0 & 0 \\ 0 & 0 & 0 & 1 \end{bmatrix}$$

is an example of a 4 by 4 *permutation matrix*. To create a matrix that swaps the second and third coefficient of a ket (or entries in a column vector), begin with I_4 and interchange the second and third columns. This is M.

For a general vector,

$$M \begin{bmatrix} v_1 \\ v_2 \\ v_3 \\ v_4 \end{bmatrix} = \begin{bmatrix} v_1 \\ v_3 \\ v_2 \\ v_4 \end{bmatrix}$$

Therefore $M\left(|\psi\rangle_1 \otimes |\psi\rangle_2\right) = |\psi\rangle_2 \otimes |\psi\rangle_1$.

When used this way, we call the quantum gate with this matrix in the standard ket basis the **SWAP** gate.

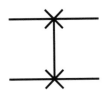

When I include the **SWAP** gate in a circuit it spans two wires. Remember the ×s!

8.3.3 The quantum CNOT / CX gate

The **CNOT** gate is one of the most important gates in quantum computing. It's used to create entangled qubits. It's not the only kind of gate that can do it, but it's simple and very commonly used.

The "C" in **CNOT** is for "controlled." Unlike the 1-qubit **X** gate, which unconditionally flips $|0\rangle$ to $|1\rangle$ and vice versa, **CNOT** has two qubit inputs and two outputs. Remember that quantum gates must be reversible. For this reason we must have the same number of inputs as outputs. We call the qubits q_1 and q_2 and their states $|\psi\rangle_1$ and $|\psi\rangle_2$, respectively.

This is the way **CNOT** works: it takes two inputs, $|\psi\rangle_1$ and $|\psi\rangle_2$.

- If $|\psi\rangle_1$ is $|1\rangle$, then the state of q_1 remains $|\psi\rangle_1$ but $|\psi\rangle_2$ becomes $\mathbf{X}|\psi\rangle_2$.
- Otherwise the states of q_1 and q_2 are not changed.

Put another way, **CNOT** always operates like $\mathbf{ID}|\psi\rangle_1$ for q_1. When $|\psi\rangle_1 = |1\rangle$, **CNOT** acts as $\mathbf{X}|\psi\rangle_2$ for q_2. Otherwise, it acts as $\mathbf{ID}|\psi\rangle_2$. The **CNOT** is a conditional bit flip.

In the classical case, we create a controlled **not** from a **xor** with this circuit:

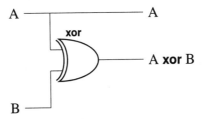

The matrix for **CNOT** is

$$\begin{bmatrix} 1 & 0 & 0 & 0 \\ 0 & 1 & 0 & 0 \\ 0 & 0 & 0 & 1 \\ 0 & 0 & 1 & 0 \end{bmatrix}.$$

It is a permutation matrix that swaps the third and fourth coefficients of $|\psi\rangle_1 \otimes |\psi\rangle_2$. Note that the upper left 2 by 2 submatrix is I_2 and the lower right 2 by 2 submatrix is the **X** matrix. It is more obvious if we rewrite the matrix in block form as

$$
\left[
\begin{array}{cc|cc}
1 & 0 & 0 & 0 \\
0 & 1 & 0 & 0 \\
\hline
0 & 0 & 0 & 1 \\
0 & 0 & 1 & 0
\end{array}
\right].
$$

When I include the **CNOT** gate in a circuit it spans two wires. The top line is the control qubit.

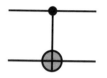

CNOT operates on the standard $\mathbb{C}^2 \otimes \mathbb{C}^2$ basis as

$$
\begin{aligned}
\textbf{CNOT}|00\rangle &= |00\rangle \\
\textbf{CNOT}|01\rangle &= |01\rangle \\
\textbf{CNOT}|10\rangle &= |11\rangle \\
\textbf{CNOT}|11\rangle &= |10\rangle.
\end{aligned}
$$

By linearity,

$$
\begin{aligned}
\textbf{CNOT}\,(a_{00}|00\rangle &+a_{01}|01\rangle + a_{10}|10\rangle + a_{11}|11\rangle) \\
&= a_{00}\textbf{CNOT}|00\rangle + a_{01}\textbf{CNOT}|01\rangle + a_{10}\textbf{CNOT}|10\rangle + a_{11}\textbf{CNOT}|11\rangle \\
&= a_{00}|00\rangle + a_{01}|01\rangle + a_{10}|11\rangle + a_{11}|10\rangle \\
&= a_{00}|00\rangle + a_{01}|01\rangle + a_{11}|10\rangle + a_{10}|11\rangle.
\end{aligned}
$$

Applying the change-of-basis Hadamard **H** gates before and after **CNOT** illustrates an interesting property of **CNOT**. The matrix form of $\textbf{H}^{\otimes 2} \circ \textbf{CNOT} \circ \textbf{H}^{\otimes 2}$ is

$$
M =
\begin{bmatrix}
\frac{1}{2} & \frac{1}{2} & \frac{1}{2} & \frac{1}{2} \\
\frac{1}{2} & -\frac{1}{2} & \frac{1}{2} & -\frac{1}{2} \\
\frac{1}{2} & \frac{1}{2} & -\frac{1}{2} & -\frac{1}{2} \\
\frac{1}{2} & -\frac{1}{2} & -\frac{1}{2} & \frac{1}{2}
\end{bmatrix}
\begin{bmatrix}
1 & 0 & 0 & 0 \\
0 & 1 & 0 & 0 \\
0 & 0 & 0 & 1 \\
0 & 0 & 1 & 0
\end{bmatrix}
\begin{bmatrix}
\frac{1}{2} & \frac{1}{2} & \frac{1}{2} & \frac{1}{2} \\
\frac{1}{2} & -\frac{1}{2} & \frac{1}{2} & -\frac{1}{2} \\
\frac{1}{2} & \frac{1}{2} & -\frac{1}{2} & -\frac{1}{2} \\
\frac{1}{2} & -\frac{1}{2} & -\frac{1}{2} & \frac{1}{2}
\end{bmatrix}.
$$

This reduces to the simpler

$$M = \begin{bmatrix} 1 & 0 & 0 & 0 \\ 0 & 0 & 0 & 1 \\ 0 & 0 & 1 & 0 \\ 0 & 1 & 0 & 0 \end{bmatrix}.$$

What can we tell about this? By inspection we can see it is a permutation operation that swaps the second and fourth coefficients of the standard ket expression in $\mathbb{C}^2 \otimes \mathbb{C}^2$.

The effect of M on the standard basis kets is

$$M|00\rangle = |00\rangle \qquad M|01\rangle = |11\rangle \qquad M|10\rangle = |10\rangle \qquad M|11\rangle = |01\rangle.$$

Examine what happens by looking at the second qubit. If it is $|1\rangle$ then the first qubit flips. If it is $|0\rangle$ then the first qubit remains the same.

This is the opposite behavior of **CNOT** yet it is constructed from it with Hadamard operations before and after. With **CNOT** it appeared that the state of the second qubit was controlled by the first. In this construction, it is the opposite. By doing the change of basis to and from $|+\rangle$ and $|-\rangle$ we've gotten evidence that **CNOT** is doing more than we perhaps expected.

If we wanted the control qubit to be the second one in this way, we would draw it with the ● on the bottom. This is sometimes called a **reverse CNOT**.

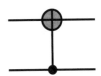

CNOT is used to create the entangled Bell state vectors. We save the construction until subsection 9.3.2 when we have more of the circuit machinery in hand.

8.3.4 The quantum CY and CZ gates

The **CNOT** gate is the same as the **CX** gate. We can also create controlled 2-qubit gates for other 1-qubit gates. In block matrix form,

$$\left[\begin{array}{cc|cc} 1 & 0 & 0 & 0 \\ 0 & 1 & 0 & 0 \\ \hline 0 & 0 & 0 & -i \\ 0 & 0 & i & 0 \end{array}\right] \quad \text{and} \quad \left[\begin{array}{cc|cc} 1 & 0 & 0 & 0 \\ 0 & 1 & 0 & 0 \\ \hline 0 & 0 & 1 & 0 \\ 0 & 0 & 0 & -1 \end{array}\right]$$

are the matrices in the standard basis for the **CY** and **CZ** gates, respectively. The **CZ** is a conditional sign flip.

Question 8.3.3

What are the matrices for the **CS** and **CH** gates?

These kinds of gates are shown in the following circuit,

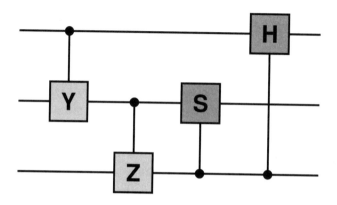

with the last two reversed to show that they can operate between arbitrary wires.

In general, if

$$\mathbf{U} = \begin{bmatrix} a & b \\ c & d \end{bmatrix}$$

is a unitary matrix, then the matrix for controlled-**U** is

$$\begin{bmatrix} 1 & 0 & 0 & 0 \\ 0 & 1 & 0 & 0 \\ 0 & 0 & a & b \\ 0 & 0 & c & d \end{bmatrix}$$

Question 8.3.4

What is the matrix for the controlled-$\sqrt{\mathbf{NOT}}$, where $\sqrt{\mathbf{NOT}}$ is defined in subsection 7.6.12?

8.3.5 The quantum $\mathbf{CR}_{\varphi}^{z}$ gate

Another useful set of controlled gates are the ones where the action is $\mathbf{R}_{\varphi}^{z}$. The first qubit controls whether the phase change, or rotation around the Bloch sphere z axis, should happen.

The general matrix for $\mathbf{CR}_{\varphi}^{z}$ is

$$\begin{bmatrix} 1 & 0 & 0 & 0 \\ 0 & 1 & 0 & 0 \\ 0 & 0 & 1 & 0 \\ 0 & 0 & 0 & e^{\varphi i} \end{bmatrix} = \begin{bmatrix} 1 & 0 & 0 & 0 \\ 0 & 1 & 0 & 0 \\ 0 & 0 & 1 & 0 \\ 0 & 0 & 0 & \cos(\varphi) + \sin(\varphi)i \end{bmatrix}.$$

When I include the $\mathbf{CR}_{\varphi}^{z}$ gate in a circuit it has what should now be a familiar form. I specify a particular radian value of φ such as $\mathbf{CR}_{\frac{\pi}{8}}^{z}$.

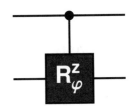

8.3.6 The quantum Toffoli CCNOT gate

The quantum Toffoli **CCNOT** gate is a double control gate operating on three qubits. If the states of the first two qubits are $|1\rangle$ then it applies **X** to the third. Otherwise it is **ID** on the third. In all cases it is **ID** for the first two qubits.

Its matrix is an 8 by 8 permutation matrix which swaps the last two coefficients, like **CNOT**.

$$\begin{bmatrix} 1 & 0 & 0 & 0 & 0 & 0 & 0 & 0 \\ 0 & 1 & 0 & 0 & 0 & 0 & 0 & 0 \\ 0 & 0 & 1 & 0 & 0 & 0 & 0 & 0 \\ 0 & 0 & 0 & 1 & 0 & 0 & 0 & 0 \\ 0 & 0 & 0 & 0 & 1 & 0 & 0 & 0 \\ 0 & 0 & 0 & 0 & 0 & 1 & 0 & 0 \\ 0 & 0 & 0 & 0 & 0 & 0 & 0 & 1 \\ 0 & 0 & 0 & 0 & 0 & 0 & 1 & 0 \end{bmatrix}$$

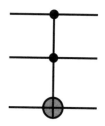

The **CCNOT** gate spans three wires in a circuit. The top two lines are the control qubits.

The Toffoli gate is also known as the **CCX** gate.

8.3.7   The quantum Fredkin CSWAP gate

The quantum Fredkin **CSWAP** gate is a control gate operating on three qubits. If the state of the first qubit is $|1\rangle$ then the states of the second and third qubits are swapped, as in **SWAP**. If it is $|0\rangle$, nothing is changed.

Its matrix is an 8 by 8 permutation matrix.

$$\begin{bmatrix} 1 & 0 & 0 & 0 & 0 & 0 & 0 & 0 \\ 0 & 1 & 0 & 0 & 0 & 0 & 0 & 0 \\ 0 & 0 & 1 & 0 & 0 & 0 & 0 & 0 \\ 0 & 0 & 0 & 1 & 0 & 0 & 0 & 0 \\ 0 & 0 & 0 & 0 & 1 & 0 & 0 & 0 \\ 0 & 0 & 0 & 0 & 0 & 0 & 1 & 0 \\ 0 & 0 & 0 & 0 & 0 & 1 & 0 & 0 \\ 0 & 0 & 0 & 0 & 0 & 0 & 0 & 1 \end{bmatrix}$$

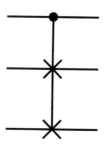

Like the **CCNOT**, the **CSWAP** gate spans three wires. The top line is the control qubit.

8.4 Summary

In this chapter we introduced the standard 2-qubit and 3-qubit gate operations to complement the classical forms from section 2.4. The **CNOT** gate allows us to entangle qubits. Entanglement, along with superposition and interference, is a key differentiator in quantum computing.

Now that we have a collection of gates, it's time to put them into circuits and implement algorithms. We begin doing that in the next chapter.

References

[1] Paul R. Halmos. *Finite-Dimensional Vector Spaces*. 1st ed. Undergraduate Texts in Mathematics. Springer Publishing Company, Incorporated, 1993.

[2] W. Heisenberg. *Across the frontiers*. Ox Bow Press, 1990.

[3] S. Lang. *Algebra*. 3rd ed. Graduate Texts in Mathematics 211. Springer-Verlag, 2002.

[4] Saunders Mac Lane. *Categories for the Working Mathematician*. 2nd ed. Graduate Texts in Mathematics 5. Springer New York, 1998.

9

Wiring Up the Circuits

The world is all gates, all opportunities,
strings of tension waiting to be struck.

Ralph Waldo Emerson

Now that we understand qubits and the operations we can apply to one or more of them, it's time to string together the actions to do something useful. In this chapter we build out circuits and discuss their properties. From there we survey basic algorithms such as those that involve oracles and searches. Through this you gain a better understanding the core programming idioms in quantum computing.

Non-trivial quantum algorithms take advantage of qubit entanglement, the graceful way in which qubits work together and interact until we get our answer. I think of this scripted interplay among qubits as an elegant dance, and that's how this book got its title.

Topics covered in this chapter

9.1 So many gates …

In practice, a hardware quantum computer implements a core set of primitive gates and the others are built from them using circuits. These core operations may be among the ones we saw in the last chapter or they may be much stranger looking: any 2 by 2 unitary matrix can be considered a 1-qubit gate.

The primitive gates depend on the technology used to create the physical quantum computer. More advanced gates are then built from these primitive gates. The *Qiskit open source quantum computing framework*, for example, provides a large selection of gates you can use, many of them built from the core ones. [16]

At the hardware level, experimental physicists and engineers work to optimize the core gates. Above that, other physicists and computer scientists try to create the best performing higher-level gates.

In the classical case, machine code is extremely low level and directly instructs the processor. Above that is assembly code, which abstracts the machine code a bit and makes certain common operations easier. The C programming language is above assembly code but still gives you very fine control over how you use memory. From there you get the very high-level languages like Python, Go, and Swift. At this level you need to know very little if anything about the hardware on which you are running.

In this chapter I stick to the most commonly available gates as we build out the circuits. These are only "book circuits": when you want to do quantum coding, you may have to adapt to the ones available.

9.2 From gates to circuits

A *quantum register* is a collection of qubits we use for computation. By convention, we number the qubits in the register as $q_0, q_1, \ldots, q_n$. All qubits in a register are initialized to state $|0\rangle$.

A *quantum circuit* is a sequence of gates applied to one or more qubits in a quantum register.

In some algorithms we group qubits into one or more labeled registers to better delineate their roles. It's common to have an "upper register" and a "lower register," for example. [20, Chapter 2]

Let's look at some simple example circuits to see how they are put together and what the components are called.

9.2.1 Constructing a circuit

The simplest circuit is

$$q_0\colon\ |0\rangle \quad\boxed{\nearrow}\quad |m_0\rangle \ = |0\rangle$$

Here qubit q_0 is initialized to $|0\rangle$ and we do an immediate measurement. $|m_0\rangle$ holds its measurement state which, of course, must be $|0\rangle$. Each horizontal application of gates for a qubit is called a *wire* or a *line*.

Another "do nothing" circuit invokes the **ID** gate and then gets measured.

$$q_0:\ |0\rangle\!-\!\boxed{\textbf{ID}}\!-\!\boxed{\nearrow}\!-\!|m_0\rangle\ =|0\rangle$$

The result is the same. The wire has depth 1: we count the gates but do not include the final measurement. The depth of a circuit is the maximum of the depths of its wires. We often omit the **ID** gate in circuits that have multiple wires. You'll see the bare wire where the gate would have been.

To exercise a gate which changes state, we use **X** to flip from $|0\rangle$ to $|1\rangle$.

$$q_0:\ |0\rangle\!-\!\boxed{\textbf{X}}\!-\!\boxed{\nearrow}\!-\!|m_0\rangle\ =|1\rangle$$

The circuit has width 1 because it involves 1 qubit in a non-trivial way. The next circuit operates on a 3-qubit quantum register but it also only has width 1.

$$q_0:\ |0\rangle\!-\!\!-\!\!-\!\!-\!\!-\!|m_0\rangle\ =|0\rangle$$

$$q_1:\ |0\rangle\!-\!\boxed{\textbf{X}}\!-\!\boxed{\nearrow}\!-\!|m_1\rangle\ =|1\rangle$$

$$q_2:\ |0\rangle\!-\!\!-\!\!-\!\!-\!\!-\!|m_2\rangle\ =|0\rangle$$

If there is no measurement shown you can assume there is one at the end of the wire. I prefer to make mine explicit.

Two consecutive **X** gates return us back to $|0\rangle$.

$$q_0:\ |0\rangle\!-\!\boxed{\textbf{X}}\!-\!\boxed{\textbf{X}}\!-\!\boxed{\nearrow}\!-\!|m_0\rangle\ =|0\rangle$$

This is a circuit of depth 2.

Software development frameworks for classical computing have been available since at least the 1990s. These provide the tools and libraries you need to create applications. As an industry, we have learned the best practices for giving coders what they need for efficient software creation.

When you use a quantum software development framework like Qiskit, it may optimize your circuit. [16] One technique is to remove unneeded gates. In the current example, since nothing is happening been the two **X** gates, we can remove both.

$$q_0: |0\rangle \quad\quad\quad \boxed{\text{measure}} \quad |m_0\rangle \quad = |0\rangle$$

The Hadamard **H** gate puts the qubit into a non-trivial superposition. Upon measurement, it collapses to either $|0\rangle$ or $|1\rangle$ with equal probability.

$$q_0: |0\rangle \quad \boxed{H} \quad \boxed{\text{measure}} \quad |m_0\rangle \quad = |0\rangle \text{ or } |1\rangle$$

The combination of two consecutive **H** gates with no measurement in between is the same as an **ID** gate. You can optimize the circuit by removing two consecutive **H** gates. (Recall the matrix for **H** is its own inverse.)

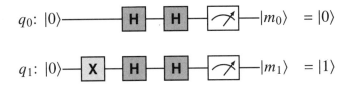

$$q_0: |0\rangle \quad\quad \boxed{H} \boxed{H} \quad \boxed{\text{measure}} \quad |m_0\rangle \quad = |0\rangle$$

$$q_1: |0\rangle \quad \boxed{X} \boxed{H} \boxed{H} \quad \boxed{\text{measure}} \quad |m_1\rangle \quad = |1\rangle$$

The first wire has depth 2 and the second has depth 3. The circuit depth is 3, the maximum of the wire depths. Next up, consider

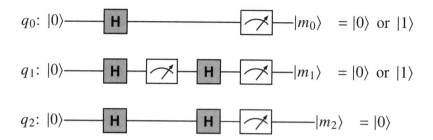

This last circuit is a prime example of quantum randomness. In the first wire we put q_0 into superposition. When measured it is either $|0\rangle$ or $|1\rangle$ with equal probability.

In the second wire, we do an **H** followed by a measurement. The state is then $50 - 50$ randomly $|0\rangle$ or $|1\rangle$. We do another **H** followed by a measurement. Again, the state is $50 - 50$ $|0\rangle$ or $|1\rangle$.

If we omit the measurement in the middle, the qubit stays in superposition and the two **H** gates cancel each other. We always get $|0\rangle$.

The real-life situation typically used to illustrate this involves flipping a coin, which can land heads up or down. If it starts heads up and you flip it, it lands randomly heads up or down. If you flip it again it still lands heads up or down with equal probabilities, if it is a fair coin. This happens whether or not you peek at the coin between flips. This is classical behavior.

Quantum theory instead says *if you do not peek,* the coin lands after the second flip in the same state in which it started. Weird but true.

Two qubit gates like **CNOT** affect more than one qubit. This circuit inverts the standard **CNOT** behavior so that the second qubit is the control.

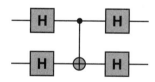

It implements a **reverse CNOT** gate.

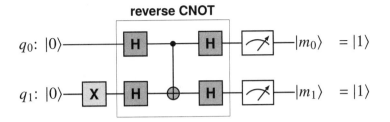

Multi-qubit gates can involve non-adjacent wires.

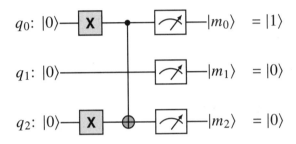

9.2.2 A note on controlled gates

In the previous chapter I noted that when we say "the quantum state was not changed," we really mean "the quantum state was not changed, except possibly being multiplied by a complex number of absolute value 1."

A **CNOT** gate is really a controlled-**X** (**CX**).

$$
\begin{array}{ccc}
\begin{array}{c}\text{---}\bullet\text{---} \\ \text{---}\oplus\text{---}\end{array} & = & \begin{array}{c}\text{---}\bullet\text{---} \\ \text{---}\boxed{\text{X}}\text{---}\end{array}
\end{array}
$$

Its matrix is

$$
\left[\begin{array}{cc|cc}
1 & 0 & 0 & 0 \\
0 & 1 & 0 & 0 \\
\hline
0 & 0 & 0 & 1 \\
0 & 0 & 1 & 0
\end{array}\right].
$$

The identity matrix is in the upper left and the **X** matrix is in the lower right. In basis kets and the computational basis, we have

$$\textbf{CNOT}|00\rangle = \begin{bmatrix} 1 & 0 & 0 & 0 \\ 0 & 1 & 0 & 0 \\ 0 & 0 & 0 & 1 \\ 0 & 0 & 1 & 0 \end{bmatrix} \begin{bmatrix} 1 \\ 0 \\ 0 \\ 0 \end{bmatrix} = \begin{bmatrix} 1 \\ 0 \\ 0 \\ 0 \end{bmatrix} = |00\rangle$$

and

$$\textbf{CNOT}|11\rangle = \begin{bmatrix} 1 & 0 & 0 & 0 \\ 0 & 1 & 0 & 0 \\ 0 & 0 & 0 & 1 \\ 0 & 0 & 1 & 0 \end{bmatrix} \begin{bmatrix} 0 \\ 0 \\ 0 \\ 1 \end{bmatrix} = \begin{bmatrix} 0 \\ 0 \\ 1 \\ 0 \end{bmatrix} = |10\rangle.$$

The results are exactly what we expect. In these two cases, the second qubit state is flipped between 0 and 1 only if the first qubit is 1. The same happens if you do the math for $|01\rangle$ and $|10\rangle$.

But what if I were to tell you that

This doesn't seem right. I thought nothing happened to the control qubit! On the left the top qubit state should "stay the same" and ditto for the bottom qubit on the right. But if these two circuits are equivalent then something odd is going on.

For the version on the left we have

$$\begin{aligned}
|0\rangle \otimes |0\rangle &\mapsto & |0\rangle \otimes |0\rangle \\
|0\rangle \otimes |1\rangle &\mapsto & |0\rangle \otimes |1\rangle \\
|1\rangle \otimes |0\rangle &\mapsto & |1\rangle \otimes |0\rangle \\
|1\rangle \otimes |1\rangle &\mapsto & |1\rangle \otimes (-|1\rangle).
\end{aligned}$$

On the right,

$$\begin{aligned}
|0\rangle \otimes |0\rangle &\mapsto & |0\rangle \otimes |0\rangle \\
|0\rangle \otimes |1\rangle &\mapsto & |0\rangle \otimes |1\rangle \\
|1\rangle \otimes |0\rangle &\mapsto & |1\rangle \otimes |0\rangle \\
|1\rangle \otimes |1\rangle &\mapsto & (-|1\rangle) \otimes |1\rangle.
\end{aligned}$$

Look at the last lines. If they are the same, then

$$|1\rangle \otimes (-|1\rangle) = (-|1\rangle) \otimes |1\rangle.$$

This is true by the bilinearity of the tensor product: we can move the multiple of -1 from one side to the other.

9.3 Building blocks and universality

In section 2.4 we discussed classical gates and I illustrated how to create an **or** gate from **nand** gates. **nand** is universal in that we can create all the other classical logic gates from it. For example,

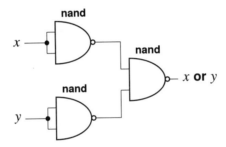

This means any software we build for classical computers could ultimately be constructed from millions of **nand** gates. That would be horribly inefficient. There are higher-level gates and circuits in modern processors that are tremendously faster.

The basic **CNOT** acts like a **xor** on the standard kets.

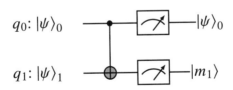

This maps the basis kets in this way:

$$|00\rangle \mapsto |00\rangle \qquad |10\rangle \mapsto |11\rangle$$
$$|01\rangle \mapsto |01\rangle \qquad |11\rangle \mapsto |10\rangle$$

The **xor** result is the second qubit state of $|m_1\rangle$. More than simply a logical operation, this implements addition mod 2. That is, this standard gate does a basic arithmetic operation "$\oplus$". For example, $|1\rangle \oplus |1\rangle = |0\rangle$ and

$$(a|0\rangle + b|1\rangle) \oplus |1\rangle = a|0\rangle \oplus |1\rangle + b|1\rangle \oplus |1\rangle = a|1\rangle + b|0\rangle.$$

If we do not want to modify one of the input qubits, we can put the value of the **xor** in a third output, or *ancilla*, qubit.

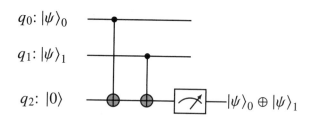

Question 9.3.1

Suppose we have three input qubits, q_0, q_1, and q_2, in states $|\psi\rangle_0$, $|\psi\rangle_1$, and $|\psi\rangle_2$, respectively. We want to put $|\psi\rangle_0 \oplus |\psi\rangle_1$ in ancilla qubit q_3 and put $|\psi\rangle_0 \oplus |\psi\rangle_2$ in ancilla qubit q_4. Draw the circuit that does this.

9.3.1 The Toffoli gate

The quantum Toffoli **CCNOT** gate operates on three qubits. If the first two qubits are $|1\rangle$ then it flips the third, otherwise it does nothing. For example,

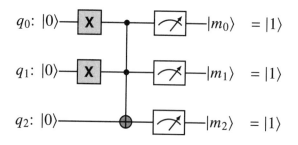

Here q_0 and q_1 are each moved into the $|1\rangle$ state. The Toffoli gate then flips the state of the third qubit from $|0\rangle$ to $|1\rangle$. Consider this particular use of the gate:

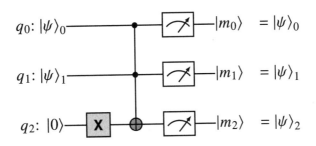

What is the value of $|\psi\rangle_2$ as $|\psi\rangle_0$ and $|\psi\rangle_1$ take on values of $|0\rangle$ and $|1\rangle$?

$|\psi\rangle_0$	$|\psi\rangle_1$	$|\psi\rangle_2$		p	q	p **nand** q
$|1\rangle$	$|1\rangle$	$|0\rangle$		true	true	false
$|1\rangle$	$|0\rangle$	$|1\rangle$		true	false	true
$|0\rangle$	$|1\rangle$	$|1\rangle$		false	true	true
$|0\rangle$	$|0\rangle$	$|1\rangle$		false	false	true

Compare the ket table on the left with the truth table for **nand** on the right. Substituting $|1\rangle$ for true and $|0\rangle$ for false, they are identical.

> The Toffoli gate can be used to create the quantum equivalent of a **nand** gate. Since all classical logic circuits can be built from **nand**s, all classical software applications could, in theory, be run on a quantum computer. [13] [17, Chapter 6]

True, but you wouldn't want to do this now because quantum gates are much slower than classical operations in modern processors. Also, as we shall discuss in section 11.1, the limited time we have to complete computations using today's qubits means that only small amounts of classical code could be run.

As we shall see starting in section 9.5, specific algorithms that take advantage of quantum properties can be significantly faster than classical alternatives. We will eventually employ classical and quantum computers in a complementary and hybrid fashion, taking advantage of their best and most powerful features.

If you set the state of the first qubit to $|1\rangle$ the Toffoli gate reduces to **CNOT**.

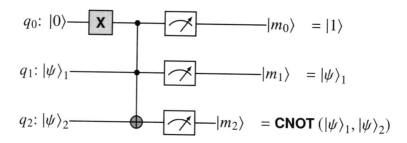

Setting the first two states to $|1\rangle$ gives us **X**.

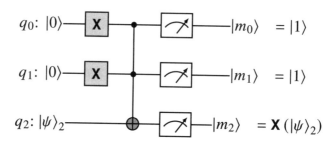

The basic Toffoli gate has the same effect as **and** on the basis kets $|0\rangle$ and $|1\rangle$.

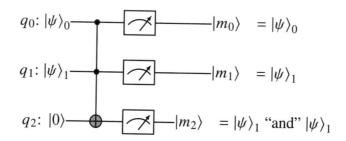

The initial values are the states for q_0 and q_1.

9.3.2   Building more complicated circuits

What are the values of $|m_0\rangle$ and $|m_1\rangle$ in this next circuit? The circuit has three 2-qubit gates: a **CNOT**, a reverse **CNOT**, and a final **CNOT**.

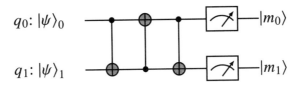

It has this effect on the standard basis kets:

$$|00\rangle \mapsto |00\rangle \qquad |01\rangle \mapsto |10\rangle$$
$$|10\rangle \mapsto |01\rangle \qquad |11\rangle \mapsto |11\rangle$$

Let's examine the states after each gate for the beginning state $|01\rangle$. After the first **CNOT**, the state is still $|01\rangle$. The **reverse CNOT** changes the state to $|11\rangle$. The final **CNOT** moves it to $|10\rangle$. It appears to swap the states of the two qubits.

If we do not have the **SWAP** available when we are coding for a particular quantum computer, we can implement it this way as a reusable circuit. We can take it a step further and use only Hadamard **H** gates and **CNOT** gates.

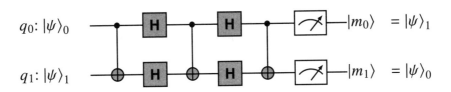

I hope you are getting the sense that some gates are especially valuable as building blocks for other gates. It appears that with some ingenuity we can do almost anything.

Question 9.3.2

Show that this circuit

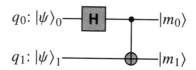

creates the four Bell states

$$|\Phi^+\rangle = \tfrac{\sqrt{2}}{2}|00\rangle + \tfrac{\sqrt{2}}{2}|11\rangle \qquad |\Psi^+\rangle = \tfrac{\sqrt{2}}{2}|01\rangle + \tfrac{\sqrt{2}}{2}|10\rangle$$

$$|\Phi^-\rangle = \tfrac{\sqrt{2}}{2}|00\rangle - \tfrac{\sqrt{2}}{2}|11\rangle \qquad |\Psi^-\rangle = \tfrac{\sqrt{2}}{2}|01\rangle - \tfrac{\sqrt{2}}{2}|10\rangle$$

from the standard basis kets $|00\rangle$, $|01\rangle$, $|10\rangle$, and $|11\rangle$.

Question 9.3.3

What do you get if you apply the same circuit again to each of the four Bell states? Use linearity. This is an example of measurement in a basis other than $|00\rangle$, $|01\rangle$, $|10\rangle$, and $|11\rangle$.

Among all the gates we have discussed, several are more fundamental than others, especially **H** and **CNOT**. Just as **nand** was universal for classical logic gates, is there a finite set of quantum gates from which we can create all the others?

No. We cannot get them all with a finite set because of the infinite number of $\mathbf{R}^z_\varphi$ gates as φ varies between 0 and 2π.

If we start with the 1-qubit gates built from all the 2 by 2 complex unitary matrices and we throw in the **CNOT** gate, we do get a universal, albeit infinite, set. We can even build all n-qubit gates from blocks of these.

It's beyond the scope of this book, but it is possible to numerically closely approximate the effect of any quantum gate using combinations of only **CNOT**, **H**, and **T**. [14, Section 4.5]

9.3.3 Copying a qubit

Let's try to put together a circuit that makes a copy of a qubit's state. We're looking for something like

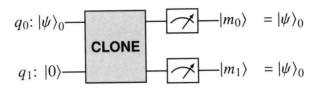

The initial state of q_1 doesn't matter since it is a placeholder we want to replace with the state of q_0. We are not looking for a gate that clones one particular qubit state but rather one that makes a copy of any arbitrary state.

If the **CLONE** gate exists, let C be its unitary matrix in the standard ket basis in $\mathbb{C}^2 \otimes \mathbb{C}^2$. As usual, we take

$$|\psi\rangle_0 = a|0\rangle + b|1\rangle.$$

The result after cloning is $|\psi\rangle_0 \otimes |\psi\rangle_0$. That is,

$$C\left(|\psi\rangle_0 \otimes |0\rangle\right) = |\psi\rangle_0 \otimes |\psi\rangle_0.$$

Are these really equal? On the left,

$$
\begin{aligned}
C\left(|\psi\rangle_0 \otimes |0\rangle\right) &= C\left((a|0\rangle + b|1\rangle) \otimes |0\rangle\right) \\
&= C\left(a|0\rangle \otimes |0\rangle + b|1\rangle \otimes |0\rangle\right) \\
&= aC\left(|0\rangle \otimes |0\rangle\right) + bC\left(|1\rangle \otimes |0\rangle\right) \text{ by linearity} \\
&= a|00\rangle + b|11\rangle \text{ by the definition of \textbf{CLONE} and } C.
\end{aligned}
$$

On the right,

$$
\begin{aligned}
|\psi\rangle_0 \otimes |\psi\rangle_0 &= (a|0\rangle + b|1\rangle) \otimes (a|0\rangle + b|1\rangle) \\
&= a^2|00\rangle + ab|01\rangle + ab|10\rangle + b^2|11\rangle
\end{aligned}
$$

For arbitrary a and b in $\mathbb{C}$ with $|a|^2 + |b|^2 = 1$,

$$a|00\rangle + b|11\rangle \neq a^2|00\rangle + ab|01\rangle + ab|10\rangle + b^2|11\rangle.$$

There is no **CLONE** gate that can duplicate the quantum state of a qubit. This is called the *No-Cloning Theorem* and has ramifications for the design of algorithms, how we might do quantum error correction, and the ultimate creation of quantum memory.

Did this result surprise you? In the classical case we can clone a bit value but this is not possible for a qubit state. It falls out naturally from the theory but it is quite a restriction.

9.3.4 Teleportation

If I can't copy the state of a qubit and give it to you, is there any way for you to get it even if mine is destroyed? The answer is yes and goes by the name of quantum teleportation. The technique was first published in 1993 by IBM Fellow Charles Bennett et al. [2]

Though the name sounds like something out of science fiction, it doesn't involve any dematerialization and rematerialization, or faster-than-light travel. In fact, it requires two classical bits of information to be transferred via traditional means.

The technique involves three qubits: M, which is my qubit; Y, which is your qubit; and Q, which is the qubit whose state I want to transfer from me to you. We use entanglement as the connection and transference mechanism.

My qubit Q is in some arbitrary quantum state $|\psi\rangle_Q = a|0\rangle + b|1\rangle$. When we are done, you will know this state but I will no longer have access to it.

We begin by entangling M and Y. There are an infinite number of ways of doing this but the usual choice is to use one of the four Bell states. I'll use $|\Phi^+\rangle = \frac{\sqrt{2}}{2}|00\rangle + \frac{\sqrt{2}}{2}|11\rangle$, but you can use any of them with appropriate changes to the algorithm and math below.

To keep track of which qubit belongs to whom, I modify the notation slightly so that

$$|\Phi^+\rangle_{MY} = \frac{\sqrt{2}}{2}|00\rangle_{MY} + \frac{\sqrt{2}}{2}|11\rangle_{MY}$$

to indicate the first qubit is mine and the second is yours.

We did this entanglement while we were close together, but now you can get on a plane and go as far away as you wish. You might even take a trip to a space station orbiting the planet. The qubits are entangled and will stay that way in this scenario.

Next, I put Q into the mix to get

$$|\psi\rangle_Q \otimes |\Phi^+\rangle_{MY} = \left(a|0\rangle_Q + b|1\rangle_Q\right) \otimes \left(\frac{\sqrt{2}}{2}|00\rangle_{MY} + \frac{\sqrt{2}}{2}|11\rangle_{MY}\right).$$

Consider the identities I first introduced in Question 8.2.2:

$$|00\rangle = \tfrac{\sqrt{2}}{2}\left(|\Phi^+\rangle + |\Phi^-\rangle\right) \qquad |11\rangle = \tfrac{\sqrt{2}}{2}\left(|\Phi^+\rangle - |\Phi^-\rangle\right)$$

$$|01\rangle = \tfrac{\sqrt{2}}{2}\left(|\Psi^+\rangle + |\Psi^-\rangle\right) \qquad |10\rangle = \tfrac{\sqrt{2}}{2}\left(|\Psi^+\rangle - |\Psi^-\rangle\right).$$

Using linearity, we rewrite the above in the following way:

$$\left(a|0\rangle_Q + b|1\rangle_Q\right) \otimes \left(\frac{\sqrt{2}}{2}|00\rangle_{MY} + \frac{\sqrt{2}}{2}|11\rangle_{MY}\right)$$

$$= \frac{\sqrt{2}}{2}\left(a|000\rangle_{QMY} + a|011\rangle_{QMY} + b|100\rangle_{QMY} + b|111\rangle_{QMY}\right)$$

$$= \frac{\sqrt{2}}{2}\left(a|00\rangle_{QM}\otimes|0\rangle_Y + a|01\rangle_{QM}\otimes|1\rangle_Y + b|10\rangle_{QM}\otimes|0\rangle_Y + b|11\rangle_{QM}\otimes|1\rangle_Y\right)$$

$$= \frac{\sqrt{2}}{2}\Big(a\frac{\sqrt{2}}{2}\left(|\Phi^+\rangle_{QM} + |\Phi^-\rangle_{QM}\right)\otimes|0\rangle_Y + a\frac{\sqrt{2}}{2}\left(|\Psi^+\rangle_{QM} + |\Psi^-\rangle_{QM}\right)\otimes|1\rangle_Y$$

$$+ b\frac{\sqrt{2}}{2}\left(|\Psi^+\rangle_{QM} - |\Psi^-\rangle_{QM}\right)\otimes|0\rangle_Y + b\frac{\sqrt{2}}{2}\left(|\Phi^+\rangle_{QM} - |\Phi^-\rangle_{QM}\right)\otimes|1\rangle_Y\Big)$$

$$= \frac{1}{2}\Big(a\left(|\Phi^+\rangle_{QM} + |\Phi^-\rangle_{QM}\right)\otimes|0\rangle_Y + a\left(|\Psi^+\rangle_{QM} + |\Psi^-\rangle_{QM}\right)\otimes|1\rangle_Y$$

$$+ b\left(|\Psi^+\rangle_{QM} - |\Psi^-\rangle_{QM}\right)\otimes|0\rangle_Y + b\left(|\Phi^+\rangle_{QM} - |\Phi^-\rangle_{QM}\right)\otimes|1\rangle_Y\Big)$$

$$= \frac{1}{2}\Big(\left(|\Phi^+\rangle_{QM} + |\Phi^-\rangle_{QM}\right)\otimes a|0\rangle_Y + \left(|\Psi^+\rangle_{QM} + |\Psi^-\rangle_{QM}\right)\otimes a|1\rangle_Y$$

$$+ \left(|\Psi^+\rangle_{QM} - |\Psi^-\rangle_{QM}\right)\otimes b|0\rangle_Y + \left(|\Phi^+\rangle_{QM} - |\Phi^-\rangle_{QM}\right)\otimes b|1\rangle_Y\Big)$$

$$= \frac{1}{2}\Big(|\Phi^+\rangle_{QM}\otimes a|0\rangle_Y + |\Phi^-\rangle_{QM}\otimes a|0\rangle_Y + |\Psi^+\rangle_{QM}\otimes a|1\rangle_Y + |\Psi^-\rangle_{QM}\otimes a|1\rangle_Y$$

$$+ |\Psi^+\rangle_{QM}\otimes b|0\rangle_Y - |\Psi^-\rangle_{QM}\otimes b|0\rangle_Y + |\Phi^+\rangle_{QM}\otimes b|1\rangle_Y - |\Phi^-\rangle_{QM}\otimes b|1\rangle_Y\Big)$$

$$= \frac{1}{2}\Big(|\Phi^+\rangle_{QM}\otimes(a|0\rangle_Y + b|1\rangle_Y) + |\Phi^-\rangle_{QM}\otimes(a|0\rangle_Y - b|1\rangle_Y)$$

$$+ |\Psi^+\rangle_{QM}\otimes(b|0\rangle_Y + a|1\rangle_Y) + |\Psi^-\rangle_{QM}\otimes(-b|0\rangle_Y + a|1\rangle_Y)\Big)$$

Note the coefficients and signs in the last expression.

That was a lot of ket manipulation! Note what happened: the a and the b, which started out in the state of Q, which I own, is now on Y, which you own. Other than entangling qubits we did not do any sort of measurement, we just rewrote the ket and tensor formulas.

Now I measure. I don't do it in the $|00\rangle$, $|01\rangle$, $|10\rangle$, and $|11\rangle$ basis, I do it in the $|\Phi^+\rangle$, $|\Phi^-\rangle$,

$|\Psi^+\rangle$, and $|\Psi^-\rangle$ basis. After measurement I have exactly one of the expressions

$$|\Phi^+\rangle_{QM} \otimes (a|0\rangle_Y + b|1\rangle_Y) \qquad\qquad |\Phi^-\rangle_{QM} \otimes (a|0\rangle_Y - b|1\rangle_Y)$$

$$|\Psi^+\rangle_{QM} \otimes (b|0\rangle_Y + a|1\rangle_Y) \qquad\qquad |\Psi^-\rangle_{QM} \otimes (-b|0\rangle_Y + a|1\rangle_Y)$$

The probability of getting any one of them is 0.25 **and I know which one I have** by looking at Q and M! My measurement did not affect Y other than breaking the entanglement. The original quantum state of Q was destroyed.

Now I call you up or text you or email you or send you a letter through the post and tell you which of the basis vectors I observed. This information is represented using two bits and is sent using a classical communication channel.

Question 9.3.4

You can get these bits by reversing the circuit shown in Question 9.3.2: apply a **CNOT** gate to QM and then an **H** gate to Q. Which 2-bit strings correspond to each of $|\Phi^+\rangle_{QM}$, $|\Phi^-\rangle_{QM}$, $|\Psi^+\rangle_{QM}$, and $|\Psi^-\rangle_{QM}$?

If I saw $|\Phi^+\rangle_{QM}$ then the quantum state of Q was successfully teleported to Y and you have nothing else to do.

If I saw $|\Phi^-\rangle_{QM}$ then the quantum state of Y has the sign of b wrong. You apply a **Z** gate to do the phase flip and Y now has the original state of Q.

If I saw $|\Psi^+\rangle_{QM}$ then the quantum state of Y has a and b reversed. You apply an **X** gate to do the bit flip and Y now has the original state of Q.

If I saw $|\Psi^-\rangle_{QM}$ then the quantum state of Y has a and b reversed with the wrong signs. You apply an **X** gate and then a **Z** gate.

Question 9.3.5

Do the math and the post-measurement analysis for using $|\Psi^-\rangle_{QM}$ instead of $|\Phi^+\rangle_{QM}$.

Question 9.3.6

Create the quantum circuit for teleportation as I have described it here.

Rather than use "me" and "you" and M and Y, it's common to see "Alice" and "Bob" and "A" and "B" in the literature. I made this more personal so you would have a better feeling for what is being done.

9.4 Arithmetic

In section 2.5 we looked at the rudimentary ideas of doing binary addition via logic gates. Now we revisit that, but see how to do it using quantum gates. Like most such algorithms, many papers have been published on how to optimize the circuits using methods such as the Quantum Fourier Transform we cover in section 10.1.

I'm going to keep to a straightforward approach to help bridge the gap between classical and quantum approaches. The gates we use are simple, and we replace bits with qubits. That is, instead of 0 and 1 we use $|0\rangle$ and $|1\rangle$, respectively. The data input qubits are called $|x\rangle$ and $|y\rangle$, and each are in the state $|0\rangle$ or $|1\rangle$ at any given time. We are essentially mimicking what we would do in the classical case.

If we do not worry about carry-in and carry-out qubits, our circuit looks like

$$q_0: |x\rangle \quad\quad\quad |x\rangle$$
$$q_1: |y\rangle \quad\quad\quad |x\rangle \oplus |y\rangle$$

where "$\oplus$" is addition modulo 2. This is implemented as a **CNOT** gate acting as an **xor**. We use q_1 to store this output as well as the $|y\rangle$ input.

To include a carry-out state $|c_{out}\rangle$ we employ a Toffoli **CCNOT** gate and use a third qubit, q_2, to hold the value.

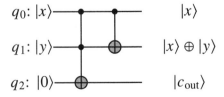

$$q_0: |x\rangle \quad\quad\quad |x\rangle$$
$$q_1: |y\rangle \quad\quad\quad |x\rangle \oplus |y\rangle$$
$$q_2: |0\rangle \quad\quad\quad |c_{out}\rangle$$

Question 9.4.1

Why do we place the Toffoli **CCNOT** gate before the **CNOT** gate? Are they interchangeable?

The only thing remaining to take into account is a carry-in state $|c_{in}\rangle$. We put this in a new qubit and rearrange the circuit slightly.

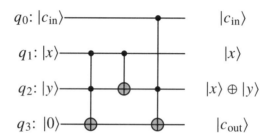

Note how the first two qubits on which the final Toffoli gate operates are non-adjacent.

Our inputs are held in qubits 0, 1, and 2, and the outputs are in qubits 2 and 3. This is called a **CARRY** gate. Rather than spell out all the individual operations, we can write it as "subroutine" gate like the one to the right.

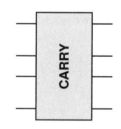

Since this is a quantum gate, it is reversible. We call the circuit subroutine we get by running the **CARRY** gate backwards the **CARRY**$^{-1}$ gate. It is shown on the left.

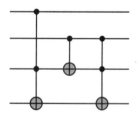

In general purpose gates we do not always specify whether any particular qubit should be $|0\rangle$ or $|1\rangle$.

The same is true of the final gate subroutine we need, the **SUM** gate. It takes the inputs from the first two qubits and adds the result to what was in the third qubit. It does not worry about any carry qubits.

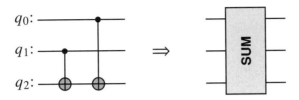

With this background, we can combine the pieces to create a circuit that adds two binary numbers of three bits each. We represent the first number as $x = x_2 x_1 x_0$ and the second as $y = y_2 y_1 y_0$. Because of carries, the result can have four bits. We might have $x = 011_2$ and $y = 101_2$, for example. It's important to pay attention to the bit/qubit ordering in algorithms as it is easy to use them in the reverse and therefore incorrect direction.

x_0 and y_0 are the *least* significant bits and x_2 and y_2 are the *most* significant bits. By analogy, in the decimal real number 247, 2 is the most significant digit and 7 is the least.

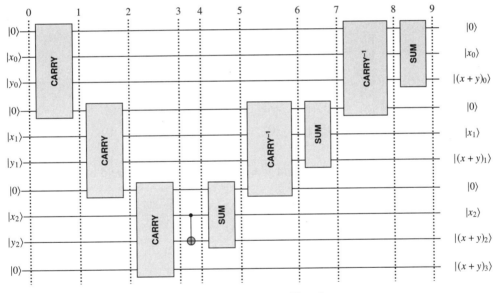

The 3 Qubit Adder Circuit

In this circuit, I've labeled the steps with vertical lines with numbers. From least significant qubit to the most, the sum will be in qubits 2, 5, 8, and 9.

When the **CARRY**, **CARRY**$^{-1}$, and **SUM** are given $|0\rangle$ inputs, they produce $|0\rangle$ outputs. Thus the circuit above successfully adds $|000\rangle$ to $|000\rangle$ to produce $|000\rangle$. Now let's try $1 + 1$. In this case, $x = y = |001\rangle$. We start with the least significant bits as the uppermost inputs and work downwards. This means

$$q_0 = |0\rangle \quad q_1 = |1\rangle \quad q_2 = |1\rangle \quad q_3 = |0\rangle \quad q_4 = |0\rangle$$
$$q_5 = |0\rangle \quad q_6 = |0\rangle \quad q_7 = |0\rangle \quad q_8 = |0\rangle \quad q_9 = |0\rangle$$

Now let's trace the evolution of the qubit states as we move from left to right in the circuit. The top line in the following table labels the steps and the first column labels the qubits. The bulk of the table gives the quantum state values of the qubits before a step with the given number.

We expect to see $|1\rangle$ in qubits 1 and 5, and $|0\rangle$ elsewhere.

Qubit	Step									
	0	1	2	3	4	5	6	7	8	9
0	$|0\rangle$	$|0\rangle$	$|0\rangle$	$|0\rangle$	$|0\rangle$	$|0\rangle$	$|0\rangle$	$|0\rangle$	$|0\rangle$	$|0\rangle$
1	$|1\rangle$	$|1\rangle$	$|1\rangle$	$|1\rangle$	$|1\rangle$	$|1\rangle$	$|1\rangle$	$|1\rangle$	$|1\rangle$	$|1\rangle$
2	$|1\rangle$	$|0\rangle$	$|0\rangle$	$|0\rangle$	$|0\rangle$	$|0\rangle$	$|0\rangle$	$|0\rangle$	$|1\rangle$	$|0\rangle$
3	$|0\rangle$	$|1\rangle$	$|1\rangle$	$|1\rangle$	$|1\rangle$	$|1\rangle$	$|1\rangle$	$|1\rangle$	$|0\rangle$	$|0\rangle$
4	$|0\rangle$	$|0\rangle$	$|0\rangle$	$|0\rangle$	$|0\rangle$	$|0\rangle$	$|0\rangle$	$|0\rangle$	$|0\rangle$	$|0\rangle$
5	$|0\rangle$	$|0\rangle$	$|0\rangle$	$|0\rangle$	$|0\rangle$	$|0\rangle$	$|0\rangle$	$|1\rangle$	$|1\rangle$	$|1\rangle$
6	$|0\rangle$	$|0\rangle$	$|0\rangle$	$|0\rangle$	$|0\rangle$	$|0\rangle$	$|0\rangle$	$|0\rangle$	$|0\rangle$	$|0\rangle$
7	$|0\rangle$	$|0\rangle$	$|0\rangle$	$|0\rangle$	$|0\rangle$	$|0\rangle$	$|0\rangle$	$|0\rangle$	$|0\rangle$	$|0\rangle$
8	$|0\rangle$	$|0\rangle$	$|0\rangle$	$|0\rangle$	$|0\rangle$	$|0\rangle$	$|0\rangle$	$|0\rangle$	$|0\rangle$	$|0\rangle$
9	$|0\rangle$	$|0\rangle$	$|0\rangle$	$|0\rangle$	$|0\rangle$	$|0\rangle$	$|0\rangle$	$|0\rangle$	$|0\rangle$	$|0\rangle$

Success! Now let's add $x = 111_2$ and $y = 101_2$. The answer should be 1100_2. The qubit inputs are

$$q_0 = |0\rangle \quad q_1 = |1\rangle \quad q_2 = |1\rangle \quad q_3 = |0\rangle \quad q_4 = |1\rangle$$
$$q_5 = |0\rangle \quad q_6 = |0\rangle \quad q_7 = |1\rangle \quad q_8 = |1\rangle \quad q_9 = |0\rangle$$

When we run it through the circuit we expect to see the output

$$q_0 = |0\rangle \quad q_1 = |1\rangle \quad q_2 = |0\rangle \quad q_3 = |0\rangle \quad q_4 = |1\rangle$$
$$q_5 = |0\rangle \quad q_6 = |0\rangle \quad q_7 = |1\rangle \quad q_8 = |1\rangle \quad q_9 = |1\rangle$$

The evolution is

Qubit	Step									
	0	1	2	3	4	5	6	7	8	9
0	$|0\rangle$	$|0\rangle$	$|0\rangle$	$|0\rangle$	$|0\rangle$	$|0\rangle$	$|0\rangle$	$|0\rangle$	$|0\rangle$	$|0\rangle$
1	$|1\rangle$	$|1\rangle$	$|1\rangle$	$|1\rangle$	$|1\rangle$	$|1\rangle$	$|1\rangle$	$|1\rangle$	$|1\rangle$	$|1\rangle$
2	$|1\rangle$	$|0\rangle$	$|0\rangle$	$|0\rangle$	$|0\rangle$	$|0\rangle$	$|0\rangle$	$|0\rangle$	$|1\rangle$	$|0\rangle$
3	$|0\rangle$	$|1\rangle$	$|1\rangle$	$|1\rangle$	$|1\rangle$	$|1\rangle$	$|1\rangle$	$|1\rangle$	$|0\rangle$	$|0\rangle$
4	$|1\rangle$	$|1\rangle$	$|1\rangle$	$|1\rangle$	$|1\rangle$	$|1\rangle$	$|1\rangle$	$|1\rangle$	$|1\rangle$	$|1\rangle$
5	$|0\rangle$	$|0\rangle$	$|1\rangle$	$|1\rangle$	$|1\rangle$	$|1\rangle$	$|0\rangle$	$|0\rangle$	$|0\rangle$	$|0\rangle$
6	$|0\rangle$	$|0\rangle$	$|1\rangle$	$|1\rangle$	$|1\rangle$	$|1\rangle$	$|0\rangle$	$|0\rangle$	$|0\rangle$	$|0\rangle$
7	$|1\rangle$	$|1\rangle$	$|1\rangle$	$|1\rangle$	$|1\rangle$	$|1\rangle$	$|1\rangle$	$|1\rangle$	$|1\rangle$	$|1\rangle$
8	$|1\rangle$	$|1\rangle$	$|1\rangle$	$|1\rangle$	$|1\rangle$	$|1\rangle$	$|1\rangle$	$|1\rangle$	$|1\rangle$	$|1\rangle$
9	$|0\rangle$	$|0\rangle$	$|0\rangle$	$|1\rangle$	$|1\rangle$	$|1\rangle$	$|1\rangle$	$|1\rangle$	$|1\rangle$	$|1\rangle$

and we again get the correct answer.

This algorithm as implemented in the circuit is slightly more complicated than it might seem it needs to be because we want to reset the carry qubits 0, 3, and 6 back to their initial values when we are done. There is no reversible gate operation that sets the state of a qubit absolutely to $|0\rangle$ or $|1\rangle$. Therefore, to get back to the initial value we must reverse the steps that got us there. Setting them to their known initial values allows us to reuse the qubits later, if we wish.

To use this circuit to add 2 numbers, each represented by 3 qubits, we needed 10 qubits. Generally, to add n qubits we need $3(n + 1) + 1$ qubits if we do addition this way. It's possible to similarly translate other classical algorithms for arithmetic into quantum versions. The number of qubits we need is $O(n)$ in the number of qubits we need to add two n bit/qubit numbers.

> To add these two n bit/qubit numbers, we need n **CARRY** gates, n **SUM** gates, $n - 1$ **CARRY**$^{-1}$ gates, and one **CNOT** gate. Since the number of included gates is fixed, this tells us we need $O(n)$ gates.

Question 9.4.2

Counting all the steps in the **CARRY**, **CARRY**$^{-1}$, and **SUM** circuit subroutines, what is the depth of the 3 qubit adder circuit? What is the depth of an n qubit adder circuit?

As I mentioned at the beginning of the section, researchers have developed much more efficient quantum algorithms for arithmetic operations. However, if you can perform such arithmetic on a classical processor, it will be much faster than quantum alternatives today. You would typically do it on a quantum computing system only if it was needed in the middle of a larger quantum algorithm.

Question 9.4.3

What do you get when you run a reversible addition circuit backwards? A subtraction circuit! Examine the full adder circuit and determine where you would place the bits/qubits of x and y in the input to the reverse circuit to compute $x - y$. Is this a full subtraction circuit or are there restrictions on x and y?

Rather than add $x = 111_2$ and $y = 101_2$, let's multiply them by the basic method you learned for multiplication of integers.

$$
\begin{array}{rccccc}
 & & & 1 & 1 & 1 \\
\times & & & 1 & 0 & 1 \\
\hline
 & & & 1 & 1 & 1 \\
+ & & 0 & 0 & 0 \\
+ & 1 & 1 & 1 \\
\hline
1 & 0 & 0 & 0 & 1 & 1 \\
\end{array}
$$

For each of the 3 qubits, we produce 3 partial sums, which we then add. The qubit-wise multiplication can be done by Toffoli gates.

Question 9.4.4

Write a reversible circuit to create the three partial sums for multiplying two 3 qubit numbers.

More generally, we need $O(n)$ gates to produce n partial sums. We then do $n - 1$ additions.

Multiplication of two n bit/qubit numbers requires $O(n^2)$ gates.

Next up is exponentiation. Given an integer x, say you want to compute x^9. You can do 8 multiplications by

$$xxxxxxxxx,$$

or you can note that $x^9 = x((x^2)^2)^2$. This requires only 4 multiplications:

- Set $a_1 = xx$.
- Set $a_2 = a_1 a_1$.

- Set $a_3 = a_2a_2$.
- Set $a_4 = a_3x$.

Then $x^9 = a_4$. This is the technique of *repeated squaring*. Note that $9 = 1001_2$.

For $x^{15} = x(x^7)^2 = x(xx^6)^2 = x(x(x^3)^2)^2 = x(x(xxx)^2)^2$, we need 7 multiplications and $15 = 1111_2$. For $x^{32} = ((((xx)^2)^2)^2)^2$, we need 5 multiplications and $32 = 100000_2$.

In general, if we need n bits to represent the exponent b, we can compute x^b with no more that $2n$ multiplications. That is, exponentiation is $O(n)$ in the number of multiplications.

If the binary representation of b is $b_{n-1}b_{n-2}\cdots b_1b_0$ with b_0 the least significant bit, then

$$x^b = x^{b_{n-1}2^{n-1}} \times x^{b_{n-2}2^{n-2}} \times \cdots \times x^{b_12^1} \times x^{b_02^0}.$$

For example, for $b = 13 = 1101_2$,

$$x^{13} = x^{1\times2^3} \times x^{1\times2^2} \times x^{0\times2^1} \times x^{1\times2^0}.$$

Raising an n bit/qubit number to an n bit/qubit exponent is $O(n) \times O(n) \times O(n) = O(n^3)$ in the number of gates.

Modular arithmetic can be done either by subtraction or division with quotient and remainder. For example,

$$17 \bmod 11 \equiv (17 - 11) \bmod 11 \equiv 6 \bmod 11.$$

Alternatively, $17 \div 11 = 1$ with *remainder* 6. The 1 is called the *quotient*.

The Python `divmod()` function takes two numbers and returns the (quotient, remainder) pair.

To learn more

The ways that you learned to do arithmetic are not always the most efficient methods to do it on a computer. [6, Chapter 9] Those techniques do form a good starting place to think about how you might implement and modify them for a quantum computing system. [10] [15, Section 2.4] From there you can investigate how to optimize them using quantum gates and circuits. [8] [9]

9.5 Welcome to Delphi

In ancient Greece, the Oracle at Delphi was a high priestess at the Temple of Apollo who, under the right conditions and during warm weather, issued prophecies. Less elaborate functions were served by other priestesses who would sometimes issue "yes" or "no" answers to questions put to them.

It's in this second sense that we introduce the notion of oracle for quantum computing. An oracle is a function which we supply with data and it responds with a 1 for yes and a 0 for no. The oracles we use cannot answer arbitrary questions but are instead built to respond to a specific query. For the algorithms that use them, two things are significant:

1. The implementation of the oracle must be as fast and efficient as possible.
2. We want to call the oracle the fewest number of times as possible to minimize the complexity of the algorithm.

An oracle is often called a *black box*, meaning we understand its behavior but not how it does what it does. Its function of answer 1/yes/true or 0/no/false also means it acts like a predicate.

Since all data can ultimately be represented by bits, we express the inputs to the oracle function as strings of 0s and 1s. If we call the function f, which is traditional, we can express it as

$$f : \{0,1\}^n \rightarrow \{0,1\}$$

This is an exceptionally terse way of saying:

Suppose we have a string s of n 0s and 1s, for example $s = 10110010$ for $n = 8$. When we apply f to s, we get back a 1 if the criteria of the oracle is met, 0 otherwise.

It might help you to think of 1 as true and 0 as false.

In section 2.1 we introduced the 7-bit ASCII character set. Let's define an example f by

$$f(\mathbf{x}) = f(x_0x_1x_2x_3x_4x_5x_6) = \begin{cases} 1 & \text{if } x_0x_1x_2x_3x_4x_5x_6 = 1000010 = \text{`B'} \\ 0 & \text{otherwise} \end{cases}$$

where each x_i is either 0 or 1. Then $f(1101010) = 0$ but $f(1000010) = 1$.

Question 9.5.1

Can you think of a Hamlet joke about f?

Though the test above is a simple equality, I don't state how we implement it. For example, suppose

$$f(\mathbf{x}) = \begin{cases} 1 & \text{if } \mathbf{x} = 1 \\ 0 & \text{otherwise .} \end{cases}$$

and I pass it the mathematical expression $\sin^2(z) + \cos^2(z)$ encoded as 0s and 1s. The oracle would have to understand trigonometry to answer correctly.

For quantum computing, we would call the oracle with a ket.

$$f(|\psi\rangle) = \begin{cases} |1\rangle & \text{if } |\psi\rangle = |1000010\rangle = |66\rangle_7 \\ |0\rangle & \text{otherwise .} \end{cases}$$

We are primarily interested in the action of the oracle on the standard basis kets.

Here's another oracle:

$$f(|x_0 x_1 x_2 x_3 x_4 x_5 x_6 x_7\rangle) = \begin{cases} |1\rangle & \text{if } (x_0 x_1 x_2 x_3) \times (x_4 x_5 x_6 x_7) = 1111_2 \\ |0\rangle & \text{otherwise .} \end{cases}$$

This returns $|1\rangle$ if the product of the two binary numbers encoded in the two halves of the input is 1111_2 (= 15 decimal). The oracle doesn't need to know how to factor, it needs to know how to multiply and test for equality. With many qubits and a lot of time, we could use the oracle to identify the factors of a large integer.

In practice, for quantum computing we need to encapsulate the function of the oracle within a unitary matrix and gate $\mathbf{U}_f$ somehow. (The $\mathbf{U}$ in $\mathbf{U}_f$ stands for "unitary.") One way to do it is to adjust the sign of a ket that meets the oracle's criteria for success.

1. Encode the data we are searching through as standard basis kets $|x\rangle$. For example, if we have 100 data items then use 7 qubits (because $2^7 = 128$) and assign each data element to a basis ket $|x\rangle_7$.
2. Let our oracle f produce $f(|y\rangle) = 1$ for one special encoded piece of data $|y\rangle$, 0 otherwise.
3. Find a unitary (hence reversible) matrix $\mathbf{U}_f$ such that

$$\mathbf{U}_f |x\rangle = \begin{cases} -|y\rangle & \text{if } |x\rangle = |y\rangle \text{ (which means } f(|y\rangle) = 1) \\ |x\rangle & \text{otherwise.} \end{cases}$$

This means $\mathbf{U}_f$ inverts the sign of the ket input that satisfies the condition of the oracle and leaves the rest alone.

I think most mathematical/quantum expressions of how oracles and their corresponding unitary matrices work look very complicated. I don't think the last is an exception. Let's translate this via a simple example.

We have two qubits and we want the oracle to return $|1\rangle$ if we give it $|01\rangle$, $|0\rangle$ otherwise. For $|\psi\rangle$ being one of the standard basis kets $|00\rangle$, $|01\rangle$, $|10\rangle$, or $|11\rangle$,

$$f(|\psi\rangle) = \begin{cases} |1\rangle & \text{if } |\psi\rangle = |01\rangle \\ |0\rangle & \text{if } |\psi\rangle = |00\rangle, |10\rangle, \text{ or } |11\rangle. \end{cases}$$

Then $\mathbf{U}_f$ should behave as

$$\mathbf{U}_f |\psi\rangle = \begin{cases} -|01\rangle & \text{if } |\psi\rangle = |01\rangle \\ |\psi\rangle & \text{if } |\psi\rangle = |00\rangle, |10\rangle, \text{ or } |11\rangle. \end{cases}$$

In standard vector notation, recall we have

$$|00\rangle = \begin{bmatrix} 1 \\ 0 \\ 0 \\ 0 \end{bmatrix} \quad |01\rangle = \begin{bmatrix} 0 \\ 1 \\ 0 \\ 0 \end{bmatrix} \quad |10\rangle = \begin{bmatrix} 0 \\ 0 \\ 1 \\ 0 \end{bmatrix} \quad |11\rangle = \begin{bmatrix} 0 \\ 0 \\ 0 \\ 1 \end{bmatrix}.$$

Then the matrix for $\mathbf{U}_f$

$$\begin{bmatrix} 1 & 0 & 0 & 0 \\ 0 & -1 & 0 & 0 \\ 0 & 0 & 1 & 0 \\ 0 & 0 & 0 & 1 \end{bmatrix}$$

does what we need.

Question 9.5.2

What's the determinant of this matrix? Is the matrix unitary?

Depending on the algorithm and how we use the oracle, we might create a different $\mathbf{U}_f$ but it must still reflect what f tells us. While we think of $\mathbf{U}_f$ as a matrix, we implement it as a circuit. The trick is to construct it to be extremely computationally efficient.

9.6 Amplitude amplification

Suppose we have three qubits and the standard basis kets for them each correspond to a possible solution to some problem. We want to devise an algorithm that chooses the best solution among them. I'm purposely not telling you what the problem is or how the kets map to the data and solution. Just assume we want to identify one of them that the algorithm can determine as best.

The first question is how to see that this best ket stands out from another. The general form for a 3-qubit register state is

$$\sum_{j=0}^{7} a_j |j\rangle_3 = a_0|000\rangle + a_1|001\rangle + a_2|010\rangle + a_3|011\rangle +$$

$$a_4|100\rangle + a_5|101\rangle + a_6|110\rangle + a_7|111\rangle$$

with

$$1 = \sum_{j=0}^{7} |a_j|^2.$$

If we initialize each qubit to $|0\rangle$ and then apply $\mathbf{H}^{\otimes 3}$, we get a balanced superposition via a change of basis

$$|\varphi\rangle = \mathbf{H}^{\otimes 3}|000\rangle = \sum_{j=0}^{7} a_j|j\rangle = \sum_{j=0}^{7} \frac{1}{\sqrt{8}}|j\rangle = \frac{1}{\sqrt{8}} \sum_{j=0}^{7} |j\rangle.$$

All the coefficients are equal and the square of each absolute value is $\frac{1}{8}$. If we measure the qubits now, we have an equal chance of getting any one of the eight basis kets.

At the qubit level, we have done a change of basis from the computational basis $\{|0\rangle, |1\rangle\}$ to $\{|+\rangle, |-\rangle\}$ via $\mathbf{H}$.

Through a process called *amplitude amplification,* we manipulate the qubit states so that the basis ket which represents the best solution has the coefficient a_j with the largest probability $|a_j|^2$.

When we measure, that standard basis ket has the highest chance of being observed. We want to make $|a_j|^2$ as large as possible, ideally equal to 1.0.

Ultimately, this is the goal of every quantum algorithm: have the results of the final qubit measurements correspond with high probability to the best solution.

In practice, we often create a reusable portion of a circuit, what we call a *circuit subroutine,* which we call several times. Each time it gets us closer to what we hope is the ideal result by increasing its corresponding probability. We also decrease the probabilities of the "bad" results.

9.6.1 Flipping the sign

Let me show how this might evolve. In the following graph, the vertical axis shows the probability amplitudes. That is, we map the coefficients a_j on this. Normally these are general values in $\mathbb{C}$, but we assume they are real in this example.

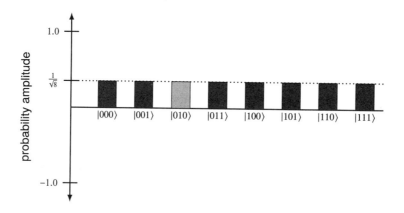

I've highlighted one ket, $|010\rangle$, that I want to modify over several interactions to increase its probability. I've chosen it at random to show the process. The dotted line shows the average of the probability amplitudes.

In the last section we saw how to negate the sign for a given ket in a balanced superposition. Let's do that now for $|010\rangle$. Create an 8 by 8 matrix $U_{|010\rangle}$, which is the identity matrix, except for the $(3, 3)$ entry, which we change to -1. After applying this, the above changes to

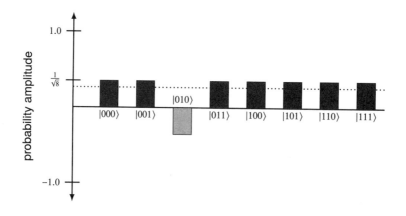

Note how the average amplitude has dropped. The sum of the squares of the absolute values of the amplitudes is still 1. All basis kets have the same probability of being the result when measured.

> **Question 9.6.1**
>
> What is the new average amplitude?

In this example with three qubits we had one particular computational basis ket whose sign we wanted to flip, the third. In general, it is the oracle f that completely determines which basis ket we sign-flip.

Now for the really clever trick.

9.6.2 Inversion about the mean

Let $S = \{s_j\}$ be a finite collection of numbers in $\mathbb{R}$ and let m be the mean (average) of the numbers. If we create a new collection T containing the numbers $\{t_j = 2m - s_j\}$, T has the following properties:

1. The average of the numbers in T is still m.
2. If $s_j = m$ then $t_j = s_j$.
3. $t_j - m = m - s_j$ and so $|t_j - m| = |s_j - m|$.
4. If $s_j < m$ then $t_j > m$ and if $s_j > m$ then $t_j < m$.

This is called *inversion about the mean*.

Let's see what this does when our collection is $S = \{1, -2, 3, 4\}$. The average, or mean, is $\frac{3}{2}$ and so $T = \{2, 5, 0, -1\}$, as you can see on the right.

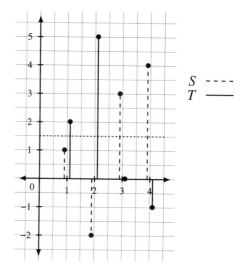

Note how much higher above the mean is the result in T for the single negative value -2 in S.

For a second example, below, let

$$S = \{0.25, -0.25, 0.25, 0.25\}.$$

The mean $m = 0.125$ and $T = \{0, 0.5, 0, 0\}$.

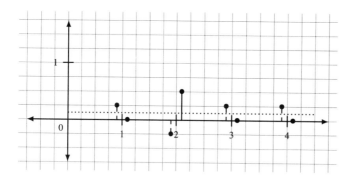

After this transformation, the lone negative value stands out significantly once inverted about the mean.

Consider

$$\mathbf{U}_\varphi = 2|\varphi\rangle\langle\varphi| - I_8$$

for

$$|\varphi\rangle = \sum_{j=0}^{7} \frac{1}{\sqrt{8}}|j\rangle$$

and where I_8 is the 8 by 8 identity matrix. The form should be familiar to you from what we discussed above. This is a unitary matrix that accomplishes an "inversion about the mean" where by "mean" we are talking about the balanced superposition ket $|\varphi\rangle = \mathbf{H}^{\otimes 2}|000\rangle\mathbf{H}^{\otimes 2}$.

Question 9.6.2

Show that the matrix for $\mathbf{U}_\varphi$ is

$$\begin{bmatrix} \frac{2}{8}-1 & \frac{2}{8} & \frac{2}{8} & \frac{2}{8} & \frac{2}{8} & \frac{2}{8} & \frac{2}{8} & \frac{2}{8} \\ \frac{2}{8} & \frac{2}{8}-1 & \frac{2}{8} & \frac{2}{8} & \frac{2}{8} & \frac{2}{8} & \frac{2}{8} & \frac{2}{8} \\ \frac{2}{8} & \frac{2}{8} & \frac{2}{8}-1 & \frac{2}{8} & \frac{2}{8} & \frac{2}{8} & \frac{2}{8} & \frac{2}{8} \\ \frac{2}{8} & \frac{2}{8} & \frac{2}{8} & \frac{2}{8}-1 & \frac{2}{8} & \frac{2}{8} & \frac{2}{8} & \frac{2}{8} \\ \frac{2}{8} & \frac{2}{8} & \frac{2}{8} & \frac{2}{8} & \frac{2}{8}-1 & \frac{2}{8} & \frac{2}{8} & \frac{2}{8} \\ \frac{2}{8} & \frac{2}{8} & \frac{2}{8} & \frac{2}{8} & \frac{2}{8} & \frac{2}{8}-1 & \frac{2}{8} & \frac{2}{8} \\ \frac{2}{8} & \frac{2}{8} & \frac{2}{8} & \frac{2}{8} & \frac{2}{8} & \frac{2}{8} & \frac{2}{8}-1 & \frac{2}{8} \\ \frac{2}{8} & \frac{2}{8} & \frac{2}{8} & \frac{2}{8} & \frac{2}{8} & \frac{2}{8} & \frac{2}{8} & \frac{2}{8}-1 \end{bmatrix}$$

Simplify this and show it is a unitary matrix. Start by computing the first few entries in the outer product matrix $|\varphi\rangle\langle\varphi|$.

Now we can take our amplitudes in the 3-qubit example and invert about the mean.

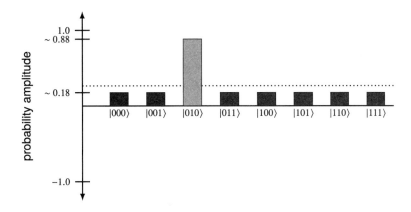

In general, for n qubits this inversion about the mean is accomplished by

$$\mathbf{U}_\varphi = 2|\varphi\rangle\langle\varphi| - I_{2^n}$$

---

where

$$|\varphi\rangle = \mathbf{H}^{\otimes n}|0\rangle^{\otimes n} = \sum_{j=0}^{2^n-1} \frac{1}{\sqrt{2^n}}|j\rangle = \frac{1}{\sqrt{2^n}} \sum_{j=0}^{2^n-1} |j\rangle$$

and I_{2^n} is the 2^n by 2^n identity matrix. Using linearity, we can re-express it using gate notation as

$$\mathbf{U}_\varphi = 2|\varphi\rangle\langle\varphi| - I_{2^n} = \mathbf{H}^{\otimes n}\left(2|0\rangle^{\otimes n}\langle 0|^{\otimes n} - \mathbf{ID}^{\otimes n}\right)\mathbf{H}^{\otimes n}$$

This is called the *Grover diffusion operator*. [11] [3] The 3-qubit circuit is

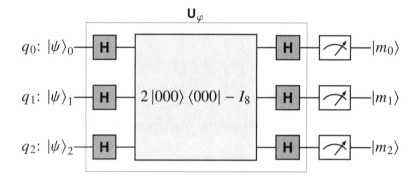

This is meant to be employed as a reusable circuit subroutine. The inputs vary based on where in the full circuit it is used.

Question 9.6.3

Apply $U_{|010\rangle}$ followed by $\mathbf{U}_\varphi$ one more time and draw a graph of the probability amplitudes.

Inversion about the mean is an example of using interference to find the best answer. We manipulate the probability amplitudes to make the "good" result more likely to be seen when we measure while we simultaneously make the "bad" ones highly unlikely. We boost the good one via *constructive interference* and reduce the bad ones by *destructive interference*.

9.7 Searching

We just saw how if we have one standard basis ket in mind, we flip the probability amplitude and then amplify the amplitude for that ket. When we repeat the process enough times, we are likely to measure the right ket with high probability.

In the last section I showed that the ket $|010\rangle$, which I knew about, can be picked out of all the kets. So I found what I knew was there, and I even knew where it was beforehand. Here we put everything together to describe the famous quantum search algorithm discovered by Lov Kumar Grover, a computer scientist.

9.7.1 Grover's search algorithm

Instead of using the magic gate matrix $U_{|010\rangle}$, which flips the sign of the amplitude of the given ket, we instead employ $\mathbf{U}_f$, which is related to the oracle f.

In essence, I have an oracle which I can call but I cannot see. I create $\mathbf{U}_f$ and then by repeating $\mathbf{U}_f \mathbf{U}_\varphi$ enough times, I can find the special element for which f returns 1. How many times is enough?

In subsection 2.8.2 we saw that in the worse case we have to look through every item in an unstructured collection of size N to see if something we want to locate is present. If we add random access and pre-sort, we can locate it in $O(\log(N))$ time via a binary search. Grover search on unstructured data can significantly decrease the time required.

To make it concrete, suppose you say "I'm thinking of a number between 1 and 100" and then you give me an oracle f that can identify the number. I can find the number you are thinking of in approximately $10 = \sqrt{100}$ iterations of $\mathbf{U}_f \mathbf{U}_\varphi$. Classically, it can require 99 calls to the oracle.

Question 9.7.1

Why 99 and not 100?

We go from $O(N)$ to $O\left(\sqrt{N}\right)$ for $N = 100$. This is a quadratic improvement. It doesn't look like much of a gain for small numbers but going from $10000 = 10^4$ to $100 = 10^2$ is significant.

Figure 9.1 shows the circuit for performing the search.

We can be even more precise: it takes approximately $\frac{\pi}{4}\sqrt{N} \approx 0.7854\sqrt{N}$ iterations to maximize the probability of getting the object for which you are searching.

Here are the steps for using the Grover search algorithm:

1. Identify the data you want to search. Let N be the number of items.
2. Find the smallest positive n in $\mathbb{Z}$ such that $N \leq 2^n$. You need n qubits for the search algorithm.

3. Figure out a way to map uniquely from the data items to search to basis kets like $|j\rangle_n$ for $j < N$. This means we have an easy to go from the data object to the basis ket and back again.

4. Construct an oracle f and gate/matrix $\mathbf{U}_f$ that flips the sign of the object for which you are searching. You do not know its basis ket ahead of time or you would otherwise extract it from the database.

5. Run the $\mathbf{U}_f \mathbf{U}_\varphi$ circuit $\sqrt{N}$ times, rounding down.

6. Measure and read off the ket which corresponds to the sought item in the data. Map this back to the item within the data collection.

7. If the answer is not correct, repeat the above. The chance of error is $O\left(\frac{1}{N}\right)$.

Some implementations of the oracle require $n + 1$ qubits.

The oracles we have discussed return 1 for one and only one input but some oracles return 1 for several possibilities. Grover's algorithm can still be used and finds one of the items even faster: if there are t possible matches then the number of iterations is $O\left(\sqrt{\frac{N}{t}}\right)$. More precisely, it is approximately

$$\frac{\pi}{4}\sqrt{\frac{N}{t}}.$$

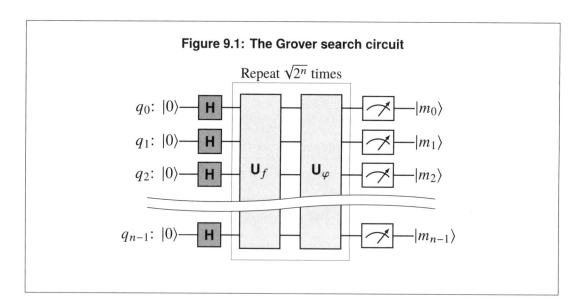

Figure 9.1: The Grover search circuit

9.7.2 Using the oracle

Just as there are different circuits that do the same thing more or less efficiently, there are different ways of coding the Grover search circuit. Here is an explicit search as you might do it in the IBM Q Experience, though the qubits are shown in reverse order there. [18]

Though the oracle identifies one basis ket uniquely, I'm not going to tell you ahead of time which one it is.

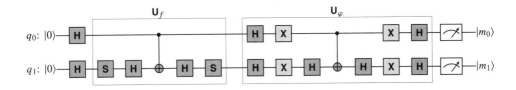

The circuit has depth 13 and width 2.

You really wouldn't go to all this trouble to find one object out of four. Nevertheless, it shows many of the characteristic aspects of larger examples.

Circuit section 1: Balanced superposition

After initializing the two qubits to $|0\rangle$, we place the entire quantum register in a balanced superposition.

The vertical dashed line allows me to describe the steps in the circuit and have no effect on computation. The step is to the right of the labelled dashed line. Within each step are the gate steps numbered from 1 going left to right.

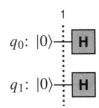

The state of the register after step 1 is

Step	State
1	$\frac{1}{2}(|00\rangle + |01\rangle + |10\rangle + |11\rangle)$

Circuit section 2: $\mathbf{U}_f$

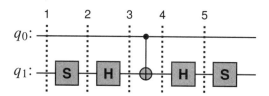

The states of the quantum register after each step are

Step	State
1	$\frac{1}{2}(|00\rangle + i|01\rangle + |10\rangle + i|11\rangle)$
2	$\frac{\sqrt{2}}{4}((1+i)|00\rangle + (1-i)|01\rangle + (1+i)|10\rangle + (1-i)|11\rangle)$
3	$\frac{\sqrt{2}}{4}((1+i)|00\rangle + (1-i)|01\rangle + (1-i)|10\rangle + (1+i)|11\rangle)$
4	$\frac{1}{2}(|00\rangle + i|01\rangle + |10\rangle - i|11\rangle)$
5	$\frac{1}{2}(|00\rangle - |01\rangle + |10\rangle + |11\rangle)$

Circuit section 3: $\mathbf{U}_\varphi$

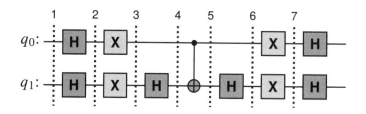

The states of the quantum register after each step are

Step	State
1	$\frac{1}{2}(|00\rangle + |01\rangle - |10\rangle + |11\rangle)$
2	$\frac{1}{2}(|00\rangle - |01\rangle + |10\rangle + |11\rangle)$
3	$\frac{\sqrt{2}}{2}(|01\rangle + |10\rangle)$
4	$\frac{\sqrt{2}}{2}(|01\rangle + |11\rangle)$
5	$\frac{1}{2}(|00\rangle - |01\rangle + |10\rangle - |11\rangle)$
6	$\frac{1}{2}(-|00\rangle + |01\rangle - |10\rangle + |11\rangle)$
7	$-|01\rangle$

Circuit section 4: Measurement

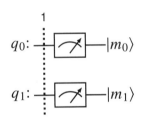

The states of the qubit measurement after the final step is

Step	State
1	$-|01\rangle$

With 100% probability, $|01\rangle$ is the answer.

It's rare to get such a high percentage but it always happens with this algorithm in one iteration for two qubits.

9.7.3 Understanding the oracle

We know that the circuit section

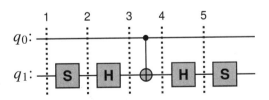

implements the oracle and the sign flip for $|01\rangle$ in the balance superposition. How and why does it work?

Let's work from the inside out and examine what

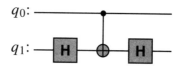

does to the standard basis kets.

When the state of q_0 is $|0\rangle$, the **CNOT** does nothing and q_1 remains the same. The two **H** gates cancel each other. So $|00\rangle$ and $|01\rangle$ are untouched.

When the state of q_0 is $|1\rangle$, the **CNOT** flips between $|0\rangle$ and $|1\rangle$. The first **H** gate takes

$$|0\rangle \mapsto \tfrac{\sqrt{2}}{2}|0\rangle + \tfrac{\sqrt{2}}{2}|1\rangle = |+\rangle \quad \text{and} \quad |1\rangle \mapsto \tfrac{\sqrt{2}}{2}|0\rangle - \tfrac{\sqrt{2}}{2}|1\rangle = |-\rangle.$$

The **CNOT** interchanges $|00\rangle$ and $|01\rangle$, yielding

$$\tfrac{\sqrt{2}}{2}|0\rangle + \tfrac{\sqrt{2}}{2}|1\rangle = |+\rangle \quad \text{and} \quad -\tfrac{\sqrt{2}}{2}|0\rangle + \tfrac{\sqrt{2}}{2}|1\rangle = -|-\rangle.$$

The final **H** undoes the superposition and we get

$$|0\rangle \quad \text{and} \quad -|1\rangle.$$

All together we have

$$
\begin{array}{ll}
|00\rangle \mapsto |00\rangle & |10\rangle \mapsto |10\rangle \\
|01\rangle \mapsto |01\rangle & |11\rangle \mapsto -|11\rangle
\end{array}
$$

> This is a good idiom to remember to flip the sign of $|11\rangle$ only. Had we wanted our oracle to identify $|11\rangle$, this would have been the $\mathbf{U}_f$ subcircuit we would use for $|11\rangle$.

We haven't figured out $|01\rangle$ yet, but we have another $\mathbf{U}_f$ in hand!

What we know so far: if we do something, then flip the sign of $|11\rangle$, then do the something again, we flip the sign of $|01\rangle$.

The **S** gate leaves $|0\rangle$ alone but takes $|1\rangle$ to $i|1\rangle$. Applied a second time it gives us the state $i\,i\,|1\rangle = -|1\rangle$. Two applications are the same as the **Z** gate. Applying the **S** gate on the second qubit produces

$$
\begin{array}{ll}
|00\rangle \mapsto |00\rangle & |10\rangle \mapsto |10\rangle \\
|01\rangle \mapsto i\,|01\rangle & |11\rangle \mapsto i\,|11\rangle
\end{array}
$$

Now we apply the $|11\rangle$ sign flip

$$|00\rangle \mapsto |00\rangle \qquad |10\rangle \mapsto |10\rangle$$
$$|01\rangle \mapsto i\,|01\rangle \qquad |11\rangle \mapsto -i\,|11\rangle$$

and **S** one more time

$$|00\rangle \mapsto |00\rangle \qquad |10\rangle \mapsto |10\rangle$$
$$|01\rangle \mapsto i\,i\,|01\rangle \qquad |11\rangle \mapsto -i\,i\,|11\rangle$$

which is

$$|00\rangle \mapsto |00\rangle \qquad |10\rangle \mapsto |10\rangle$$
$$|01\rangle \mapsto -|01\rangle \qquad |11\rangle \mapsto |11\rangle$$

and it does what we want! Complex numbers are your friend.

Question 9.7.2

Construct the subcircuit for flipping the sign of $|10\rangle$.

Question 9.7.3

Construct the subcircuit for flipping the sign of $|00\rangle$.

You construct circuits for algorithms by learning or devising idioms, implementing them into subcircuits, and then reusing them to complete your task. You should look for symmetry and ask yourself why it is present and what function it performs.

To learn more

Grover's algorithm can be extended in different ways. Suppose we have N items and instead of trying to find one distinguished item among them, our oracle will somehow identify K of them. We do not know what K is *a priori*. Given Grover, we would expect to be able to find K in something like $O\left(\sqrt{\frac{N}{K}}\right)$ queries to the oracle, and this is almost true up to an ϵ that shows up in both how closely we approximate K and within the $O()$ expression.

The original 1998 proof of this involved the Quantum Fourier Transform (QFT), which we work through in section 10.1. [4] A more recent proof in 2019 dispenses with the need for the QFT. [1] Both techniques involve amplitude amplification.

9.7.4 The data problem

A quantum computer is not a big data machine. There is no way to quickly input large amounts of data before the loading operations exhaust the physical qubit coherence times.

In Grover's algorithm, we do not have to pre-load all the data as long as the oracle can correctly identify the object we need.

How do we construct the oracle? If the data is complicated then we might need a very sophisticated oracle subroutine. This is likely to involve many steps and, again, our coherence time might limit what we can compute. If the oracle involved multiplication of numbers with a few dozen digits we could easily require hundreds of qubits and circuit steps.

Grover's algorithm works best as a subroutine for information that is already represented within the states of the qubits or implicitly usable. If another quantum algorithm requires a fast search of data to which it has direct access or a very fast oracle, Grover's algorithm could be the ideal choice.

9.8 The Deutsch-Jozsa algorithm

Before we leave oracles, I want to walk you through one of the other early quantum algorithms that employs them. It also shows us another form in which oracles are expressed in quantum circuits.

Let's begin with an example. Suppose I buy two standard decks of 52 playing cards. In a separate room where you cannot see me, I create a single deck of 52 cards where one of the following is true:

1. All the cards are red or all the cards are black.
2. Half the cards (26) are black and half are red.

The first option is called "constant" and the second is "balanced."

I now go to you and give you the problem of finding out which of the two possibilities is the case for the deck I am holding. You do so by looking at and then discarding the card at the top of the deck.

In the best case, the first card is one color and the second is the other. Therefore the deck is balanced. In the worst case you must examine $27 = 1 + 52/2$ cards. The first 26 cards might be black, say. If the next is black, then all are black. If it is red, the deck is balanced.

Regarding an oracle, we are asking "is the card at the top of the deck black?". It returns 1 if it is, 0 if it is red. As I stated, we must consult the oracle 27 times in the worst case to get the correct answer.

When we first saw the definition of an oracle, it was expressed as a function f operating on strings of n bits

$$f : \{0,1\}^n \rightarrow \{0,1\}.$$

When we translate this to quantum computing, we consider standard basis kets instead of bit strings.

Our new problem is this: for all possible bit strings of length n or standard basis kets for n qubits, the oracle f either always returns 0, always returns 1, or is balanced. How many calls to the oracle do we need to do to determine whether it is constant or balanced?

There are 2^n bit strings of length n and the brute force classical approach could require our looking (that is, calling the oracle) at one more than half of them, which is $2^{n-1} + 1$.

The quantum solution to this, with all the improvements to the original, is called the *Deutsch-Jozsa algorithm*, which we cover in the next several sections. [5]

David Deutsch in 2017 after receiving the Dirac Medal. Photo subject to use via the Creative Commons Attribution 3.0 Unported license ©①

It was discovered by physicist David Deutsch and mathematician Richard Jozsa, and was based on earlier work by Deutsch. [7]

9.8.1 More Hadamard math

When we first saw the single qubit Hadamard gate, I remarked there was a nice relationship between the 0s and 1s and exponents

$$\mathsf{H}|u\rangle = \frac{\sqrt{2}}{2} \left(|0\rangle + (-1)^u |1\rangle \right)$$

when u is one of $\{0,1\}$. With the summation notation, we can make this even more concise.

$$\mathsf{H}|u\rangle = \frac{\sqrt{2}}{2} \sum_{v \text{ in } \{0,1\}} (-1)^{uv} |v\rangle.$$

Please take a moment to verify that this is true and really quite clever. By the way, note how we use the powers of -1 to change the signs of kets. For a physicist, this is a *phase change*. For a regular person, this is a *change of sign*.

The cleverness continues as we look at more qubits. [19] [20]

For bit strings u of length 2, that is, of the form $\{0, 1\}^2$, we have the corresponding standard basis kets $|00\rangle$, $|01\rangle$, $|10\rangle$, and $|00\rangle$. In this case, when we say $|u\rangle$ for u in $\{0, 1\}^2$, we mean these basis kets. I use subscripts to label the individual bits. Therefore

$$|u\rangle = |u_1 u_2\rangle$$

where u_1 and u_2 are 0 or 1. Using this convention,

$$(\mathbf{H} \otimes \mathbf{H})\,|u\rangle = \mathbf{H}|u_1\rangle \otimes \mathbf{H}|u_2\rangle$$

$$= \left(\frac{\sqrt{2}}{2} \sum_{v_1 \text{ in } \{0,1\}} (-1)^{u_1 v_1}|v_1\rangle \right) \otimes \left(\frac{\sqrt{2}}{2} \sum_{v_2 \text{ in } \{0,1\}} (-1)^{u_2 y_2}|v_2\rangle \right)$$

$$= \frac{1}{2} \sum_{v \text{ in } \{0,1\}^2} (-1)^{u_1 v_1 + u_2 v_2}|v\rangle.$$

Question 9.8.1

Confirm

$$(\mathbf{H} \otimes \mathbf{H})\,|00\rangle = \frac{1}{2}(|00\rangle + |01\rangle + |10\rangle + |11\rangle).$$

The exponent $u_1 v_1 + u_2 v_2$ for -1 looks like a dot product. We rewrite the last line above to conclude

$$(\mathbf{H} \otimes \mathbf{H})\,|u\rangle = \frac{1}{2} \sum_{v \text{ in } \{0,1\}^n} (-1)^{u \cdot v}|v\rangle.$$

The final generalization is to n qubits. After calculations like the above, the formula becomes

$$\mathbf{H}^{\otimes n}|u\rangle = \frac{1}{\sqrt{2^n}} \sum_{v \text{ in } \{0,1\}^n} (-1)^{u \cdot v}|v\rangle$$

for u a bit string of length n. This shows how Hadamard gates transform standard kets in a particularly nice way. We can simplify this even more by computing the dot product mod 2. It makes no difference as an exponent for -1.

Question 9.8.2

Using this formula, fully write out the ket expressions for $\mathbf{H}^{\otimes 3}|000\rangle$, $\mathbf{H}^{\otimes 3}|001\rangle$, and $\mathbf{H}^{\otimes 3}|110\rangle$.

Let's compute a special case which we need in the Deutsch-Jozsa algorithm. For u we take a bit string of n 0s with a single 1 tacked on at the end. So $|u\rangle = |0\rangle^{\otimes n} \otimes |1\rangle$. In this case, $u \cdot v = v_{n+1}$ because $u_j = 0$ for $1 \le j \le n$ and $u_{n+1} = 1$.

$$\mathbf{H}^{\otimes n+1}|0\cdots 01\rangle = \frac{1}{\sqrt{2^{n+1}}} \sum_{v \text{ in } \{0,1\}^{n+1}} (-1)^{v_{n+1}}|v\rangle$$

That last bit of v is controlling the sign. Let's isolate that by rewriting $|v\rangle = |x\rangle \otimes |y\rangle$ where y is now the last bit. The formula becomes

$$\mathbf{H}^{\otimes n+1}|0\cdots 01\rangle = \frac{1}{\sqrt{2^{n+1}}} \sum_{x \text{ in } \{0,1\}^n} \sum_{y \text{ in } \{0,1\}} (-1)^{y}|x\rangle \otimes |y\rangle$$

$$= \frac{1}{\sqrt{2^{n+1}}} \sum_{x \text{ in } \{0,1\}^n} \left(|x\rangle \otimes |0\rangle - |x\rangle \otimes |1\rangle\right)$$

$$= \frac{1}{\sqrt{2^{n+1}}} \sum_{x \text{ in } \{0,1\}^n} |x\rangle \otimes \left(|0\rangle - |1\rangle\right)$$

$$= \frac{1}{\sqrt{2^n}} \sum_{x \text{ in } \{0,1\}^n} |x\rangle \otimes \left(\frac{\sqrt{2}}{2}|0\rangle - \frac{\sqrt{2}}{2}|1\rangle\right)$$

$$= \left(\frac{1}{\sqrt{2^n}} \sum_{x \text{ in } \{0,1\}^n} |x\rangle\right) \otimes \left(\frac{\sqrt{2}}{2}|0\rangle - \frac{\sqrt{2}}{2}|1\rangle\right)$$

$$= \mathbf{H}^{\otimes n}|0\rangle^{\otimes n} \otimes \mathbf{H}|1\rangle$$

We already knew the last equality, but I worked through the computation so you could see common arithmetic methods for working with Hadamard gates.

9.8.2 Another oracle circuit

The oracle for Grover search in subsection 9.7.1 was incorporated into the circuit and modified only the input states. We here look at another way of constructing a circuit with an oracle.

There are $n + 1$ wires in the circuit in Figure 9.2. n of those represent the bit strings, and we initialize those to $|0\rangle$. The extra wire, at the bottom of the diagram, is for a "work" or "scratchpad" qubit. It's not part of the input data but we need it to do our computation. More formally, it is called an *ancilla qubit*.

Aside from giving us an extra place to put information, we can also initialize ancilla qubits to known states. Algorithmically, we know the state of the ancilla but we likely don't precisely know the states of the others. In our circuit, q_n starts in state $|0\rangle$ but we use an **X** to flip it to $|1\rangle$.

Is there a gate that turns an arbitrary state directly into either $|0\rangle$ or $|1\rangle$? If there were, such a gate would not be reversible, and so not unitary. With an ancilla qubit, we can set the initial value to a precise value.

Circuit section 1: Superposition

This part of the circuit implements the superposition from subsection 9.8.1.

$$\mathbf{H}^{\otimes n+1}|0\cdots01\rangle = \left(\frac{1}{\sqrt{2^n}} \sum_{x \text{ in } \{0,1\}^n} |x\rangle\right) \otimes \left(\frac{\sqrt{2}}{2}|0\rangle - \frac{\sqrt{2}}{2}|1\rangle\right)$$

$$= \mathbf{H}^{\otimes n}|0\rangle^{\otimes n} \otimes \mathbf{H}|1\rangle$$

Circuit section 2: $\mathbf{U}_f$

In the inputs to $\mathbf{U}_f$, we label the collective states for the n "data" qubits q_0 through q_{n-1} as $|x\rangle$. We label the state for the ancilla qubit q_n as $|y\rangle$. All together, the input to $\mathbf{U}_f$ is $|x\rangle \otimes |y\rangle$.

The output is $|x\rangle \otimes |y \oplus f(x)\rangle$. Remember, $f(x)$ always returns 0 or 1, and "$\oplus$" is addition mod 2.

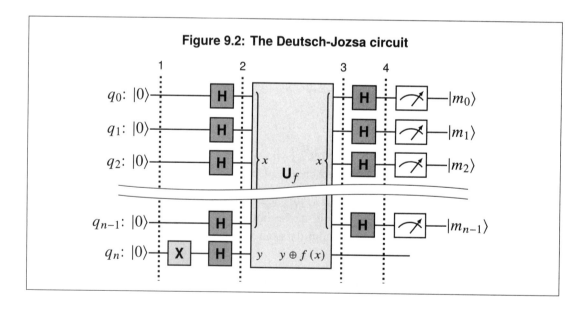

Figure 9.2: The Deutsch-Jozsa circuit

Question 9.8.3

Show that the transformation

$$|x\rangle \otimes |y\rangle \mapsto |x\rangle \otimes |y \oplus f(x)\rangle$$

is invertible and unitary.

Applying $\mathbf{U}_f$ to

$$\left(\frac{1}{\sqrt{2^n}} \sum_{x \text{ in } \{0,1\}^n} |x\rangle \right) \otimes \left(\frac{\sqrt{2}}{2}|0\rangle - \frac{\sqrt{2}}{2}|1\rangle \right)$$

yields

$$\left(\frac{1}{\sqrt{2^n}} \sum_{x \text{ in } \{0,1\}^n} |x\rangle \right) \otimes \left(\frac{\sqrt{2}}{2}|0 \oplus f(x)\rangle - \frac{\sqrt{2}}{2}|1 \oplus f(x)\rangle \right)$$

by linearity, which is

$$\left(\frac{1}{\sqrt{2^n}} \sum_{x \text{ in } \{0,1\}^n} |x\rangle \right) \otimes \left(\frac{\sqrt{2}}{2}|f(x)\rangle - \frac{\sqrt{2}}{2}|1 \oplus f(x)\rangle \right).$$

If $f(x) = 0$ then the right-hand side is

$$\frac{\sqrt{2}}{2}|f(x)\rangle - \frac{\sqrt{2}}{2}|1 \oplus f(x)\rangle = \frac{\sqrt{2}}{2}|0\rangle - \frac{\sqrt{2}}{2}|1\rangle.$$

When $f(x) = 1$, the only other possibility, we have

$$\frac{\sqrt{2}}{2}|1\rangle - \frac{\sqrt{2}}{2}|0\rangle.$$

Note how the how the signs reverse. We can combine both these equations into

$$\frac{\sqrt{2}}{2}|f(x)\rangle - \frac{\sqrt{2}}{2}|1 \oplus f(x)\rangle = (-1)^{f(x)}(|0\rangle - |1\rangle).$$

The full result of $\mathbf{U}_f$ is

$$\left(\frac{1}{\sqrt{2^n}} \sum_{x \text{ in } \{0,1\}^n} |x\rangle \right) \otimes \left(\frac{\sqrt{2}}{2} |0 \oplus f(x)\rangle - \frac{\sqrt{2}}{2} |1 \oplus f(x)\rangle \right)$$

$$= \left(\frac{1}{\sqrt{2^n}} \sum_{x \text{ in } \{0,1\}^n} |x\rangle \right) \otimes \left((-1)^{f(x)} \frac{\sqrt{2}}{2} (|0\rangle - |1\rangle) \right)$$

$$= \left(\frac{1}{\sqrt{2^n}} \sum_{x \text{ in } \{0,1\}^n} (-1)^{f(x)} |x\rangle \right) \otimes \left(\frac{\sqrt{2}}{2} (|0\rangle - |1\rangle) \right)$$

The expression to the right of "$\otimes$" is constant and we no longer need it. Remember that this is the state of the ancilla qubit q_n. I've indicated that we can discard it by not measuring it in the circuit diagram.

The output of the $\mathbf{U}_f$ gate for the n data qubits q_0 to q_{n-1} is

$$\frac{1}{\sqrt{2^n}} \sum_{x \text{ in } \{0,1\}^n} (-1)^{f(x)} |x\rangle.$$

We have encoded the effect of the oracle into the phase of each $|x\rangle$. We've done this by multiplying $|x\rangle$ by a value of absolute value 1, namely $(-1)^{f(x)}$. This is called *phase kickback*.

Circuit section 3: Final H gates

We apply the $\mathbf{H}^{\otimes n}$ gates to the above expression using the formula

$$\mathbf{H}^{\otimes n} |u\rangle = \frac{1}{\sqrt{2^n}} \sum_{v \text{ in } \{0,1\}^n} (-1)^{u \cdot v} |v\rangle$$

from subsection 9.8.1. This produces

$$\frac{1}{2^n} \sum_{x \text{ in } \{0,1\}^n} (-1)^{f(x)} \left(\sum_{v \text{ in } \{0,1\}^n} (-1)^{x \cdot v} |v\rangle \right)$$

By rearranging terms, we can move the inner sum to the outside:

$$\frac{1}{2^n} \sum_{v \text{ in } \{0,1\}^n} \left(\sum_{x \text{ in } \{0,1\}^n} (-1)^{f(x)} (-1)^{x \cdot v} \right) |v\rangle.$$

Before we turn to measurement, let me say that I know this section is very intensive with all the arithmetic with the Σ summations. We compress a lot of information about the kets and their amplitudes into formulas and then manipulate and simplify them. It is worth your time to work through each until you are comfortable with what is happening in each.

Circuit section 4: Measurement

Given

$$\frac{1}{2^n} \sum_{v \text{ in } \{0,1\}^n} \left(\sum_{x \text{ in } \{0,1\}^n} (-1)^{f(x)}(-1)^{x \cdot v} \right) |v\rangle,$$

if f is constant then all the $f(x)$ are the same and equal to $f(0)$. We can rewrite the above as

$$\frac{1}{2^n}(-1)^{f(0)} \sum_{v \text{ in } \{0,1\}^n} \left(\sum_{x \text{ in } \{0,1\}^n} (-1)^{x \cdot v} \right) |v\rangle.$$

The inner sum

$$\sum_{x \text{ in } \{0,1\}^n} (-1)^{x \cdot v}$$

is a very special form. It is equal to 2^n if v is the zero bit string and is 0 otherwise.

If v is the zero bit string, the sum is 2^n copies of $(-1)^0 = 1$ added together, which is 2^n. If v is not all zeroes, half the $(-1)^{x \cdot v}$ are 1 and the other half -1. They cancel each out when added together. [17, Section 7.1]

Question 9.8.4

Confirm this for $n = 2$.

This means

$$\frac{1}{2^n}(-1)^{f(0)} \sum_{v \text{ in } \{0,1\}^n} \left(\sum_{x \text{ in } \{0,1\}^n} (-1)^{x \cdot v} \right) |v\rangle = \frac{1}{2^n}(-1)^{f(0)} 2^n |0\rangle^{\otimes n}$$

$$= (-1)^{f(0)} |0\rangle^{\otimes n}.$$

The amplitude $(-1)^{f(0)}$ has absolute value 1 and so has no effect on the measurement.

If f is constant then the result upon measurement is $|0\rangle^{\otimes n}$ with 100% probability.

We can glean more by looking at

$$\frac{1}{2^n} \sum_{v \text{ in } \{0,1\}^n} \left(\sum_{x \text{ in } \{0,1\}^n} (-1)^{f(x)} (-1)^{x \cdot v} \right) |v\rangle.$$

When $|v\rangle = |0\rangle^{\otimes n}$, this expression reduces to

$$\frac{1}{2^n} \left(\sum_{x \text{ in } \{0,1\}^n} (-1)^{f(x)} \right) |0\rangle^{\otimes n}.$$

This is the amplitude for $|0\rangle^{\otimes n}$. It is equal to 1 when f is constant and 0 when f is balanced. The first case is constructive interference and the second is destructive interference.

When we run the Deutsch-Jozsa algorithm, if we get $|0\rangle^{\otimes n}$ after measurement, then the oracle is constant. Otherwise it is balanced.

9.9 Simon's algorithm

There is one more algorithm using an oracle that we now cover before we leave this chapter and that is *Simon's algorithm*. It may seem like an odd use of an oracle but the techniques are further used in section 10.5 when we develop function period finding before tackling Shor's factoring algorithm.

9.9.1 The problem

Our problem is to find how the values of a function on whole numbers repeat themselves. Recall that by $\{0, 1\}^n$ we mean a string or sequence of n 0s and 1s. For example, if $n = 2$ then

$$\{0, 1\}^2 = \{00, 01, 10, 11\}.$$

A function

$$f : \{0, 1\}^n \to \{0, 1\}^n$$

takes such binary strings of length n to other such binary strings. For any two x and y in $\{0, 1\}^n$, when does $f(x) = f(y)$?

In general, we don't know, but we could require that there exists some r in $\{0, 1\}^n$ such that $f(x) = f(y)$ if and only if $x \oplus y$ is in $\{0^n, r\}$.

Here we have "$\oplus$" being the binary **xor** function, which is addition modulo 2. (We covered modular arithmetic in section 3.7.)

By 0^n I mean the string of length n containing only 0s:

$$0^n = \underbrace{000\ldots00}_{n}.$$

What does it mean if $x \oplus y = 0^n$? For single bits b_1 and b_2, $b_1 \oplus b_2 = 0$ if and only if b_1 and b_2 are both 0 or both 1. That is, they are equal. So $x \oplus y = 0^n$ means that x and y are the same in each of the n bit positions, and so $x = y$. It's then obvious that $f(x) = f(y)$.

For this case where $x \oplus y = 0^n$, the condition on f says $f(x) = f(y)$ only if $x = y$. f is *one-to-one*, meaning each element in its domain maps to a unique element in its range.

Question 9.9.1

If we further label f as

$$f : A = \{0, 1\}^n \to B = \{0, 1\}^n,$$

show that for every binary n-string z in B there is a binary n-string x in A such that $f(x) = z$.

The other possibility is that there is a non-zero bit string r of length so that $x \oplus y = r$ **for every** x and y if $f(x) = f(y)$. Moreover, if this is the case then $f(x) = f(y)$.

What is r?

Question 9.9.2

Let $f : \{0, 1\}^3 \to \{0, 1\}^3$ be defined by

$$
\begin{array}{llll}
000 \mapsto 101 & 001 \mapsto 010 & 010 \mapsto 000 & 011 \mapsto 110 \\
100 \mapsto 000 & 101 \mapsto 110 & 110 \mapsto 101 & 111 \mapsto 010
\end{array}
$$

What is r?

In subsection 2.8.1 I introduced the "big O" notation. The number of interesting operations used to solve a problem on n objects is $O(f(n))$ if there is a positive real number c and an integer m such that

$$\text{number of operations} \leq cf(n)$$

once $n \geq m$, for some function f.

This means that the problem is **at most as hard** as $cf(n)$.

The number of interesting operations used to solve a problem on n objects is $\Omega(f(n))$ if there is a positive real number c and an integer m such that

$$\text{number of operations} \geq cf(n)$$

once $n \geq m$, for some function f.

This means that the problem is **at least as hard** as $cf(n)$.

If we were to proceed classically by brute force, we would have to call f many times to ensure we find the correct r. How many? Well, there are 2^n possible r values to consider. Even with refinements and optimizations, the problem is $\Omega\left(\sqrt{2^n}\right)$, which is exponentially hard.

In case this value $\sqrt{2^n}$ seems odd to you, note that it is the common probability amplitude after applying an **H** gate to n qubits initialized to $|0\rangle$.

With Simon's quantum algorithm, we can find r with $O(n)$ calls to the oracle f.

Question 9.9.3

Continuing as before, for

$$f : A = \{0, 1\}^n \rightarrow B = \{0, 1\}^n,$$

with r non-zero, show that f is *two-to-one*. That is, if $f(x) = z$ with z in B, then there is one and only one other $y \neq x$ such that we also have $f(y) = z$. Both x and y are in A.

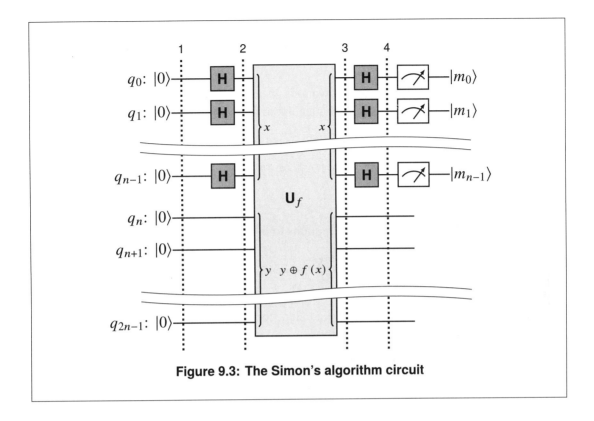

Figure 9.3: The Simon's algorithm circuit

9.9.2 The circuit

From this point on, our task is to find r and we will assume $r \neq 0^n$. Note that since r has n bits, we need at least n qubits to represent this answer. For example, if we are working with integers with 2048 bits, we need at least 2048 qubits.

In Figure 9.3 I show the circuit for Simon's algorithm.

This looks familiar. We apply **H** gates to several qubits, do something amazing and mysterious with $\mathbf{U}_f$, apply more **H** gates, and then interpret what is output.

We need and use the n qubits at the top because $f : \{0,1\}^n \to \{0,1\}^n$. We also use n ancilla qubits at the bottom. All told, we need $2n$ qubits and we may need to iterate the algorithm multiple times, as we work through below.

In the Deutsch-Jozsa circuit in Figure 9.2,

$$\mathbf{U}_f : |x\rangle \otimes |y\rangle \mapsto |x\rangle \otimes |y \oplus f(x)\rangle$$

where $|x\rangle$ is the quantum state of the top n qubits but $|y\rangle$ is the state of a single ancilla qubit. The same expression holds here as well, except that $|x\rangle$ and $|y\rangle$ each encompass the quantum state of n qubits.

By computations similar to what we worked through in subsection 9.8.2 for the Deutsch-Jozsa algorithm, the states of the qubits change in the following way in the circuit.

As we begin circuit section 1, the full state of the qubits is

$$|0\rangle^{\otimes n} \otimes |0\rangle^{\otimes n}.$$

This is the initial state.

We then apply the **H** gates to the first n qubits to have

$$\frac{1}{\sqrt{2^n}} \sum_{x \text{ in } \{0,1\}^n} |x\rangle \otimes |0\rangle^{\otimes n}$$

at the end of this section. That is, we have a balanced superposition of the first n qubits but we are still in the initial state for the last n qubits.

Next, apply $\mathbf{U}_f$ in circuit section 2. Since

$$|x\rangle \otimes |y\rangle \mapsto |x\rangle \otimes |y \oplus f(x)\rangle$$

under this transformation, we now have

$$|x\rangle \otimes |f(x)\rangle$$

because $|y\rangle = |0\rangle^{\otimes n}$.

We conclude the circuit by applying **H** gates again to the first n qubits in section 3. Given the identities for the Hadamard transformation on multiple qubits from subsection 9.8.1, we end up with

$$\frac{1}{2^n} \sum_{x \text{ in } \{0,1\}^n} \sum_{u \text{ in } \{0,1\}^n} (-1)^{x \cdot u} |u\rangle \otimes |f(x)\rangle$$

which is the same as

$$\sum_{u \text{ in } \{0,1\}^n} |u\rangle \otimes \left(\frac{1}{2^n} \sum_{x \text{ in } \{0,1\}^n} (-1)^{x \cdot u} |f(x)\rangle \right).$$

As before, "$\cdot$" is a dot product where we do the addition modulo 2.

9.9.3 Analysis of the circuit results

The above represents a full quantum state and so the sum of the squares of the absolute values of the coefficients of the $|u\rangle \otimes |f(x)\rangle$ is 1:

$$\sum_{u \text{ in } \{0,1\}^n} \sum_{x \text{ in } \{0,1\}^n} \left| \frac{1}{2^n} (-1)^{x \cdot u} \right|^2 = 1.$$

In the expression

$$\sum_{u \text{ in } \{0,1\}^n} |u\rangle \otimes \left(\frac{1}{2^n} \sum_{x \text{ in } \{0,1\}^n} (-1)^{x \cdot u} |f(x)\rangle \right),$$

the probability of observing a particular $|u\rangle$ is the square of the magnitude of the coefficient. That is, we will see $|u\rangle$ with probability

$$\left\| \frac{1}{2^n} \sum_{x \text{ in } \{0,1\}^n} (-1)^{x \cdot u} |f(x)\rangle \right\|^2.$$

The case $r = 0$

In the case of $r = 0$, f is one-to-one. Because there are a finite number of values, when we iterate over all the x values we are also iterating over all the $f(x)$, just in a different order. Taking the sum of the squares of the coefficients for all the $|x\rangle$ to get the square of the length yields the same value as the sum of the squares of the coefficients when we use the $|x\rangle$.

Put another way, two vectors that differ only by a permutation of their coefficients have the same length.

For each $|f(x)\rangle$ we square each $\dfrac{(-1)^{x \cdot u}}{2^n}$ to get $\dfrac{1}{2^{2n}}$ and then add up for all 2^n of the $|f(x)\rangle$. Hence

$$\left\| \frac{1}{2^n} \sum_{x \text{ in } \{0,1\}^n} (-1)^{x \cdot u} |f(x)\rangle \right\|^2 = 2^n \frac{1}{2^{2n}} = \frac{1}{2^n}.$$

With this, we conclude that when $r = 0$ the probability of seeing any of the 2^n values of $|u\rangle$ when we measure is 2^{-n}. If we run the circuit many times and see this uniform distribution of the $|u\rangle$, we might suspect $r = 0$.

Though it is trivial in this case, note that because $r = 0$, this uniform distribution of kets we observe is equal to all $|u\rangle$ such that $u \cdot r = 0$.

The case $r \neq 0$

Now we explicitly assume that $r \neq 0$ and therefore f is two-to-one. While x can take on any of the 2^n values in $\{0, 1\}^n$, $f(x)$ will only land on half of them. I call this subset $\{0, 1\}_f^n$. It is the range of f.

For a particular z in $\{0, 1\}_f^n$, there are exactly two binary n-strings that map to it: some x_z and its companion $x_z \oplus r$.

The probability to observe a particular $|u\rangle$ when we measure is still

$$\left\| \frac{1}{2^n} \sum_{x \text{ in } \{0,1\}^n} (-1)^{x \cdot u} |f(x)\rangle \right\|^2,$$

but we can rewrite this as

$$\left\| \frac{1}{2^n} \sum_{z \text{ in } \{0,1\}_f^n} \left((-1)^{x_z \cdot u} + (-1)^{(x_z \oplus r) \cdot u} \right) |z\rangle \right\|^2.$$

Via some arithmetic on the exponents, this is further equal to

$$\left\| \frac{1}{2^n} \sum_{z \text{ in } \{0,1\}_f^n} (-1)^{x_z \cdot u} \left(1 + (-1)^{r \cdot u} \right) |z\rangle \right\|^2 = \begin{cases} 0 & \text{if } r \cdot u = 1 \\ \dfrac{1}{2^{n-1}} & \text{if } r \cdot u = 0. \end{cases}$$

Question 9.9.4

Show that $(x_z \oplus r) \cdot u = (x_z \cdot u) \oplus (r \cdot u)$.

Less trivially than in the case $r = 0$, the uniform distribution of kets we observe here is on all $|u\rangle$ such that $u \cdot r = 0$.

But what is r ?

Since this is a quantum algorithm, randomness is involved. When we run the circuit several times we should see various values when we measure $|u\rangle$. Suppose we execute it $n - 2$ more times and we get $n - 1$ binary strings $u_1, \ldots, u_{n-1}$. Now assume the u_k are different and linearly independent vectors. This is a big assumption, but let's go with it for a while.

This means that

$$u_1 \cdot r = 0$$
$$u_2 \cdot r = 0$$
$$\vdots$$
$$u_{n-1} \cdot r = 0$$

or

$$\left(u_{1,1}r_1 + u_{1,2}r_2 + \cdots + u_{1,n}r_n\right) \bmod 2 = 0$$
$$\left(u_{2,1}r_1 + u_{2,2}r_2 + \cdots + u_{2,n}r_n\right) \bmod 2 = 0$$
$$\vdots$$
$$\left(u_{n-1,1}r_1 + u_{n-1,2}r_2 + \cdots + u_{n-1,n}r_n\right) \bmod 2 = 0$$

Since the u_k are linearly independent, we can uniquely solve for $(r_1, \ldots, r_n) = r$ via a binary form of Gaussian elimination.

Now we test to see if $f(0^n) = f(0^n \oplus r) = f(r)$. If this is true, then we are assured that r is our answer. If not, then we must have $r = 0^n$.

Question 9.9.5

Why must $r = 0^n$ in this last case?

The only remaining question is whether we can find the n linearly independent u_k. Obviously, they exist among the 2^n binary n-strings but if we must search all those strings our algorithm will be exponential.

Via the analysis in [12, Appendix G], the chance of success for finding a linearly independent set of n binary n-strings within a collection of $n + k$ non-zero binary n-strings that are produced by our circuit is

$$1 - \frac{1}{2^{k+1}}.$$

We can find such a set, and hence determine r, with a chance of failure less than one in a billion by choosing $k = 30$. This means we only need to run the circuit $n + 30$ times to have such a small probability of not correctly determining r.

Question 9.9.6

What parts of the overall algorithm involve quantum processing and what parts involve classical?

The expression $f(x) = f(x \oplus r)$ looks a lot like $f(x) = f(x + r)$, which looks like f starts repeating itself after r values. This would make f a periodic function and r would be its period.

We are familiar with periodic functions on the real numbers from trigonometry. Both the sine and the cosine are such because $\cos(x) = \cos(x + r)$ and $\sin(x) = \sin(x + r)$ with $r = 2\pi$.

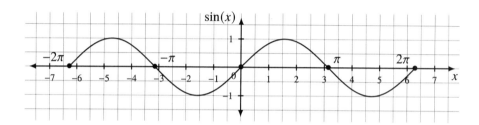

Simon's algorithm was the inspiration for Peter Shor when he developed a more general quantum period finding routine as part of his famous integer factorization algorithm. We cover the former in section 10.5 and the latter in section 10.6.

9.10 Summary

In this chapter we examined how to link gates together for multiple qubits to create circuits. Circuits implement algorithms and these are the building blocks for solutions. After all, we're not only interested in the theory of how one might do quantum computing, we want to accomplish real work.

We looked at some well known basic algorithms for quantum computing, including Simon's, Deutsch-Jozsa, amplitude amplification, and Grover's search.

Quantum computing will show its advantage when it can perform calculations that are intractable today. To be really valuable, quadratic or exponential speed increases over classical methods will be required.

References

[1]   Scott Aaronson and Patrick Rall. *Quantum Approximate Counting, Simplified.* 2019. URL: https://arxiv.org/abs/1908.10846.

[2]   Charles H. Bennett et al. "Teleporting an unknown quantum state via dual classical and Einstein-Podolsky-Rosen channels". In: *Physical Review Letters* 70 (1993), pp. 1895–1899.

[3]   Gilles Brassard and Peter F. Høyer. "An Exact Quantum Polynomial-Time Algorithm for Simon's Problem". In: ISTCS '97 (1997).

[4]   Gilles Brassard et al. "Quantum Amplitude Amplification and Estimation". In: *Quantum Computation and Information.* Ed. by S. J. Lomonaco and H. E. Brandt. Contemporary Mathematics Series. AMS, 2000.

[5]   Creative Commons. *Attribution 3.0 Unported (CC BY 3.0)* ⓒ①. URL: https://creativecommons.org/licenses/by/3.0/legalcode.

[6]   R. Crandall and C. Pomerance. *Prime Numbers: a Computational Approach.* 2nd ed. Springer, 2005.

[7]   D Deutsch and R Jozsa. "Rapid Solution of Problems by Quantum Computation". English. In: *Proceedings of the Royal Society A: Mathematical, Physical and Engineering Sciences* 439 (1992), pp. 553–558.

[8]   Craig Gidney. *Asymptotically Efficient Quantum Karatsuba Multiplication.* 2019. URL: https://arxiv.org/abs/1904.07356.

[9]   Craig Gidney. "Halving the cost of quantum addition". In: *Quantum* 2 (June 2018), p. 74.

[10]   Phil Gossett. *Quantum Carry-Save Arithmetic.* 1998. URL: https://arxiv.org/abs/quant-ph/9808061.

[11]   Lov K. Grover. "A Fast Quantum Mechanical Algorithm for Database Search". In: *Proceedings of the Twenty-Eighth Annual ACM Symposium on the Theory of Computing, Philadelphia, Pennsylvania, USA, May 22-24, 1996.* 1996, pp. 212–219.

[12]   N. David Mermin. *Quantum Computer Science: An Introduction.* Cambridge University Press, 2007.

[13]   Ashok Muthukrishnan. *Classical and Quantum Logic Gates: An Introduction to Quantum Computing.* URL: http://www2.optics.rochester.edu/users/stroud/presentations/muthukrishnan991/LogicGates.pdf.

[14]   Michael A. Nielsen and Isaac L. Chuang. *Quantum Computation and Quantum Information.* 10th ed. Cambridge University Press, 2011.

[15]   A.O. Pittenger. *An Introduction to Quantum Computing Algorithms.* Progress in Computer Science and Applied Logic. Birkhäuser Boston, 2012.

[16]   Qiskit.org. *Qiskit: An Open-source Framework for Quantum Computing.* URL: https://qiskit.org/documentation/.

[17] Eleanor Rieffel and Wolfgang Polak. *Quantum Computing: A Gentle Introduction*. 1st ed. The MIT Press, 2011.

[18] The IBM Q Experience team. *Grover's Algorithm*. URL: https://quantumexperience.ng.bluemix.net/proxy/tutorial/full-user-guide/004-Quantum_Algorithms/070-Grover's_Algorithm.html.

[19] John Watrous. *CPSC 519/619: Introduction to Quantum Computing*. URL: https://cs.uwaterloo.ca/~watrous/LectureNotes/CPSC519.Winter2006/all.pdf.

[20] John Watrous. *The Theory of Quantum Information*. 1st ed. Cambridge University Press, 2018.

10

From Circuits to Algorithms

I am among those who think that science has great beauty.

Marie Curie

In the last chapter we became comfortable with putting together gates to create circuits for simple algorithms. We're now ready to look at more advanced quantum algorithms and considerations on how and when to use them.

Our target in this chapter is Peter Shor's 1995 algorithm for factoring large integers almost exponentially faster than classical methods. To get there we need more tools, such as phase estimation, the Quantum Fourier Transform, and function period finding. These are important techniques in their own rights, but are necessary in combination for quantum factoring.

We also return to the idea of complexity that we first saw for classical algorithms in section 2.8. This allows us to understand what "almost exponentially faster" means.

This chapter contains more mathematics and equations for quantum computing than we have encountered previously. I recommend that you take the time to understand the linear algebra and complex number computations throughout. While they may appear daunting at first, the techniques are used frequently in quantum computing algorithms.

Topics covered in this chapter

10.1 Quantum Fourier Transform

The Quantum Fourier Transform (QFT) is widely used in quantum computing, notably in Shor's factorization algorithm in section 10.6. If that weren't enough, the Hadamard **H** is the 1-qubit QFT and we've seen many examples of its use.

Most treatments of the QFT start by comparing it to the classical Discrete Fourier Transform and then the Fast Fourier Transform. If you don't know either of these, don't worry. I'm presenting the QFT in detail for its own sake in quantum computing. Should you know about or read up about the classical analogs, the similarities should be clear.

10.1.1 Roots of unity

We are all familiar with square roots. For example, $\sqrt{4}$ is equal to either 2 or -2. We can also write $\sqrt{2} = 2^{\frac{1}{2}}$ and say that there are two "2nd-roots of 2." Similarly, 5 is a cube root, or

"3rd-root," of 125. In general, we talk about an "Nth-root" for some natural number N. When we consider the complex numbers, we have a rich collection of such Nth-roots for 1.

Let N be in $\mathbb{N}$. An Nth *root of unity* is a complex number ω such that $\omega^N = 1$. "ω" is the lowercase Greek letter "omega." There are N Nth roots of unity and 1 is always one of those.

If every other Nth root of unity can be expressed as a natural number power of ω, then ω is a *primitive Nth root of unity*. If N is prime then every Nth root of unity is primitive except 1.

For $N = 1$, there is only one first root of unity, and that is 1 itself.

When $N = 2$, there are two second roots of unity: 1 and -1. -1 is primitive.

With $N = 3$, we can start to see a pattern. For this, remember Euler's formula

$$e^{\varphi i} = \cos(\varphi) + \sin(\varphi) i$$

and $\left| e^{\varphi i} \right| = 1$. If $\varphi = 2\pi$, we go all the way around the unit circle and back to 1. If $\varphi = \frac{2\pi}{3}$ then we go only one-third of the way. Rotating another $\frac{2\pi}{3}$ radians we get to two-thirds around. The third roots of unity are

$$\omega_0 = e^{\frac{0 \times 2\pi}{3} i} = 1$$

$$\omega_1 = e^{\frac{1 \times 2\pi}{3} i} = \cos\left(\frac{2\pi}{3}\right) + \sin\left(\frac{2\pi}{3}\right) i = -\frac{1}{2} + \frac{\sqrt{3}}{2} i$$

$$\omega_2 = e^{\frac{2 \times 2\pi}{3} i} = \cos\left(\frac{4\pi}{3}\right) + \sin\left(\frac{4\pi}{3}\right) i = -\frac{1}{2} - \frac{\sqrt{3}}{2} i$$

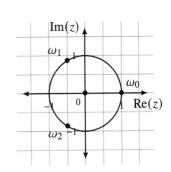

and ω_1 and ω_2 are primitive.

For $N = 4$ we can do something similar, though the situation is much simpler.

$$\omega_0 = e^{\frac{0 \times 2\pi}{4} i} = 1$$

$$\omega_1 = e^{\frac{1 \times 2\pi}{4} i} = \cos\left(\frac{\pi}{2}\right) + \sin\left(\frac{\pi}{2}\right) i = i$$

$$\omega_2 = e^{\frac{2 \times 2\pi}{4} i} = \cos\left(\pi\right) + \sin\left(\pi\right) i = -1$$

$$\omega_3 = e^{\frac{3 \times 2\pi}{4} i} = \cos\left(\frac{3\pi}{2}\right) + \sin\left(\frac{3\pi}{2}\right) i = -i$$

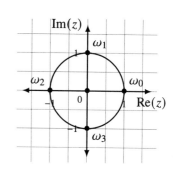

Question 10.1.1

What are the primitive fourth roots of unity?

Notice something else here: for the first time, we see overlaps in the collection of roots of unity for different N. This happens when the greatest common divisor of the different N is greater than 1.

For $N = 5$ and above we continue in this way.

$\omega = e^{\frac{2\pi i}{N}}$ is a primitive Nth root of unity.

When there is a fraction in the exponent of such expressions, it is common to pull the i into the numerator, as I have done here.

For some N as we have considered, let ω be an Nth root of unity. What is ω^{-1}? This one is easy because

$$1 = \omega \overline{\omega} = \omega \omega^{-1}.$$

Complex conjugation to the rescue again!

If $\omega_k = e^{\frac{2\pi k i}{N}}$ is a root of unity then

$$\overline{\omega_k} = \omega_k^{-1} = e^{-\frac{2\pi k i}{N}}.$$

Question 10.1.2

Show that if ω_k is a primitive Nth root of unity, so is $\overline{\omega_k}$. Start by assuming that $\overline{\omega_k}$ is not primitive and see if that leads to a contradiction.

The polynomial $x^N - 1$ can be factored as

$$x^N - 1 = (x - 1)\left(x^{N-1} + x^{N-2} + \cdots + x + 1\right).$$

Every Nth root of unity satisfies $x^N - 1 = 0$. The first factor $(x - 1)$ is 0 only for $\omega = 1$. Therefore, every other Nth root of unity $\omega \neq 1$ satisfies

$$0 = \omega^{N-1} + \omega^{N-2} + \cdots + \omega + 1.$$

We can re-express this as

$$0 = \sum_{k=0}^{N-1} \omega^k \text{ with } \omega \neq 1.$$

Taking it a step further, let $\omega = e^{\frac{2\pi i}{N}}$. Then

$$0 = \sum_{k=0}^{N-1} e^{\frac{2\pi k i}{N}} = \sum_{k=0}^{N-1} e^{-\frac{2\pi k i}{N}}$$

with the second equality holding because we are using $\overline{\omega}$ instead of ω as the Nth root of unity. Thought of another way, we are adding together the same set of roots of unity, just in a different order. In these equivalent versions, this is called the *summation formula*.

For j and k in $\mathbb{Z}$, the *Kronecker delta function* $\delta_{j,k}$ is defined by

$$\delta_{j,k} = \begin{cases} 0 & \text{if } j \neq k \\ 1 & \text{if } j = k. \end{cases}$$

For $0 \leq j < N$, the *extended summation formula* is

$$\sum_{k=0}^{N-1} e^{\frac{2\pi j k i}{N}} = \sum_{k=0}^{N-1} e^{-\frac{2\pi j k i}{N}} = N\delta_{j,0} = \begin{cases} 0 & \text{if } \delta_{j,0} = 0 \\ N & \text{if } \delta_{j,0} = 1. \end{cases}$$

To see this, let

$$\omega_1 = e^{\frac{2\pi i}{N}}.$$

This is a primitive Nth root of unity and for any $0 < j < N$,

$$\omega_2 = \omega_1^j = e^{\frac{2\pi j i}{N}}$$

is another Nth root of unity $\neq 1$. From the summation formula,

$$0 = \sum_{k=0}^{N-1} \omega_2^k = \sum_{k=0}^{N-1} e^{\frac{2\pi j k i}{N}}$$

When $j = 0$, we are simply adding 1 to itself N times.

Question 10.1.3

For a given N, let $\{\omega_0, \omega_1, \ldots, \omega_{N-2}, \omega_{N-1}\}$ be the Nth roots of unity. What is their product

$$\omega_0 \, \omega_1 \, \cdots \, \omega_{N-1}?$$

To start thinking about this, look at

$$(x - \omega_0)\,(x - \omega_1) \cdots (x - \omega_{N-1}).$$

To learn more

Roots of unity are an important concept and tool in several parts of mathematics, especially algebra [16] [8] and algebraic number theory [12] [19].

10.1.2 The formula

As we have seen, a general quantum state $|\varphi\rangle$ on n qubits can be written

$$|\varphi\rangle = \sum_{j=0}^{2^n-1} a_j |j\rangle_n = \sum_{j=0}^{N-1} a_j |j\rangle_n$$

for $N = 2^n$. There are N amplitudes a_j corresponding to the N standard basis kets $|j\rangle$. For a fixed $|\varphi\rangle$, we get a complex-valued function

$$a : \{0, 1, 2, \ldots, N - 1\} \to a_j$$

where $a(j) = a_j$. We also know $1 = \sum_{j=0}^{N-1} |a_j|^2$.

Definition

The quantum Fourier Transform of $|\varphi\rangle$ is

$$\mathbf{QFT}_n : |\varphi\rangle = \sum_{j=0}^{N-1} a_j |j\rangle_n \rightarrow \sum_{j=0}^{N-1} b_j |j\rangle_n$$

where

$$b_j = \frac{1}{\sqrt{N}} \sum_{k=0}^{N-1} a_k e^{\frac{2\pi jki}{N}}.$$

We can simplify this by letting $\omega = e^{\frac{2\pi i}{N}}$, which is a primitive Nth root of unity.

$$b_j = \frac{1}{\sqrt{N}} \sum_{k=0}^{N-1} a_k \omega^{jk}$$

For $N = 2^n$ and

$$|\varphi\rangle = \sum_{j=0}^{N-1} a_j |j\rangle_n,$$

its Quantum Fourier Transform is

$$\mathbf{QFT}_n\left(|\varphi\rangle\right) = \frac{1}{\sqrt{N}} \sum_{j=0}^{N-1} \sum_{k=0}^{N-1} a_k \omega^{jk} |j\rangle_n$$

for $\omega = e^{\frac{2\pi i}{N}}$, a primitive Nth root of unity.

Question 10.1.4

Show that $\mathbf{QFT}_n$ is a linear transformation.

Question 10.1.5

Show that

$$\mathbf{QFT}_n\left(|0\rangle^{\otimes n}\right) = \frac{1}{\sqrt{N}} \sum_{j=0}^{N-1} |j\rangle_n.$$

When we have 1 qubit, $N = 2$ and $\omega = -1$. So

$$b_j = \frac{1}{\sqrt{2}}\left(a_0(-1)^{-0j} + a_1(-1)^{1j}\right) = \frac{\sqrt{2}}{2}\left(a_0 + a_1(-1)^j\right)$$

For $|\psi\rangle = |0\rangle$, $a_0 = 1$ and $a_1 = 0$. So $b_0 = \frac{\sqrt{2}}{2}$ and $b_1 = \frac{\sqrt{2}}{2}$. For $|\psi\rangle = |1\rangle$, $a_0 = 0$ and $a_1 = 1$. So $b_0 = \frac{\sqrt{2}}{2}$ and $b_1 = -\frac{\sqrt{2}}{2}$.

All together,

$$\textbf{QFT}_1|0\rangle = \frac{\sqrt{2}}{2}\left(|0\rangle + |1\rangle\right) = |+\rangle$$
$$\textbf{QFT}_1|1\rangle = \frac{\sqrt{2}}{2}\left(|0\rangle - |1\rangle\right) = |-\rangle$$

This is none other than **H**!

So $\textbf{QFT}_1 = \textbf{H}$. Is $\textbf{QFT}_n = \textbf{H}^{\otimes n}$?

No.

Matrix

From its definition, you can see that $\textbf{QFT}_n$ has matrix

$$\textbf{QFT}_n = \frac{1}{\sqrt{N}}\begin{bmatrix} 1 & 1 & 1 & 1 & \cdots & 1 \\ 1 & \omega & \omega^2 & \omega^3 & \cdots & \omega^{N-1} \\ 1 & \omega^2 & \omega^4 & \omega^6 & \cdots & \omega^{2(N-1)} \\ 1 & \omega^3 & \omega^6 & \omega^9 & \cdots & \omega^{3(N-1)} \\ \vdots & \vdots & \vdots & \vdots & \ddots & \vdots \\ 1 & \omega^{N-1} & \omega^{2(N-1)} & \omega^{3(N-1)} & \cdots & \omega^{(N-1)(N-1)} \end{bmatrix}.$$

Since ω is an Nth root of unity, we can simplify several of the exponents. For example,

- $\omega^{(N-1)(N-1)} = \omega^{N^2-2N+1} = \omega$ because $\omega^{N^2} = \left(\omega^N\right)^N = 1^N = 1$.
- $\omega^{kN} = \omega^N = 1$ for k in $\mathbb{Z}$.
- $\omega^{2(N-1)} = \omega^{N-2}$.
- $\omega^{-1} = \omega^{N-1}$.

Applying rules such as these, we get

$$
\mathbf{QFT}_n = \frac{1}{\sqrt{N}}
\begin{bmatrix}
1 & 1 & 1 & 1 & \cdots & 1 \\
1 & \omega & \omega^2 & \omega^3 & \cdots & \omega^{N-1} \\
1 & \omega^2 & \omega^4 & \omega^6 & \cdots & \omega^{N-2} \\
1 & \omega^3 & \omega^6 & \omega^9 & \cdots & \omega^{N-3} \\
\vdots & \vdots & \vdots & \vdots & \ddots & \vdots \\
1 & \omega^{N-1} & \omega^{N-2} & \omega^{N-3} & \cdots & \omega
\end{bmatrix}.
$$

Question 10.1.6

Show that this is the matrix of $\mathbf{QFT}_n$ and $\mathbf{QFT}_n$ is unitary. For the latter, what is $\mathbf{QFT}_n \times \mathbf{QFT}_n^{\dagger}$?

For $n = 1$, this is

$$
\mathbf{QFT}_1 = \frac{\sqrt{2}}{2}
\begin{bmatrix}
1 & 1 \\
1 & (-1)^1
\end{bmatrix}
= \frac{\sqrt{2}}{2}
\begin{bmatrix}
1 & 1 \\
1 & -1
\end{bmatrix}
= \mathbf{H}
$$

as previously noted. But when $n = 2$, $N = 4$ and $\omega = i$,

$$
\mathbf{QFT}_2 = \frac{1}{2}
\begin{bmatrix}
1 & 1 & 1 & 1 \\
1 & i & -1 & -i \\
1 & -1 & 1 & -1 \\
1 & -i & -1 & i
\end{bmatrix}
\quad \text{and} \quad
\mathbf{H}^{\otimes 2} = \frac{1}{2}
\begin{bmatrix}
1 & 1 & 1 & 1 \\
1 & -1 & 1 & -1 \\
1 & 1 & -1 & -1 \\
1 & -1 & -1 & 1
\end{bmatrix}.
$$

So $\mathbf{QFT}_n \neq \mathbf{H}^{\otimes n}$, in general.

Recursive matrix

The higher Hadamard matrices $\mathbf{H}^{\otimes n}$ are defined recursively in terms of the lower ones by

$$
\mathbf{H}^{\otimes n+1} = \frac{\sqrt{2}}{2}
\begin{bmatrix}
\mathbf{H}^{\otimes n} & \mathbf{H}^{\otimes n} \\
\mathbf{H}^{\otimes n} & -\mathbf{H}^{\otimes n}
\end{bmatrix}.
$$

Is there a similar decomposition for $\mathbf{QFT}_n$?

Yes, but there are other matrix factors. Namely,

$$
\mathbf{QFT}_{n+1} =
\begin{bmatrix}
I_N & \Omega_N \\
I_N & -\Omega_N
\end{bmatrix}
\begin{bmatrix}
\mathbf{QFT}_n & 0 \\
0 & \mathbf{QFT}_n
\end{bmatrix}
P_{2^{n+1}}
$$

where

- $N = 2^n$,
- I_N is the N by N identity matrix,
- ω is the primitive Nth root of unity $e^{\frac{2\pi i}{N}}$,
- Ω_N is the diagonal matrix

$$
\Omega_N = \begin{bmatrix}
1 & 0 & 0 & 0 & \cdots & 0 \\
0 & \omega & 0 & 0 & \cdots & 0 \\
0 & 0 & \omega^2 & 0 & \cdots & 0 \\
\vdots & \vdots & \vdots & \vdots & \ddots & \vdots \\
0 & 0 & 0 & 0 & \cdots & \omega^{N-1}
\end{bmatrix},
$$

  and
- $P_{2^{n+1}}$ is a *shuffle transform* defined by setting its j, k entry via the formula

$$
(P_{2^{n+1}})_{j,k} = \begin{cases}
1 & \text{if } 2(j-1) = k-1 \\
1 & \text{if } 2(j-1-2^n)+1 = k-1 \\
0 & \text{otherwise.}
\end{cases}
$$

Remember that the first row and column index in the matrix is 1.

We do not derive this here but its effect is to move the vector entries with odd numbered indices to the front, followed by the even ones. For example,

$$
P_{2^2} \begin{bmatrix} v_1 \\ v_2 \\ v_3 \\ v_4 \end{bmatrix} = \begin{bmatrix} v_1 \\ v_3 \\ v_2 \\ v_4 \end{bmatrix}
$$

where

$$
P_{2^2} = \begin{bmatrix}
\mathbf{1} & 0 & 0 & 0 \\
0 & 0 & \mathbf{1} & 0 \\
0 & \mathbf{1} & 0 & 0 \\
0 & 0 & 0 & \mathbf{1}
\end{bmatrix}.
$$

Question 10.1.7

Compute P_{2^3}.

Given this recursive decomposition, we can break a QFT down into smaller and smaller gates. This is seen in the relatively simple structure of the circuit.

10.1.3 The circuit

The **QFT**$_3$ circuit can be constructed entirely from **H**, **S**, **T**, and **SWAP** gates. [20, Section 5.1]

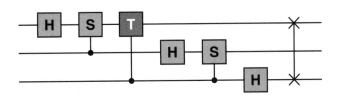

The last **SWAP** is necessary to reverse the order of the qubit quantum states. We can rewrite this with explicit **R**$_\varphi^z$ gates.

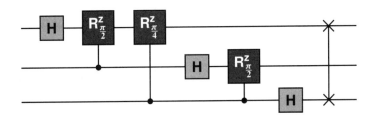

An **R**$_{\frac{\pi}{2}}^z$ = **S** gate is a rotation given by the matrix

$$\begin{bmatrix} 1 & 0 \\ 1 & e^{\frac{\pi i}{2}} \end{bmatrix} = \begin{bmatrix} 1 & 0 \\ 1 & e^{\frac{2\pi i}{2^2}} \end{bmatrix}.$$

An **R**$_{\frac{\pi}{4}}^z$ = **T** gate is a rotation given by the matrix

$$\begin{bmatrix} 1 & 0 \\ 1 & e^{\frac{\pi i}{4}} \end{bmatrix} = \begin{bmatrix} 1 & 0 \\ 1 & e^{\frac{2\pi i}{2^3}} \end{bmatrix}.$$

Define **ROT**$_k$ = **R**$_{\frac{2\pi}{2^k}}^z$. So **R**$_{\frac{\pi}{2}}^z$ = **ROT**$_2$ and **R**$_{\frac{\pi}{4}}^z$ = **ROT**$_3$.

Rewriting the above circuit in terms of **ROT**$_k$ gives

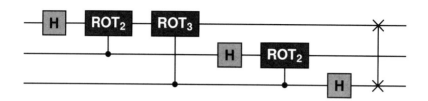

The pattern becomes clear when we extend this to 4 qubits and **QFT**$_4$.

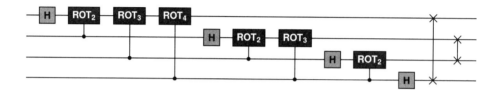

Question 10.1.8

Draw the circuit for 1 and 2 qubits. Using lots of "$\cdots$", can you sketch the circuit for **QFT**$_n$?

In this section we have discussed the Quantum Fourier Transform **QFT**$_n$ on n qubits. In some circuits we need the inverse Quantum Fourier Transform **QFT**$_n^{-1}$. We get this by running **QFT**$_n$ *backwards*. This is possible because all gates ultimately correspond to unitary operations, which are invertible and hence reversible.

Question 10.1.9

What is the defining matrix for **QFT**$_n^{-1}$?

For $N = 2^n$ and

$$|\varphi\rangle = \sum_{j=0}^{N-1} a_j |j\rangle_n,$$

its inverse Quantum Fourier Transform is

$$\text{QFT}_n^{-1}\left(|\varphi\rangle\right) = \frac{1}{\sqrt{N}} \sum_{j=0}^{N-1} \sum_{k=0}^{N-1} a_k \overline{\omega}^{jk} |j\rangle_n = \frac{1}{\sqrt{N}} \sum_{j=0}^{N-1} \sum_{k=0}^{N-1} a_k e^{-\frac{2\pi i j k}{N}} |j\rangle_n$$

for $\omega = e^{\frac{2\pi i}{N}}$, a primitive Nth root of unity.

To learn more

There is extensive literature about classical Fourier Transforms, particularly in engineering and signal processing texts. [4, Chapter 30] [10]

Quantum Fourier Transforms are used in many optimized quantum algorithms for arithmetic operations. [7] [25]

10.2 Factoring

It is hard to do a web search and find "practical" applications of factoring integers. Many of the results say things like "factoring integers is a tool in for factoring polynomials," and "factoring integers is useful for solving some differential equations." These examples are fine, but it seems like math to do more math.

One area where factoring comes into play is cryptography. There are cryptographic protocols that involve *not* being able to easily factor large numbers, in order to keep your information safe while it travels across the Internet or is stored in databases.

In this section, we examine some of the mathematics of factoring and focus in on one famous algorithm. This is Shor's algorithm, named after its discoverer, mathematician Peter Shor. It can factor large integers almost exponentially faster than classical methods *given a sufficiently large and powerful quantum computer.*

If one does not understand what "sufficiently large and powerful" means, it is easy to make poor predictions of whether, or when, quantum computing can cause problems for encryption.

10.2.1 The factoring problem

Most people don't worry about factoring integers once they graduate from high school or are no longer teenagers. In this book, we first encountered factoring when we discussed integers. In

$$-60 = (-1)\,2^2\,3\,5$$

the −1 is a unit and 2, 3, and 5 are examples of prime numbers. They are irreducible in the sense that their only factors are 1 and themselves.

As of this writing, the largest confirmed prime number has 24,862,048 digits. [21] Determining whether a number is prime or factoring that number are different but related problems. The part of mathematics that studies these topics, as well as many others, is called number theory.

While it can be hard to factor large integers, it is not simply the size that makes it difficult. The number

$$10^{25000000} = 2^{25000000} 5^{25000000}$$

is bigger than the above prime but is trivially factorable.

When we factor an N in $\mathbb{Z}$ greater than 1 we express it as a product

$$N = p_1 p_2 \cdots p_{n-1} p_n$$

for some integer $n \geq 1$ with all the p_j prime numbers. We sort the primes from smallest to largest, which means $p_j \leq p_k$ if $j < k$. Some primes may be repeated more than once. If, for example, 11 appears in the factorization five times then we say 11 has multiplicity 5. The notation we use to show that a p_j is a factor of N is $p_j | N$. Read this as "p_j divides N."

If $n = 1$ then N itself is prime. While the goal is to find all the p_j factors, we really want to start by finding one p_j and then continue by factoring the smaller integer N/p_j. As we explore factoring, we focus on that first prime factor and then re-use our methods, basic or advanced, to do what's left.

10.2.2 Big integers

On a 64 bit classical processor, the largest natural number you can represent is

$$2^{64} - 1 = 18,446,744,073,709,551,615$$

For positive and negative integers, one bit is reserved for the sign and so the largest number we can represent is

$$2^{63} - 1 = 9,223,372,036,854,775,807$$

These may look big, but from a factoring perspective they are small. We know all the primes that go well above these values.

What would you do if you wanted to factor 18,446,744,073,709,551,617, which is $(2^{64} - 1) + 2$? The processor cannot hold the number directly in its hardware.

We fix this by creating "bignum" or "big integer" objects in software. These can be as large as computer memory permits. Essentially, we divide up the big integer across multiple hardware integers and then write software routines for negation, addition, subtraction, multiplication, division, and (related) quotients and remainders. Of these, the division operations are the trickiest to code correctly, though it is well known how to do it.

Scientific applications often support big integers. These include Maple, Mathematica, MATLAB®, and cryptographic software. Python, Erlang, Haskell, Julia, Ruby, and many variants of Lisp and Scheme are programming languages which provide built-in big integers. Other languages provide the functionality via library extensions.

The GNU Multiple Precision Arithmetic Library implements highly optimized low-level routines for arbitrary precision integers and other number types. It is distributed under the dual GNU LGPL v3 and GNU GPL v2 licenses. [9]

Let's do a quick review of classical ways of factoring integers and then move on to Shor's approach.

10.2.3 Classical factoring: basic methods

Classical factoring techniques can be broken down into basic attempts to break a number N down into prime factors that require arithmetic, and more advanced versions that require sophisticated mathematics. We go into detail on the former and survey the latter.

If $N = 1$ or $N = -1$, we are done. If N is negative, we note the unit is -1 and assume that N is positive.

There is one and only one even prime number and that is 2. If N is even, meaning that the last digit is one of 0, 2, 4, 6, or 8, then $p_1 = 2$. Continue with $N = N/2$. If this N is even, then $p_2 = 2$, and reset $N = N/2$. Keep going until N is not prime. If N ever becomes 1 in the factoring processes, we are done. At this point N is odd and we may have some initial p_j factors equal to 2. To see if 3 is a factor we use a trick you may have learned while young.

Write N as a sequence of digits

$$d_t d_{t-1} \ldots d_2 d_1 d_0$$

where each d_j is one of 0, 1, 2, 3, 4, 5, 6, 7, 8, or 9. For example, if $N = 475$ then $d_2 = 4$, $d_1 = 7$, and $d_0 = 5$.

Then N is divisible by 3 if and only if $\sum_{j=0}^{t} d_j$ is.

We can tell if 3 is a factor of N if 3 divides the much smaller integer that is the sum of the digits. To prove this, let N be as above. Then

$$N = \sum_{j=0}^{t} d_j 10^j.$$

Convince yourself of this. If $N \equiv 0 \bmod 3$ then N is divisible by 3.

By modular arithmetic, which we discussed in section 3.7,

$$N \bmod 3 \equiv \left(\sum_{j=0}^{t} d_j 10^j \right) \bmod 3$$

$$\equiv \sum_{j=0}^{t} (d_j 10^j \bmod 3) \equiv \sum_{j=0}^{t} (d_j \bmod 3)(10^j \bmod 3)$$

$$\equiv \sum_{j=0}^{t} (d_j \bmod 3) \equiv \left(\sum_{j=0}^{t} d_j \right) \bmod 3$$

because every $10^j \equiv 1 \bmod 3$!

Question 10.2.1

Clearly $10^0 \equiv 1 \bmod 3$. Also, $10^1 \equiv 10 \equiv 3^2 + 1 \equiv 1 \bmod 3$. Since $a\,b \bmod 3 \equiv (a \bmod 3)(b \bmod 3)$, show that $10^2 \equiv 1 \bmod 3$. How would you demonstrate that $10^j \equiv 1 \bmod 3$?

There are no special tricks for 4 because we already factored out any 2s. $5|N$ if the last digit is 0 or 5. Because 6 is 2×3, there is nothing to do.

For the prime 7 we can do trial division: divide N by 7 and if there is no remainder, 7 is a factor. There is nothing to do for 8, 9, or 10. For the prime 11 it is trial division again.

What I have described here is one mode of attack: start with a list of sorted primes and see if N is divisible by each in turn. You don't need to look at all primes $< N$ or even $\leq N/2$, you need to look at those $\leq \sqrt{N}$.

Question 10.2.2

If the prime $p|N$, $p \neq N$, and $p \geq \sqrt{N}$, why is there another prime $q \leq \sqrt{N}$ where $q|N$?

This is not a practical method of finding factors once N gets larger because division is computationally expensive and you have to do many of them. It's not difficult to find what is called the *integer square root* of N, the largest integer s such that $s^2 \leq N$. This is $\lfloor \sqrt{N} \rfloor$. If $s^2 = N$ then N is a square.

How would we go about creating a list of primes to use in trial division? A classical way (I really mean "classical" – it's from ancient Greece) is the *Sieve of Eratosthenes*. Let's see how this works to find all primes less than or equal to 30. We "sift" through the numbers and eliminate the composite values.

We begin by creating a list of all numbers in $\mathbb{N}$ less than or equal to 30. We shade all the boxes and selectively remove the shading for the non-primes. We "mark" a box by removing its shading. When we are done, only the primes are in shaded boxes. Since 1 is not a prime, we mark it.

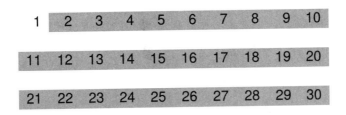

The first number that is not marked is 2. It must be prime. Now mark each of $2+2$, $2+2+2$, $2+2+2+2$ and so on. The numbers we are marking are all multiples of 2 and so are not prime. We get:

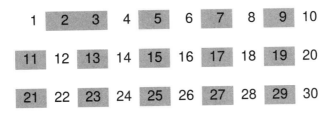

After 2, the next number that is not marked is 3. It must be prime. We mark $3+3$, $3+3+3$, etc.

Question 10.2.3

Why would it be fine to start marking at 3^2 and then $3^2 + 3$ instead of $3+3$ and then $3+3+3$?

It is efficient to simply add 3 to the last number we marked. We can use multiplication, but that's more computation than we need.

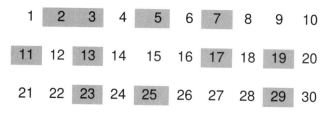

After continuing in this manner, we end up with

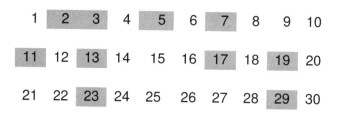

The integers that remain are prime. This is not a terrible way to create a list of a few hundred or thousand primes, but you would not want to do this to find primes with millions of digits.

A simple Python implementation of the sieve is in Listing 10.1. The output for the example above and a larger set of primes is

```
print(list_of_primes(30))

[2, 3, 5, 7, 11, 13, 17, 19, 23, 29]

print(list_of_primes(300))

[2, 3, 5, 7, 11, 13, 17, 19, 23, 29, 31, 37, 41, 43, 47, 53, 59, 61, 67,
71, 73, 79, 83, 89, 97, 101, 103, 107, 109, 113, 127, 131, 137, 139,
149, 151, 157, 163, 167, 173, 179, 181, 191, 193, 197, 199, 211, 223,
227, 229, 233, 239, 241, 251, 257, 263, 269, 271, 277, 281, 283, 293]
```

For a given number N, there are number theoretic algorithms such as the Miller-Rabin test if N is prime, or at least prime with high probability.

A further technique we can try is testing whether N is a power of another integer. Using an efficient algorithm called Newton's method, we can determine whether N is an mth power. The most efficient method to do this test does so in $O(\log(N))$ time. [1]

Which m do we need to try?

If $2^m > N$, then any other positive integer raised to the mth power is also too big. Therefore we need $2^m \leq N$. By taking base 2 logarithms and noting that both sides are greater than one, we need to look at all m in $\mathbb{Z}$ such that

$$m \leq \log_2(N).$$

For $N = 10^{50}$, for example, a maximum of $m = 167$ suffices.

Another basic method is due to mathematician Pierre de Fermat. We can always represent N as the difference of two squares $N = u^2 - v^2$. The importance of this is that we then have

$$N = u^2 - v^2 = (u + v)(u - v) \text{ where } u \neq \pm v.$$

If we can find choices of u and v so that neither $u + v$ nor $u - v$ are 1, then we have a factorization.

Pierre de Fermat, 1601-1665. Image is in the public domain. ⓟ

Another way of saying the above is

$$u^2 \equiv v^2 \bmod N \text{ where } u \neq \pm v.$$

Since N is odd, $N - 1$ and $N + 1$ are both even and

$$N = \left(\frac{N + 1}{2}\right)^2 - \left(\frac{N - 1}{2}\right)^2$$

so one possible pair of choices is $u = \frac{N+1}{2}$ and $v = \frac{N-1}{2}$.

Suppose $N = 87$. Then setting

$$u = \frac{87+1}{2} = 44 \quad \text{and} \quad v = \frac{87-1}{2} = 43$$

yields $u + v = 87$ and $u - v = 1$. This choice of u and v does not help us. These particular choices for finding a pair never works because

$$u - v = \frac{N + 1}{2} - \frac{N - 1}{2} = \frac{N + 1 - N + 1}{2} = 1.$$

```
#!/usr/bin/env python3

def list_of_primes(n):
    # Return a list of primes <= n using the Sieve of Eratosthenes

    # Prepare the list of numbers containing [0,1,...,n]
    numbers = list(range(n+1))

    # Mark the first two numbers by setting them to 0
    numbers[0] = numbers[1] = 0

    # The first prime is 2
    p = 2

    # Cycle through the numbers, marking non-primes
    while p < n:
        if p:
            index = p + p
            while index <= n:
                numbers[index] = 0
                index += p
        p += 1

    # Return the primes left in the list
    return [i for i in numbers if i != 0]

print(list_of_primes(30))
print(list_of_primes(300))
```

Listing 10.1: Sample Python 3 code demonstrating the Sieve of Eratosthenes

In any case, since 87 is prime we cannot find a good pair which works.

So is there a better alternative or approach?

Start by computing the integer square root s of N. If $s^2 = N$ we have found a factor. Otherwise, set $u = s + 1$ and consider $u^2 - N$. If this is a square, then it is v^2. If it is not, increment u by 1 and try again. Since $N = (u + v)(u - v)$, we can stop when $u + v \geq n$. (In fact, we can stop a little sooner, but this shows there is point after which we need not continue.)

Let's try $N = 143$. $s = 11$ and since $s^2 \neq 144$, set $u = s + 1 = 12$. $u^2 - N = 144 - 143 = 1$ and so set $v = 1$. v is a perfect square and so

$$N = (u + v)(u - v) = 13 \times 11.$$

This works well when a $u + v$ and $u - v$ are both close to $\sqrt{N}$. When they are not, we need to do many iterations and computations.

Question 10.2.4

Use Fermat's method to factor 3493157. Use it again to factor 13205947.

10.2.4 Classical factoring: advanced methods

In Fermat's method, we tried to find u and v such that

$$u^2 \equiv v^2 \bmod N \text{ where } u \neq \pm v.$$

This is called a congruence of square modulo N. If we had such a congruence, we could create the factorization

$$(u + v)(u - v) \equiv 0 \bmod N.$$

This does not say $(u + v)(u - v) = N$ but rather that there is a non-zero integer c such that $(u + v)(u - v) = c\,N$.

Given this, let $g = \gcd(u + v, N)$. Then g and N/g is a possible factorization of N. If either of these is 1 then we have failed and we have to find new candidates for u and v.

The key question is: how efficiently can we find u and v that are likely to work?

In Fermat's method we work outwards from the integer square root of N, hoping to find a factor near it. Let's briefly look at more advanced number theoretic approaches.

To learn more

The details are beyond the scope of this book but are well described elsewhere. [11] [15] [5]

Let B be in $\mathbb{Z}$ and ≥ 2. A positive integer is called *B-smooth* if all its prime factors are $\leq B$. B is an upper bound on the size of the primes.

A *factor base* is a set P of prime numbers. Typically, given a bound B as above, P is the set of all primes $\leq B$. If we simply say an integer n is smooth, we mean it is n-smooth, so it means all its prime factors are in P.

The following is known as *Dixon's method* after its discoverer, mathematician John D. Dixon. For a given N to factor, choose B and hence P. Let n be the number of primes in P. We want to find $n + 1$ distinct positive integers f_j such that j^2 mod N is B-smooth. Said another way,

$$f_j^2 \bmod N = \prod_{k=1}^{n} p_k{}^{e_{j,k}}$$

for non-negative integer exponents $e_{j,k}$.

How do we find these f_j? We choose them randomly between 1+ the integer square root of N and N. Suppose $N = 17$. The the integer square root of N is 4. In Python, the code

```
#import random

random.randint(5, 17)
```

returns a random integer greater than or equal to 5 and less than 17.

Now that we have our collection of f_j, look at the product

$$\prod_{j=1}^{n+1} f_j^2 \equiv \prod_{k=1}^{n} p_k^{e_{j,1}+\cdots+e_{j,n}} \bmod N$$

By linear algebraic methods on the matrix

$$\begin{bmatrix} e_{1,1} & e_{1,2} & \cdots & e_{1,n} \\ e_{2,1} & e_{2,2} & \cdots & e_{2,n} \\ \vdots & \vdots & \ddots & \vdots \\ e_{n,1} & e_{n,2} & \cdots & e_{n,n} \end{bmatrix}$$

we can find replacement values for the exponents so that the congruence still holds and $e_{j,1} + \cdots + e_{j,n}$ is **even**. This means the right-hand side of

$$\prod_{j=1}^{n+1} f_j^2 \equiv \prod_{k=1}^{n} p_k^{e_{j,1}+\cdots+e_{j,n}} \bmod N$$

is a square, as is the left. This equation has the form

$$u^2 \equiv v^2 \bmod N \text{ where } u \neq \pm v.$$

and we can proceed to test if we have a factor. If we don't we can

- look for different f_j, or
- increase B and keep trying.

Full treatments of the algorithm put bounds on the value of B. Can we find better congruences of squares, faster? The f_j we use are smooth numbers. Can we sift through all the possibilities better so that we need not consider as many bad candidates?

To learn more

Beyond Dixon's algorithm, there are more powerful and complicated number theoretic algorithms that generalize the sieving techniques to find smooth numbers much faster. The *quadratic sieve* and the *general number field sieve* are examples, with the latter being the most efficient factoring algorithm for very large integers. [2] [11, Chapter 3]

10.3 How hard can that be, again

In section 2.8 we first saw and used the $O(\ )$ notation for sorting and searching. Bubble sort runs in $O(n^2)$ time and merge sort is $O(n \log(n))$. A brute force search is $O(n)$ but adding sorting and random access allows a binary search to be $O(\log(n))$. All algorithms are of polynomial time because we can bound them, in this case, by $O(n^2)$. More precisely, there is a hierarchy of time complexities. For the examples above,

Algorithm	Complexity	Name
Bubble sort	$O(n^2)$	quadratic time
Merge sort	$O(n \log(n))$	quasilinear time
Brute force search	$O(n)$	linear time
Binary search	$O(\log(n))$	logarithmic time

Polynomial time is higher than all of them but we can say each runs in at least polynomial time.

We are concerned with this because there is a special distinction between polynomial time and exponential time. The latter describes exponential growth and can cause problems to become quickly intractable. All these descriptions apply to large n. Even though an exponential running time sounds bad, the algorithm may still be feasible when n is small.

Something running in exponential time has its execution time proportional to $2^{f(n)}$. Here $f(n)$ is a polynomial in n such as $n^2 - n + 1$. Double exponential time is even worse: the time is proportional to $2^{2^{f(n)}}$.

Subexponential is an improvement over exponential. It means for every $b > 1$ in $\mathbb{R}$, the running time is less than b^n. Set $b = 2$ to compare with exponential time. [13]

For example, an algorithm that runs in $O\left(2^{n^{\frac{1}{4}}}\right)$ is subexponential.

There are many subtleties concerning classical and quantum complexity, and a full discussion involves traditional and quantum Turing machines. As our need is only to understand where quantum computing can be better than classical computing, I don't delve into a full and formal description of complexity theory here.

Fermat's method of factoring integers is of exponential time. The general number field sieve is subexponential. There are no known polynomial time classical algorithms for factoring.

When you learned to add and multiply, you learned *deterministic* algorithms for each. Given the same input numbers, you executed a series of steps that always led to the same answer. No randomness was involved. If you encounter random numbers as you continue through this book, chances are you are looking at a non-deterministic algorithm.

A non-deterministic algorithm may involve probability and choices that could ultimately produce different answers. Even if the answers are the same, the choices made inside the algorithm may cause the algorithm to run quite quickly or excruciatingly slowly. More formally, a bad choice can cause an exponential time run while a good one can enable polynomial time execution.

The complexity class **P** refers to the collection of problems that can be solved in polynomial time using a deterministic algorithm.

The class **NP** is the collection of problems whose solutions can be checked in polynomial time. We do not know if **P** = **NP** but you could become wealthy via industry prizes if you could prove it or its contrary. Checking a solution by asking a question like "Is $p \times q$ equal to N?" is an example of a *decision problem*.

Consider integer factorization: we do not know if there is a classical way of factoring a composite integer in polynomial time but, once done, we can easily multiply together the factors to see if the solution is correct. Otherwise said, we do not know if classical factorization is in **P** but we know it is in **NP**.

If **P** ≠ **NP** we can represent the set of problems as shown on the right. Not all **NP** problems are equally difficult. They are defined by their ability to check solutions in polynomial time.

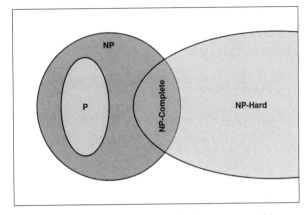

Two new terms are shown: **NP-Hard** and **NP-Complete**. The first is easier to explain: the problems in **NP-Hard** are at least as hard as the most difficult ones in **NP**.

There is a valuable notion of reducing one problem to another. If A and B are two problems, we can "reduce A to B in polynomial time" if by a series of manipulations we can express A as an instance of B.

As an analogy, suppose one problem (A) is "travel to the Empire State building in New York City." For problem B, we express it as "drive to any location in New York City." B in general is hard because of the need to get to any location and the difficulty in driving to it. A can be reduced to B because we need to get to one location somehow.

A problem is in **NP-Hard** if any **NP** problem can be reduced to it. As you can see from the diagram, the class of **NP-Hard** problems intersects **NP** but is larger. The intersection is the **NP-Complete** class, the hardest problems in **NP** to which all others can be reduced. An **NP-Hard** problem that is not in **NP** cannot be solved or verified in polynomial time.

There are many, many (many, many, ...) complexity classes that you may find interesting enough to learn more about beyond what we have discussed. **BPP** is the class of decision problems that can execute in polynomial time but with a caveat. **BPP** stands for "bounded-error probabilistic polynomial time." Since probability is involved, an algorithm that solves this problem may not always return the correct answer. The "bounded" part of the name is made concrete by insisting the correct answer is gotten at least two-thirds of the time.

The class **P** is in **BPP** but it is not known if all problems in **NP** are in **BPP**.

The above classifications refer to problems solved on a classical computer. For quantum computers, we consider the **BQP** class, those solvable in bounded-error polynomial time on a quantum computer. **BQP** is short for "bounded-error quantum polynomial time." Loosely speaking, **BQP** corresponds to quantum computers as **BPP** does for classical computers.

However, a stronger relationship exists. Any problem in **BPP** is also in **BQP**. This follows from our being able to perform classical operations on a quantum computer, though it might be extremely slow. We examined this in section 9.3 and subsection 9.3.1 when we saw that **nand** is universal for classical computing and we can create a **nand** on a quantum computer via a quantum Toffoli gate.

> **To learn more**
>
> A strong grounding in classical algorithms and complexity theory is necessary for you to become proficient in quantum algorithms. [4] [6] [26]

10.4   Phase estimation

Let U be an N by N square matrix with complex entries. From section 5.9, the solutions λ of the equation

$$\det(U - \lambda I_N) = 0$$

are the *eigenvalues* $\{\lambda_1, \lambda_2, \ldots, \lambda_N\}$ of U. Some of the λ_j may be equal. If a particular eigenvalue λ shows up k times among the N, we say λ has *multiplicity k*.

Each eigenvalue λ_j corresponds to an *eigenvector* $\mathbf{v}_j$ so that

$$U\mathbf{v}_j = \lambda_j \mathbf{v}_j$$

can take each $\mathbf{v}_j$ to be a unit vector.

When U is unitary, we can say even more: each λ_j has absolute value 1 and so can be represented as

$$\lambda_j = e^{2\pi \varphi_j i} \text{ where } 0 \leq \varphi_j < 1.$$

This is slightly different notation than we've used before, but it is common. It is the product $2\pi\varphi$ that is the full radian measure of the rotation.

Now we are ready to pose the question whose solution we outline in this section:

> Let **U** be a quantum transformation/gate on n qubits with corresponding N by N square matrix U, with $N = 2^n$. If $|\psi\rangle$ is an eigenvector of **U** (and of U, by abuse of notation), can we closely estimate φ where $e^{2\pi\varphi i}$ is the eigenvalue corresponding to $|\psi\rangle$?

Some authors call the eigenvector $|\psi\rangle$ an *eigenket*. I don't in this book.

We are looking for a fractional approximation to φ. By increasing the number of qubits we use in the circuit, we can get a closer fit to φ.

Suppose $\mathbf{U}$, $|\psi\rangle$, and $e^{2\pi\varphi i}$ are as above. By applying $\mathbf{U}$ multiple times

$$\mathbf{U}|\psi\rangle = e^{2\pi\varphi i}|\psi\rangle$$
$$\mathbf{U}^2|\psi\rangle = \mathbf{U}\left(e^{2\pi\varphi i}|\psi\rangle\right) = e^{(2+2)\pi\varphi i}$$
$$\mathbf{U}^3|\psi\rangle = \mathbf{U}\left(e^{(2+2)\pi\varphi i}|\psi\rangle\right) = e^{(2+2+2)\pi\varphi i}$$
$$\vdots$$
$$\mathbf{U}^k|\psi\rangle = e^{2k\pi\varphi i}$$

If $k = 2^j$ then $\mathbf{U}^{2^j}|\psi\rangle = e^{2^{j+1}\pi\varphi i}$.

We use this in the circuit for phase estimation in the following way.

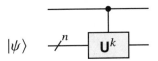

This is a controlled-$\mathbf{U}^k$ gate. The $\overset{n}{\diagup}$ indicates $|\psi\rangle$ is represented on an n qubit register and $\mathbf{U}^k$ takes n quantum state inputs and outputs. This $\mathbf{U}^k$ is only run if the input on the wire above it is $|1\rangle$.

Let $\mathbf{U}$ be a single qubit gate that operates like

$$e^{2\pi\varphi i}\begin{bmatrix} 1 & 0 \\ 0 & 1 \end{bmatrix} = \begin{bmatrix} e^{2\pi\varphi i} & 0 \\ 0 & e^{2\pi\varphi i} \end{bmatrix}$$

Then up to differing only by a global phase, these two circuits are the same:

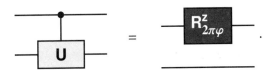

Note how the value of the control qubit is modified.

Question 10.4.1

Show this is true via matrices and the standard basis kets.

This extends to larger matrices acting as a unit multiple of the identity and is exactly the case for how **U** operates on its eigenvector $|\psi\rangle$.

Let m be the number of bits we need to get us as close as we need to the original phase. [20, Section 5.2] We use m qubits on the upper portion of the circuit, the "upper register."

When $m = 2$, our phase estimation circuit is

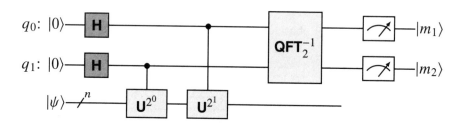

In general, we keep adding gates on the bottom up to $\mathbf{U}^{2^{m-1}}$ as is shown in Figure 10.1. By the nature of application of the phase shifts described above, these modify the quantum states in the upper register. We then take the inverse Quantum Fourier Transform.

Continuing with our 2-qubit case, the state of the upper register before applying $\mathbf{QFT}_2^{-1}$ is

$$\frac{1}{2}\left(|0\rangle + e^{2\pi 2^1 \varphi i}|1\rangle\right) \otimes \left(|0\rangle + e^{2\pi 2^0 \varphi i}|1\rangle\right)$$

$$= \frac{1}{2}\left(|00\rangle + e^{2\pi 2^0 \varphi i}|01\rangle + e^{2\pi 2^1 \varphi i}|10\rangle + e^{2\pi \boxed{3} \varphi i}|11\rangle\right)$$

$$= \frac{1}{2}\left(|00\rangle + e^{2\pi 2^0 \varphi i}|01\rangle + e^{2\pi 2^1 \varphi i}|10\rangle + e^{2\pi \boxed{(2^2 - 1)} \varphi i}|11\rangle\right)$$

$$= \frac{1}{2}\sum_{j=0}^{2^2-1} e^{2\pi j \varphi i}|j\rangle_2.$$

I've highlighted two of the terms in the exponents to show the pattern.

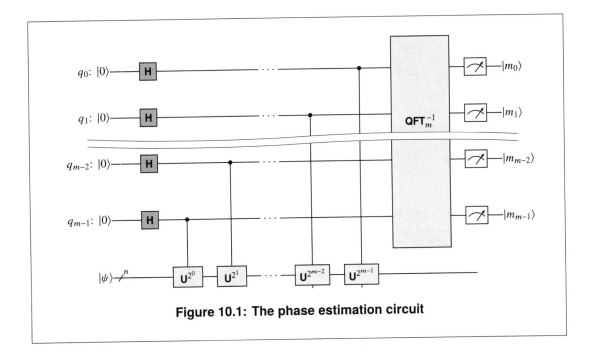

Figure 10.1: The phase estimation circuit

In the 2-qubit case,

$$\mathbf{QFT}_2 = \frac{1}{\sqrt{4}} \begin{bmatrix} 1 & 1 & 1 & 1 \\ 1 & \omega & \omega^2 & \omega^3 \\ 1 & \omega^2 & \omega^4 & \omega^6 \\ 1 & \omega^3 & \omega^6 & \omega^9 \end{bmatrix} = \frac{1}{2} \begin{bmatrix} 1 & 1 & 1 & 1 \\ 1 & i & -1 & -i \\ 1 & -1 & 1 & -1 \\ 1 & -i & -1 & i \end{bmatrix}$$

because $\omega = e^{\frac{2\pi i}{2^2}} = i$ is a primitive fourth root of unity. Since this matrix is unitary, its inverse is its conjugate transpose

$$\mathbf{QFT}_2^{-1} = \frac{1}{2} \begin{bmatrix} 1 & 1 & 1 & 1 \\ 1 & -i & -1 & i \\ 1 & -1 & 1 & -1 \\ 1 & i & -1 & -i \end{bmatrix}$$

For $N = 2^2$ and

$$|\chi\rangle = \sum_{j=0}^{2^2-1} a_j |j\rangle_2,$$

its inverse Quantum Fourier Transform is

$$\mathbf{QFT}_2^{-1}\left(|\chi\rangle\right) = \frac{1}{\sqrt{2^2}} \sum_{j=0}^{2^2-1} \sum_{k=0}^{2^2-1} a_k e^{-\frac{2\pi i j k}{2^2}} |j\rangle_2$$

for $\omega = e^{\frac{2\pi i}{N}}$, a primitive Nth root of unity. The Greek letter "χ" is pronounced "chī".

In our circuit, $a_k = \frac{1}{2}e^{2\pi\varphi k i}$. Putting all this together, the final upper register before measurement is

$$\frac{1}{2^2} \sum_{j=0}^{2^2-1} \sum_{k=0}^{2^2-1} e^{2\pi\varphi k i} e^{-\frac{2\pi j k i}{2^2}} |j\rangle_2 = \frac{1}{2^2} \sum_{j=0}^{2^2-1} \sum_{k=0}^{2^2-1} e^{2\pi\varphi k i - \frac{2\pi j k i}{2^2}} |j\rangle_2$$

$$= \frac{1}{2^2} \sum_{j=0}^{2^2-1} \sum_{k=0}^{2^2-1} e^{-\frac{2\pi k i}{2^2}\left(j - 2^2 \varphi\right)} |j\rangle_2$$

Obvious, right? Let me state why we have some of the elements in these messy expressions that we do:

- φ is the phase we want to estimate.
- The $|j\rangle_2$ are the standard basis kets.
- The sums involving the $|j\rangle_2$ are from superpositions.
- The exponential forms involving e are roots of unity or other complex numbers with absolute value 1.
- The $\mathbf{QFT}_n^{-1}$ introduces the minus sign into the exponent while using $\mathbf{QFT}_n$ would not have.

While the expressions look complicated, because they are, each part is there because of some gate action.

The general form of the last expression for m qubits instead of 2 is

$$\frac{1}{2^m} \sum_{j=0}^{2^m-1} \sum_{k=0}^{2^m-1} e^{-\frac{2\pi k i}{2^m}\left(j - 2^m \varphi\right)} |j\rangle_m.$$

Including the top register and the bottom register (which is $|\psi\rangle$), the full pre-measurement state at the end of the circuit is

$$\frac{1}{2^m} \sum_{j=0}^{2^m-1} \sum_{k=0}^{2^m-1} e^{-\frac{2\pi k i}{2^m}\left(j - 2^m \varphi\right)} |j\rangle_m \otimes |\psi\rangle.$$

The probability amplitudes are

$$\frac{1}{2^m} \sum_{k=0}^{2^m-1} e^{-\frac{2\pi ki}{2^m}(j-2^m\varphi)}$$

for the expression to the left of $\otimes |\psi\rangle$. Now let's get an estimate for φ. We do so by finding a good rational approximation between 0 and 1.

Starting with φ, multiply it by 2^m and choose the closest c in $\mathbb{Z}$ that is within $\frac{1}{2}$ of $2^m\varphi$. To be concrete, choose

$$c = \left\lfloor 2^m\varphi + \frac{1}{2} \right\rfloor.$$

Let

$$d = \left| \frac{2^m\varphi - c}{2^m} \right| = \left| \varphi - \frac{c}{2^m} \right|.$$

Then $0 \le 2^m d \le \frac{1}{2}$. We are looking for $\frac{c}{2^m}$ to be a good approximation to φ. In fact, we want the measured state of the upper register to be $|c\rangle$. By increasing m, the number of qubits in each of the registers, we can get a more accurate approximation.

This is the crux of what the algorithm does, find $|c\rangle$ as the measured standard basis ket in the upper register as a good approximation to $2^m\varphi$.

Given c and d, we rewrite the upper register quantum state as

$$\frac{1}{2^m} \sum_{j=0}^{2^m-1} \sum_{k=0}^{2^m-1} e^{-\frac{2\pi ki}{2^m}(j-2^m\varphi)} |j\rangle_m = \frac{1}{2^m} \sum_{j=0}^{2^m-1} \sum_{k=0}^{2^m-1} e^{-\frac{2\pi ki}{2^m}(j-c)} e^{2\pi dki} |j\rangle_m.$$

The probability of getting $|c\rangle$ is the square of the absolute value of the probability amplitude when $j = c$. That is,

$$P(c) = \left| \frac{1}{2^m} \sum_{k=0}^{2^m-1} e^{-\frac{2\pi ki}{2^m}(c-c)} e^{2\pi dki} \right|^2 = \left| \frac{1}{2^m} \sum_{k=0}^{2^m-1} e^{2\pi dki} \right|^2 = \frac{1}{2^{2m}} \left| \sum_{k=0}^{2^m-1} e^{2\pi dki} \right|^2.$$

If $d = 0$ the sum on the right is exactly 1 and the measured result of the upper register is $|c\rangle$ with probability 1. This happens when $\varphi = \frac{c}{2^m}$ is a rational number.

All is not lost if $d \ne 0$. In that case, the circuit returns the correct answer with probability $\frac{4}{\pi^2} \approx 0.405$. [18, Section 3.7]

This means we must run the circuit multiple times to get the correct answer. If you run the circuit 28 or more times, the probability of never having gotten the correct answer is less than 10^{-6}, as we saw in section 6.3.

10.5   Order and period finding

Consider the function a^k on whole numbers k for a fixed a in $\mathbb{N}$ greater than 1. For example, if $a = 3$ then the first 12 values are

$$3^0 = 1, \qquad 3^1 = 3, \qquad 3^2 = 9, \qquad 3^3 = 27,$$

$$3^4 = 81, \qquad 3^5 = 243 \qquad 3^6 = 729, \qquad 3^7 = 2187,$$

$$3^8 = 6561, \quad 3^9 = 19683, \quad 3^{10} = 59049, \quad 3^{11} = 177147$$

As we look at larger exponents k, the values of 3^k will just get larger and larger.

If we instead use modular arithmetic as we saw in section 3.7, 3^k cannot get arbitrarily large. For example, modulo $M = 13$, the values we get are

$$1 = 3^0 \bmod 13, \quad 3, \quad 9, \quad 1, \quad 3, \quad 9, \quad 1, \quad 3, \quad 9, \quad 1, \quad 3, \quad 9, \quad 1$$

Working modulo $M = 16$ yields

$$1, \quad 3, \quad 9, \quad 11, \quad 1, \quad 3, \quad 9, \quad 11, \quad 1, \quad 3, \quad 9, \quad 11, \quad 1, \quad 3, \quad 9, \quad 11$$

Finally, for modulo $M = 22$ we get

$$1, \quad 3, \quad 9, \quad 5, \quad 15, \quad 1, \quad 3, \quad 9, \quad 5, \quad 15, \quad 1,$$

$$3, \quad 9, \quad 5, \quad 15, \quad 1, \quad 3, \quad 9, \quad 5, \quad 15, \quad 1, \quad 3$$

In each case, the sequence starts repeating. That is, the sequences, and hence the functions, are *periodic*. If we define $f_a(x) = a^x \bmod M$ **for a coprime to M**, then the smallest positive integer r such that $f_a(x) = f_a(x + r)$ for all x is called the *period* of f_a.

Question 10.5.1

What happens when we work modulo $M = 23, 24$, and 25?

For $M = 13$ in the first example, the period $r = 3$. For $M = 16$, $r = 4$. In the final example with $M = 22$, the period $r = 5$.

Given such an a as above, we can look at all $a^1, a^2, a^3, \ldots$ modulo M and ask: what is the smallest r in $\mathbb{N}$, if it exists, such that $a^r \equiv 1 \bmod M$? If there is such an r, it is called the *order* of $a \bmod M$. Then $a^{x+r} \equiv a^x \bmod M$ by multiplication by a^x. Thus r is also the period of f_a. For this reason, the "period finding problem" is equivalent to the "order finding problem."

In the rest of this section we develop a hybrid quantum-classical algorithm to find the order of such an a, with a few conditions.

This is an important algorithm because one of its applications is integer factorization. With such an r in hand, and assuming it is even,

$$a^r \equiv 1 \bmod M \Rightarrow a^r - 1 \equiv 0 \bmod M \Rightarrow \left(a^{r/2} - 1\right)\left(a^{r/2} + 1\right) \equiv 0 \bmod M.$$

Given a good a, a good even r, and Euclid's algorithm, we might be able to find a factorization of M.

Let

$$\ell_{\text{bits}} = \lceil \log_2 M \rceil$$

be the number of bits we need to represent M. For example, if $M = 7$ then $\ell_{\text{bits}} = 3$. For $M = 64$, we use $\ell_{\text{bits}} = 6$.

Now set

$$\ell_\epsilon = 2\ell_{\text{bits}} + 1 + \left\lceil \log_2 \left(2 + \frac{1}{2\epsilon}\right) \right\rceil$$

for some very small ϵ in $\mathbb{R}$.

We use phase estimation with some post-processing using continued fractions to compute the order r of a modulo M. For $0 \leq j \leq r - 1$, we find approximations to the phases $\varphi_j = \frac{j}{r}$ accurate to $2\ell_{\text{bits}} + 1$ bits with probability $\geq \frac{1-\epsilon}{r}$. The smaller our ϵ, the larger our ℓ_ϵ is.

I've chosen to use ℓ to remind us that ℓ_{bits} and ℓ_ϵ are lengths of quantum registers.

To learn more

In what follows, I generally follow the approach in Nielsen and Chuang, though there are many variations in the literature such as in Watrous, ideas of which are reflected here. [20, Section 5.3.1] [29]

10.5.1 Modular exponentiation

Since we are finding periods of functions or orders of numbers modulo M, we need to be able to compute $a^x \bmod M$ in a quantum way.

For a binary string y of length ℓ_{bits}, that is, y is in $\{0,1\}^{\ell_{\text{bits}}}$, define

$$U|y\rangle = \begin{cases} |ay \bmod M\rangle & \text{if } 0 \le y < M \\ |y\rangle & \text{if } M \le y < 2^{\ell_{\text{bits}}}. \end{cases}$$

Remember that a is coprime to M, which means they share no non-trivial factors. This is the same as saying $\gcd(a,M)=1$ and so there exists b and c in $\mathbb{Z}$ so that $ab + Mc = 1$. Looking at this modulo M,

$$1 = ab + Mc \equiv ab \bmod M.$$

Thus a is invertible modulo M with $a^{-1} = b$.

When I write an expression like $|j \bmod M\rangle$ it is shorthand for $|j \bmod M\rangle_{\ell_{\text{bits}}}$.

The $|y\rangle$ are the computational basis vectors in a $2^{\ell_{\text{bits}}}$-dimensional vector space over $\mathbb{C}$. U is a $2^{\ell_{\text{bits}}}$ by $2^{\ell_{\text{bits}}}$ square matrix. The lower right $2^{\ell_{\text{bits}}} - M$ by $2^{\ell_{\text{bits}}} - M$ submatrix is the identity matrix. The upper left M by M submatrix is a permutation matrix because a is invertible modulo M. All other matrix entries are zero. Hence all of U is a permutation matrix consisting of 0s and 1s, and and so is unitary.

If we apply U multiple times, we get multiplication and hence exponentiation by repeated squaring:

$$U^2|y\rangle = UU|y\rangle = U|ay \bmod M\rangle = |a^2 y \bmod M\rangle.$$

For example,

$$\begin{aligned} U^{15}|y\rangle &= U(U^7)^2|y\rangle = U(UU^6)^2|y\rangle \\ &= U(U(U^3)^2)^2|y\rangle = U(U(UUU)^2)^2|y\rangle \\ &= |a^{15} y \bmod M\rangle. \end{aligned}$$

As we saw at the end of section 9.4, for any non-negative integer z with binary representation $z_{k-1}z_{k-2}\cdots z_1 z_0$ with z_0 the least significant bit,

$$a^z = a^{z_{k-1}2^{k-1}} \times a^{z_{k-2}2^{k-2}} \times \cdots \times a^{z_1 2^1} \times a^{z_0 2^0}.$$

We can define

$$\begin{aligned} |z\rangle|y\rangle &\mapsto |z\rangle U^{z_{k-1}2^{k-1}} \times U^{z_{k-2}2^{k-2}} \times \cdots \times U^{z_1 2^1} \times U^{z_0 2^0} \\ &= |z\rangle\big|a^{z_{k-1}2^{k-1}} \times a^{z_{k-2}2^{k-2}} \times \cdots \times a^{z_1 2^1} \times a^{z_0 2^0} \times y \bmod M\big\rangle \\ &= |z\rangle|a^z y \bmod M\rangle \end{aligned}$$

In practice we will allow k to be as large as ℓ_ϵ.

With these observations and calculations, we know how to do quantum modular exponentiation. Like many such quantum subroutines, we may need additional qubits for the computations outside the main circuit description. As in section 9.4, we can do modular exponentiation in $O\left(\ell_{\text{bits}}^3\right)$ gates using algorithms similar to the simple classical versions.

We have one last thing to observe about U before we move on to the circuit for order finding and its analysis. With r the order of a modulo M, the kets

$$|w_j\rangle = \frac{1}{\sqrt{r}} \sum_{k=0}^{r-1} e^{-\frac{2\pi k j i}{r}} |a^k \bmod M\rangle$$

are eigenvectors of U for $0 \le j < r$. To test this, we apply U to each of these expressions:

$$U|w_j\rangle = U\left(\frac{1}{\sqrt{r}} \sum_{k=0}^{r-1} e^{-\frac{2\pi k j i}{r}} |a^k \bmod M\rangle \right)$$

$$= \frac{1}{\sqrt{r}} \sum_{k=0}^{r-1} e^{-\frac{2\pi k j i}{r}} |a^{k+1} \bmod M\rangle$$

$$= \frac{1}{\sqrt{r}} \sum_{k=0}^{r-1} e^{-\frac{2\pi (k+1-1) j i}{r}} |a^{k+1} \bmod M\rangle$$

$$= \frac{1}{\sqrt{r}} \sum_{k=0}^{r-1} e^{-\frac{2\pi (k+1) j i}{r}} e^{-\frac{2\pi (-1) j i}{r}} |a^{k+1} \bmod M\rangle$$

$$= e^{\frac{2\pi i j}{r}} \frac{1}{\sqrt{r}} \sum_{k=0}^{r-1} e^{-\frac{2\pi (k+1) j i}{r}} |a^{k+1} \bmod M\rangle$$

$$= e^{\frac{2\pi j i}{r}} |w_j\rangle$$

Even though we use r explicitly in the above equations, remember that we do not know what it is yet! We have shown only that the eigenvalues corresponding to the eigenvectors $|w_j\rangle$ are

$$e^{\frac{2\pi j i}{r}}$$

as j goes from 0 to $r-1$. We use phase estimation to get at those eigenvalues and hence r.

Question 10.5.2

Remembering that r is the order of a modulo M, show that

$$\sum_{k=0}^{r-1} e^{-\frac{2\pi k j i}{r}} \left| a^k \bmod M \right\rangle = \sum_{k=0}^{r-1} e^{-\frac{2\pi (k+1) j i}{r}} \left| a^{k+1} \bmod M \right\rangle.$$

To learn more

Quantum Fourier Transforms are used in many optimized quantum algorithms for modular arithmetic operations. [24]

10.5.2 The circuit

To set up our circuit, we need two quantum registers. Since we use phase estimation, the first register needs enough qubits to get us the accuracy we need. This is ℓ_ϵ. The number of qubits in the second register is ℓ_{bits}.

The circuit for the quantum portion of the order finding algorithm is in Figure 10.2. The part of the circuit that I have labeled $\mathbf{U_{PE}}$ is the core of the setup for the phase estimation. Referring back to the general phase estimation circuit in Figure 10.1, $\mathbf{U_{PE}}$ needs to prepare an eigenvector in the second register and handle the controlled $\mathbf{U}^{2^j}$ gates.

Unfortunately, there is nothing that allows us to create

$$\left| w_j \right\rangle = \frac{1}{\sqrt{r}} \sum_{k=0}^{r-1} e^{-\frac{2\pi k j i}{r}} \left| a^k \bmod M \right\rangle$$

directly. The phases φ_j we wish to estimate are $\frac{j}{r}$.

When in a quandary in a quantum algorithm, create a superposition! Consider

$$\frac{1}{\sqrt{r}} \sum_{j=0}^{r-1} \left| w_j \right\rangle.$$

The $\frac{1}{\sqrt{r}}$ out front is the common probability amplitude of all the kets. Squaring this and multiplying by r gives us the required value of 1. Expanding the expression,

$$\frac{1}{\sqrt{r}} \sum_{j=0}^{r-1} \left| w_j \right\rangle = \frac{1}{\sqrt{r}} \sum_{j=0}^{r-1} \frac{1}{\sqrt{r}} \sum_{k=0}^{r-1} e^{-\frac{2\pi k j i}{r}} \left| a^k \bmod M \right\rangle$$

$$= \frac{1}{\sqrt{r}} \sum_{k=0}^{r-1} \frac{1}{\sqrt{r}} \sum_{j=0}^{r-1} e^{-\frac{2\pi k j i}{r}} \left| a^k \bmod M \right\rangle$$

$$= \frac{1}{\sqrt{r}} \sum_{k=0}^{r-1} \frac{1}{\sqrt{r}} \left[\sum_{j=0}^{r-1} e^{-\frac{2\pi k j i}{r}} \right] \left| a^k \bmod M \right\rangle$$

$$= \frac{1}{\sqrt{r}} \sum_{k=0}^{r-1} \frac{1}{\sqrt{r}} r \delta_{k,0} \left| a^k \bmod M \right\rangle$$

$$= \left| a^0 \bmod M \right\rangle$$

$$= \left| 1 \right\rangle_{\ell_{\text{bits}}}$$

where the equality involving $\delta_{k,0}$ follows from the extended summation formula for roots of unity in subsection 10.1.1.

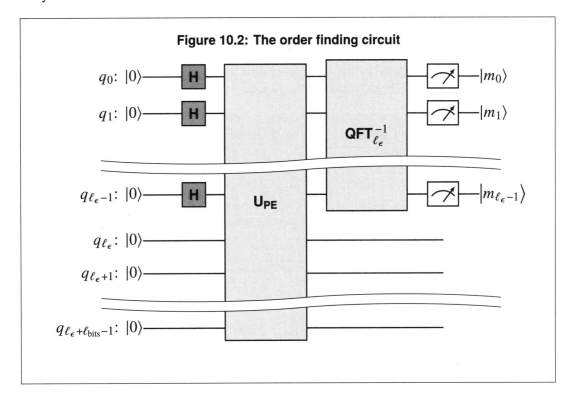

Figure 10.2: The order finding circuit

The value $|1\rangle_{\ell_{\text{bits}}}$ is **not** $|111\cdots111\rangle_{\ell_{\text{bits}}}$ but is the ket $|000\cdots001\rangle_{\ell_{\text{bits}}}$, assuming the right-most bit is the least significant.

So while we cannot prepare an individual eigenvector $|w_j\rangle$, it is simple to prepare $|1\rangle_{\ell_{\text{bits}}}$ as the circuit on the right.

$$q\ell_\epsilon: |0\rangle - \boxed{\text{X}} -$$
$$q\ell_\epsilon+1: |0\rangle -$$

$$q\ell_\epsilon+\ell_{\text{bits}}-1: |0\rangle -$$

This is the first part of $\mathbf{U_{PE}}$ for the lower quantum register. The rest of $\mathbf{U_{PE}}$ is the sequence of controlled-$\mathbf{U}^{2^j}$ gates from Figure 10.1.

Unlike the phase estimation algorithm, we are using a superposition of eigenvectors instead of a single one as the input to the lower quantum register. We're not done, because what comes out of our circuit after applying the inverse QFT requires more work to get to r. In the next section, we use continued fractions to get us closer to the factorization.

10.5.3   The continued fraction part

We chose ℓ_ϵ so that for $0 \le j \le r-1$ we can find approximations φ_j to the phases $\frac{j}{r}$ of U's eigenvalues accurate to $2\ell_{\text{bits}}+1$ bits with probability $\ge \frac{1-\epsilon}{r}$. j satisfies $0 \le j < r$ where each j is as probable as another. This is another way of saying that the j we get are uniformly distributed.

The output of the phase estimation algorithm after measurement is

$$\varphi_j = \frac{c_j}{2^\epsilon} \approx \frac{j}{r}.$$

We've done a lot of quantum work and we have in hand a rational number $\frac{c_j}{2^\epsilon}$ approximation to $\frac{j}{r}$. We want to get to $\frac{j}{r}$ and then r.

If $\gcd(c_j, 2^\epsilon) \ne 1$ then we try again because the fraction is not in reduced form. If c_j and 2^ϵ are coprime, we test if

$$a^{2^\epsilon} \equiv 1 \bmod M.$$

If this succeeds, we take $r = 2^\epsilon$ and we are done. c_j and 2^ϵ will be coprime if and only if c_j is odd, and there are many such candidates less than 2^ϵ.

We might just get a bad result with $\frac{c_j}{2^\epsilon}$ being, essentially, garbage. In this case, we can try again or decrease ϵ, thereby increasing ℓ_ϵ.

We now employ this result about approximations and continued fractions:

Let $\frac{a}{b}$ be a rational number in reduced form. Every reduced rational number $\frac{c}{d}$ that satisfies

$$\left|\frac{a}{b} - \frac{c}{d}\right| < \frac{1}{2d^2}$$

is a convergent of $\frac{a}{b}$. [14, Theorem 19]

(We discussed continued fractions and convergents in subsection 3.5.4.)

This statement is quite powerful: given two rational numbers, if the second is sufficiently close to the first, then the second is a convergent of the first. Reworking this for our situation gives the following result.

By our choice of ℓ_{bits} and ℓ_ϵ, every reduced rational number $\varphi_j = \frac{c_j}{2^\epsilon}$ that satisfies

$$\left|\varphi_j - \frac{j}{r}\right| = \left|\frac{c_j}{2^\epsilon} - \frac{j}{r}\right| < \frac{1}{2r^2}$$

is a convergent of $\frac{j}{r}$.

If $\frac{c_j}{2^\epsilon}$ is a good approximation to $\frac{j}{r}$ then it is accurate to $2\ell_{\text{bits}} + 1$ bits with probability $\geq \frac{1-\epsilon}{r}$. What does it mean for some x to be accurate to some y for some number of bits b? Simply that

$$|x - y| \leq \frac{1}{2^b} = 2^{-b}.$$

Therefore,

$$\left|\varphi_j - \frac{j}{r}\right| = \left|\frac{c_j}{2^\epsilon} - \frac{j}{r}\right| \leq \frac{1}{2^{2\ell_{\text{bits}}+1}} = \frac{1}{2 \times \left(2^{\ell_{\text{bits}}}\right)^2} \leq \frac{1}{2M^2} < \frac{1}{2r^2}$$

by our definition of ℓ_{bits} and noting that $r \leq M$.

Now we start computing the convergents for the known reduced fraction $\frac{c_j}{2^\epsilon}$. Among these will be $\frac{j}{r}$ in reduced form. We test the denominators of the convergents as candidates for r. When we find one that works, we are done.

The complexity of the overall algorithm is dominated by the quantum modular exponentiation and so is $O\left(\ell_{\text{bits}}^3\right)$.

10.6 Shor's algorithm

We now have the tools we need to sketch Shor's algorithm for factoring integers in polynomial time on a sufficiently large quantum computer.

The complete algorithm has both classical and quantum components. Work is done on both kinds of machines to get to the answer. It is the quantum portion that drops us down to polynomial complexity in the number of gates by use of phase estimation, order finding, modular exponentiation, and the Quantum Fourier Transform.

Let odd M in $\mathbb{Z}$ be greater than 3 for which you have already tried the basic tricks from subsection 10.2.3 to check it is not a multiple of 3, 5, 7, and so on. So that you don't waste your time, you should also try trial division using a small list of primes, though this is not necessary. It is necessary, however, to make sure that M is not a power of a prime number, and you can use Newton's method to test this.

So M is an odd positive number in $\mathbb{Z}$ that is not a power of a prime. It has a reasonable chance of being composite.

The following is the general approach to Shor's algorithm given M as above:

1. Choose a random number a such that $1 < a < M$. Keep track of these values since we might need to repeat this step again.
2. Check if $\gcd(a, M) = 1$. If not, we have found a factor of M and we are done. This is pretty unlikely but now we know that a and M are coprime: they have no integer factors in common.
3. Now find the non-zero order r of a mod M. This means that $a^r \equiv 1 \bmod M$. If r is odd, go back to step 1 and try again with a different a.
4. If r is even, we have

$$a^r \equiv 1 \bmod M \Rightarrow a^r - 1 \equiv 0 \bmod M$$
$$\Rightarrow \left(a^{r/2} - 1\right)\left(a^{r/2} + 1\right) \equiv 0 \bmod M$$

5. Now look at $\gcd\left(a^{r/2} - 1, M\right)$ and, if necessary, $\gcd\left(a^{r/2} + 1, M\right)$. If either of these does not equal 1, we have found a factor of M and we have succeeded.
6. If both of these greatest common divisors are 1, we repeat all the above from step 1 with another random a. We continue to do this until we find a factor.

> **To learn more**
>
> Other more advanced treatments go into the full complexity analysis and number theory behind this algorithm, starting with Shor's original paper. [27] [17, Chapter 12] [18, Chapter 3] [23, Chapter 5]

In 2001, IBM Research scientists demonstrated factoring the number 15 via Shor's algorithm on a 7-qubit NMR quantum computer. [28]

10.7 Summary

In this chapter we did some hard math to ultimately understand how to factor integers much faster than we can classically. Along the way we went deeply into several non-trivial quantum algorithms that are used in other quantum applications. These algorithms include the Quantum Fourier Transform, phase estimation, and order finding. These form a good basis for you to understand other quantum algorithms and their circuits.

We next turn our attention to the connections between the slightly abstract concepts we have seen so far and the physical quantum computers that we can build today.

> **To learn more**
>
> There are many other quantum algorithms though we have covered several of the most common ones. Techniques such as order finding are part of advanced algorithms in addition to Shor's factorization. [3] [17] [22].

References

[1]   Daniel J. Bernstein. "Detecting Perfect Powers in Essentially Linear Time". In: *Mathematics of Computation* 67.223 (July 1998), pp. 1253–1283.

[2]   David M. Bressoud. *Factorization and primality testing*. Undergraduate Texts in Mathematics. Springer-Verlag New York, 1989.

[3]   Patrick J. Coles et al. *Quantum Algorithm Implementations for Beginners*. 2018. URL: https: //arxiv.org/abs/1804.03719.

[4]   Thomas H. Cormen et al. *Introduction to Algorithms*. 3rd ed. The MIT Press, 2009.

[5]   R. Crandall and C. Pomerance. *Prime Numbers: a Computational Approach*. 2nd ed. Springer, 2005.

[6]   Sanjoy Dasgupta, Christos H. Papadimitriou, and Umesh Vazirani. *Algorithms*. McGraw-Hill, Inc., 2008.

[7]   Thomas G. Draper. *Addition on a Quantum Computer*. 2000. URL: https://arxiv.org/abs/quant-ph/0008033.

[8]   D. S. Dummit and R. M. Foote. *Abstract Algebra*. 3rd ed. Wiley, 2004.

[9]   Free Software Foundation. *The GNU Multiple Precision Arithmetic Library*. URL: https://gmplib.org/.

[10]   R.W. Hamming. *Numerical Methods for Scientists and Engineers*. Dover Books on Engineering. Dover, 1986.

[11]   Jeffrey Hoffstein, Jill Pipher, and Joseph H. Silverman. *An Introduction to Mathematical Cryptography*. 2nd ed. Undergraduate Texts in Mathematics 152. Springer Publishing Company, Incorporated, 2014.

[12]   Kenneth Ireland and Michael Rosen. *A Classical Introduction to Modern Number Theory*. 2nd ed. Graduate Texts in Mathematics 84. Springer-Verlag New York, 1990.

[13]   Burt Kaliski. "Subexponential Time". In: ed. by Henk C. A. van Tilborg and Sushil Jajodia. Springer US, 2011, pp. 1267–1267. URL: https://doi.org/10.1007/978-1-4419-5906-5_436.

[14]   A. Ya. Khinchin. *Continued Fractions*. Revised. Dover Books on Mathematics. Dover Publications, 1997.

[15]   Neal Koblitz. *A Course in Number Theory and Cryptography*. 2nd ed. Graduate Texts in Mathematics 114. Springer-Verlag, 1994.

[16]   S. Lang. *Algebra*. 3rd ed. Graduate Texts in Mathematics 211. Springer-Verlag, 2002.

[17]   Richard J. Lipton and Kenneth W. Regan. *Quantum Algorithms via Linear Algebra: A Primer*. The MIT Press, 2014.

[18]   N. David Mermin. *Quantum Computer Science: An Introduction*. Cambridge University Press, 2007.

[19]   S.J. Miller et al. *An Invitation to Modern Number Theory*. Princeton University Press, 2006.

[20]   Michael A. Nielsen and Isaac L. Chuang. *Quantum Computation and Quantum Information*. 10th ed. Cambridge University Press, 2011.

[21]   Joe Palca. *The World Has A New Largest-Known Prime Number*. URL: https://www.npr.org/2018/12/21/679207604/the-world-has-a-new-largest-known-prime-number.

[22]   A.O. Pittenger. *An Introduction to Quantum Computing Algorithms*. Progress in Computer Science and Applied Logic. Birkhäuser Boston, 2012.

[23]   Eleanor Rieffel and Wolfgang Polak. *Quantum Computing: A Gentle Introduction*. 1st ed. The MIT Press, 2011.

[24]   Rich Rines and Isaac Chuang. *High Performance Quantum Modular Multipliers*. 2018. URL: https://arxiv.org/abs/1801.01081.

[25]   Lidia Ruiz-Perez and Juan Carlos Garcia-Escartin. *Quantum arithmetic with the Quantum Fourier Transform*. May 2017. URL: https://arxiv.org/abs/1411.5949.

[26]   Robert Sedgewick and Kevin Wayne. *Algorithms*. 4th ed. Addison-Wesley Professional, 2011.

[27]   Peter W. Shor. "Polynomial-Time Algorithms for Prime Factorization and Discrete Logarithms on a Quantum Computer". In: *SIAM J. Comput.* 26.5 (Oct. 1997), pp. 1484–1509.

[28]   Lieven M. K. Vandersypen et al. "Experimental realization of Shor's quantum factoring algorithm using nuclear magnetic resonance". In: *Nature* 414.6866 (2001), pp. 883–887.

[29]   John Watrous. *CPSC 519/619: Introduction to Quantum Computing*. URL: https://cs.uwaterloo.ca/~watrous/LectureNotes/CPSC519.Winter2006/all.pdf.

11

Getting Physical

The non-physicist finds it hard to believe that really
the ordinary laws of physics, which he regards as the
prototype of inviolable precision,
should be based on the statistical tendency of matter
to go over into disorder.

Erwin Schrödinger [26]

It's now time to discuss some considerations about how we go from theoretical mathematics and physics to the applied and experimental.

The qubits we make in the lab for research and those we will create for commercial use are physical hardware devices. As such, they are subject to noise from the environment, electronic components, and manufacturing choices. Hardware improvements decrease the disturbances, but software and system ones can too. The long-term goal is to have fully error corrected, fault-tolerant quantum computing devices.

Since we're talking about physics, I explain the questionable fate of Schrödinger's cat.

Topics covered in this chapter

11.1   That's not logical

The qubits like those in the last two chapters are examples of "logical qubits." We can use them indefinitely, they never lose state when they are not used, and we can apply as many gates to them as we wish.

When you build a quantum computer, the fundamental physical implementations of qubits aren't as perfect as logical qubits. Such a qubit, called a "physical qubit," starts to lose its ability to hold onto a state after what is called its "coherence time." We also say that the qubit is *decohering*.

It's a goal of quantum computing researchers and engineers to delay the decay of a physical qubit's quantum state as long as possible. Since the decay is inevitable, a goal of fault tolerance and error correction is to handle and fix the effects of the qubits' decoherence throughout the execution of a circuit.

Is it possible to create objects that act like logical qubits from physical ones? Research today says "yes" but it will require hundreds to thousands of physical qubits to make one logical qubit. When we get there, we will have *fault tolerance* where errors are detected and corrected. We use error correcting schemes and circuits as I describe in section 11.5 to have these many physical qubits work together to create a virtual logical qubit.

Small errors creep into the state when gates are applied. After too many gates, you either pass the coherence time and the qubit becomes unreliable, or the errors accumulate to such a degree that further use and measurement is too inaccurate for useful computing.

Two simultaneous and connected goals when constructing a quantum computer are to increase coherence time and reduce errors.

You may see or hear the term "short depth circuit." How many gates are in such a circuit? There is no hard and fast number, though expect it to increase over time. A reasonable working definition of a short depth circuit is one you can run and from which you can get useful results before decoherence sets in and errors overwhelm the computation. When you are working with quantum computing hardware, make sure you can see the current operating statistics about coherence times and errors.

11.2   What does it take to be a qubit?

In his 2000 paper "The Physical Implementation of Quantum Computation," then-IBM Research Staff Member David P. DiVincenzo laid out five "requirements for the implementation of quantum computation." [10]

In his words they are:

1. *A scalable physical system with well characterized qubits*
2. *The ability to initialize the state of the qubits to a simple fiducial state, such as* $|000\ldots\rangle$
3. *Long relevant decoherence times, much longer than the gate operation time*

4. *A "universal" set of quantum gates*

5. *A qubit-specific measurement capability*

Let's discuss what each of these mean, following his lead.

Scalable physical system

In our physical system that we manufacture for quantum computing, we need to build a qubit that has two clearly delineated states, $|0\rangle$ and $|1\rangle$. If these represent energy states, other states may be possible, but we must control the qubit to keep it at either $|0\rangle$ or $|1\rangle$.

The qubit must be able to move into a true superposition of $|0\rangle$ and $|1\rangle$ that obeys the rules with amplitudes and probabilities. As we add more qubits, we must be able to entangle the qubits either directly by physical means or indirectly through a sequence of gates in a circuit.

It is not necessary to physically connect every qubit with every other qubit. The chip architecture determines the degree of connectivity so that overall performance and manufacturability is optimized. For example, on the right is the connection map for the IBM Q first generation 20-qubit chip from 2017. [16]

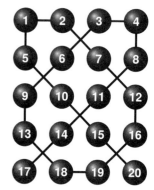

Since then, the layout of the 20-qubit chip was changed to decrease the errors coming from nearby connected qubits and prepare for eventual qubit and gate error correction.

Through the entanglement, starting with the physical connections, we must be able to see and use the doubling of the $\mathbb{C}$ (Hilbert) vector space for every additional qubit added to the system.

Since one qubit is never enough, we should be able to add, over time, sufficient additional qubits to do useful quantum calculations. The cost to add these qubits should not be prohibitive. We would not, for example, want the economic cost or the engineering complexity to scale exponentially as we increase the number of qubits.

Initializing qubits

We must be able to initialize a qubit to a known initial state with very high probability. This is called "high fidelity state preparation."

Since it is common for algorithms to begin with qubits in $|0\rangle$, this is a good choice. If $|1\rangle$ is a better choice for the technology, applying an **X** gate after initialization gives an equivalent effect.

Long decoherence

As we will see in section 11.4, decoherence causes a qubit to move from a desired quantum state to something else. If too much decoherence happens, the state is random and useless.

The qubit must have a long enough coherence time so that enough quantum gates can be executed to implement an algorithm that does something useful. If you have a long coherence time but your gates take a long time to execute, that may be equivalent to a short coherence time but with fast gates.

Long coherence, fast enough gates, and low error rates will be the key to success with NISQ quantum computers.

Universal set of gates

You can't build a house if you don't have the right tools. You can't build general quantum algorithms unless you have a complete-enough set of gates.

In section 9.3, we looked at how gates can be created from other, more primitive ones. The gates that are native to a particular qubit technology may not look anything like the ones we have seen in this book, but as long as they can be composed into a standard collection, practical algorithms can developed, implemented, and deployed.

Measurement capability

We must be able to reliably force the qubit into one or the other of two orthonormal basis states, and these are often $|0\rangle$ and $|1\rangle$. The error rate of this operation must be low enough to allow for useful computation: if we get the wrong measurement answer 50% of the time, all the previous work on executing the circuit is lost. This is called "high fidelity readout."

If we move the qubit to $|+\rangle = \frac{\sqrt{2}}{2}|0\rangle + \frac{\sqrt{2}}{2}|1\rangle$, it should measure to either $|0\rangle$ or $|1\rangle$ with 0.5 probability each. The same is true for quantum states with known probability amplitudes before measurement.

> **To learn more**
>
> In 2018, DiVincenzo published a brief retrospective on the creation of and progress in implementing quantum computers that respected his criteria. [9]

11.3 Light and photons

Light literally illuminates the things around us. It can be as dim and small as a faraway star on a clear night, or can be harsh and bright as the sun or the output of welding equipment. Understanding exactly the nature of light was a major research direction in physics in the nineteenth and early twentieth centuries.

The answers ended up being far more complicated than anyone imagined, gave birth to quantum mechanics, and involved the electromagnetic spectrum well beyond visible light.

11.3.1 Photons

Does light behave like a wave, with varying *amplitude* A (height) and *frequency* ν? The wavelength λ is the distance between two wave crests or other corresponding points. (λ is the Greek letter "lambda" and ν is the Greek letter "nu".)

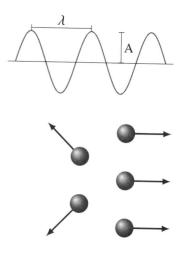

Or **does light behave like a particle** with a well-defined shape, shooting off in various directions? Can particles have different energies? Can particles have different colors?

Light has both wave-like and particle-like characteristics. The fundamental unit of light is called a *photon*. It has no charge and, according to theory, it has no mass.

A photon never sits still. The *speed of light* is how fast a photon moves in a vacuum. This is denoted by c and is 299,792,458 meters per second, which is approximately 186,282 miles per second.

When two waves take up the same space at the same time, we get *superposition*, as shown on the left-hand side in the plot below.

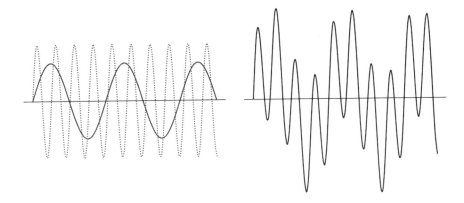

This leads to *interference*, shown on the right-hand side in the plot. At any point where the waves overlap, their amplitudes, positive or negative, add to give a new combined amplitude. If both amplitudes are the same sign, we get *constructive interference*. If they differ, we get *destructive interference*. If the new amplitude is zero, we have *complete* destructive interference.

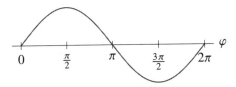

For a given wave with a repeating period as we have shown, we measure the various points along the horizontal from the beginning of the period to the end in radians and call this the *phase* φ. It takes on values from 0 to 2π.

When two waves have the same shape, the same amplitude, and the same frequency but are possibly offset horizontally, we say that the waves are *coherent*. The offset $\Delta\varphi$ is the *phase difference* or *phase offset*.

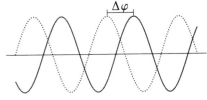

Note that some authors also call this phase difference φ, which is confusing.

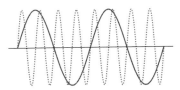

Frequency is measured in *hertz*, **Hz**, which is one full cycle per second. The solid line on the left has a frequency of 1 Hz while the dotted line has a frequency of 4 Hz.

One *gigahertz* (**GHz**) is a frequency of one billion (10^9) cycles per second. A *terahertz* (**THz**) is 10^{12} cycles per second.

To finish our discussion of photons, we need to know about a fundamental value that appears to be essential to how our universe works. That is, if we vary this number, all sort of physics that exists for us would either be different or impossible. *Planck's constant*, *h*, is an exceptionally small number and is

$$h = 6.62607004 \times 10^{-34} \text{ m}^2\text{kg/s}$$

$$= 0.000000000000000000000000000000000662607004 \text{ m}^2\text{kg/s}.$$

The unit to the right is "meters squared kilograms per second." It is related to the *Joule* (**J**), which is a unit of energy, and is 1 meters squared kilograms per second squared, or $\text{m}^2\text{kg/s}^2$. Given the abbreviations m = meter, s = second, kg = kilogram, and J = Joule, Planck's constant is

$$h = 6.62607004 \times 10^{-34}\text{J s}.$$

Thinking a of photon as a wave, let ν be its frequency measured in Hz. Its energy **E** measured in Joules is

$$E = h\nu.$$

Question 11.3.1

What is the unit we used for Hz? Determine this from the units for E and ν. One Hz is equal to how many Joules?

The higher the frequency of the photon (or the smaller its wavelength), the greater its energy.

A wave that is travelling at a constant speed has wavelength λ equal to that speed divided by the frequency ν. Wavelengths are usually measured in some variation of the meter such as the nanometer, **nm**. One nm = 10^{-9} meters, or one-billionth of a meter.

Visible light is only part of the full electromagnetic spectrum. Infrared radiation, microwaves, and radio waves all have wavelengths longer than visible light, and so less energy. Ultraviolet radiation, x-rays, and gamma (γ) rays all have shorter wavelengths, hence higher frequencies, and more energy.

γ rays have frequencies above 10^{19} Hz and wavelengths less than 10^{-11} m. On the other side, microwaves have frequencies between 1 GHz and 300 GHz, and wavelengths from 0.30 m to 0.001 m. Note that the exact wavelength boundaries between different types of electromagnetic radiation vary quite a bit in books and articles, depending on which source you consult.

Coherent light is defined as above (same amplitude, same frequency) but we require that the waves are in phase with one another. Light that varies in frequency or phase is called *incoherent light*.

Coherent light can be produced via **L**ight **A**mplification by **S**timulated **E**mission of **R**adiation, which gives rise to the word "laser." A laser can be focused to send a beam of photons a great distance or for cutting materials. They are also used in scanners at stores to read bar codes and, historically, to read and write data to compact discs, DVDs, and Blu-ray discs.

Question 11.3.2

Compact disc technology used light with a wavelength of 780 nm while Blu-ray used 405 nm. Where are these on the light spectrum?

11.3.2   The double-slit experiment

Suppose you have a device that allows you to shoot perfectly spherical pellets against a smooth, blank wall. The pellets are the same size and do not vary in their straight path or their speed once they leave the device.

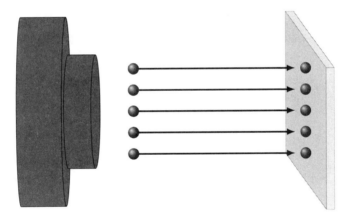

We can see where each pellet struck the wall.

Now let's insert a pellet-proof flat shield between the device and the wall. Also, we cut a horizontal slit that should allow only the center pellet to pass through.

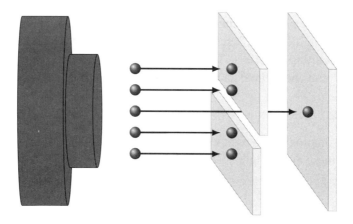

The shield blocks all the pellets except those lined up with the slit. We can repeat this with two slits.

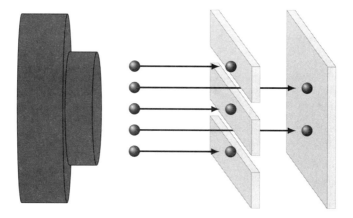

If we shoot many pellets and allow for some horizontal scattering, the impact area on the wall, shown in green, would look like the shape of the slit.

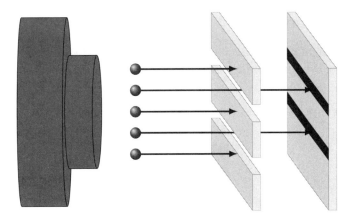

Instead of pellets, let's consider photons. Our "photon shooting" device is a laser and so the light is coherent with some constant wavelength. Thought of as a particle, a photon passes through one slot or the other but not both. However, photons also behave as waves.

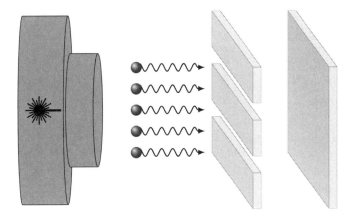

The light waves interfere with themselves after they pass through the slots. Instead of getting solid bands directly behind the slits, we get diffraction and strong bands between the slits and weakening bands to each side. This is because of constructive and destructive interference.

This ability of light to sometimes behave like a wave and sometimes behave like a particle is called the "wave-particle duality," though it is considered more of a historic physical model than modern *quantum optics*. [11] [25]

11.3.3 Polarization

Consider a general quantum state for one qubit in the computational basis: $|\psi\rangle = a|0\rangle + b|1\rangle$. We are going to pass the qubit through a series of procedures and see what comes out the other end.

The first is the $|1\rangle$ blocker.

qubit with state |1> Blocker   qubit with state $|0\rangle$ with
$|\psi\rangle = a|0\rangle + b|1\rangle$ probability $|a|^2$ or nothing

In this somewhat unnatural process, we start with a qubit in general state and pass it through the blocker. With probability $|a|^2$ the qubit will emerge on the other side and it will have state $|0\rangle$. With the complementary probability $|b|^2$, the qubit will be completely blocked and nothing will emerge. Bye bye qubit.

In a similar way we can define the $|0\rangle$ blocker.

qubit with state |0> Blocker   qubit with state $|1\rangle$ with
$|\psi\rangle = a|0\rangle + b|1\rangle$ probability $|b|^2$ or nothing

If we compose these, no qubit makes it through.

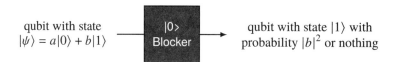

has state $|0\rangle$

Our final procedure is the $|+\rangle$ blocker.

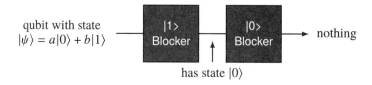

What do we get if we do this?

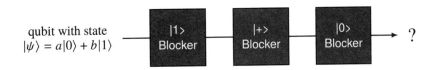

qubit with state $|\psi\rangle = a|0\rangle + b|1\rangle$ → $|1\rangle$ Blocker → $|+\rangle$ Blocker → $|0\rangle$ Blocker → ?

> Recall we have these fundamental equalities among basis kets:
>
> $$|+\rangle = \tfrac{\sqrt{2}}{2}(|0\rangle + |1\rangle) \qquad |-\rangle = \tfrac{\sqrt{2}}{2}(|0\rangle - |1\rangle)$$
>
> $$|0\rangle = \tfrac{\sqrt{2}}{2}(|+\rangle + |-\rangle) \qquad |1\rangle = \tfrac{\sqrt{2}}{2}(|+\rangle - |-\rangle)$$

After the $|1\rangle$ blocker, we have $|0\rangle$ with probability $|a|^2$, if anything. Assuming something is there, it is also equal to $\tfrac{\sqrt{2}}{2}(|+\rangle + |-\rangle)$. We run this through the $|+\rangle$ blocker and with probability $0.5 = \left(\tfrac{\sqrt{2}}{2}\right)^2$ we get $|-\rangle$ or nothing. If anything, it is also equal to $\tfrac{\sqrt{2}}{2}(|0\rangle - |1\rangle)$.

Now we pass it through the $|0\rangle$ blocker. With probability $0.5 = \left(-\tfrac{\sqrt{2}}{2}\right)^2$ we get the qubit in state $|1\rangle$.

In the case where we used only the $|1\rangle$ and $|0\rangle$, no qubit in any state made it through. Oddly enough, if we inserted the $|+\rangle$ blocker between them, that qubit in initial state $a|0\rangle + b|1\rangle$ made it through with probability $|a|^2 \times 0.5 \times 0.5 = \tfrac{|a|^2}{4}$.

This is the mathematics behind the famous three filter polarization experiment, which I now illustrate.

When you think of a photon as a wave, the wave travels in one direction but the peaks and crests of the waves are perpendicular to the direction. Consider a string on a musical instrument. If you pluck it by pulling straight up and releasing, the wave will rise and fall vertically, or $\uparrow$. If you pluck it directly on its side, the wave will rise and fall from left to right, or horizontally ($\rightarrow$).

A photon is a two state quantum system and we can use $\{|\uparrow\rangle, |\rightarrow\rangle\}$ as an orthonormal basis. Rather than thinking about a qubit making it through the above blocking processes, we consider a photon passing through or being absorbed by a polarization filter. We start with a photon in the general $a|\uparrow\rangle + b|\rightarrow\rangle$ state.

A $|\uparrow\rangle$ vertical polarization filter absorbs the photon with probability $|a|^2$ and passes it through with probability $|b|^2$. Similarly, a $|\rightarrow\rangle$ horizontal polarization filter absorbs the photon with probability $|b|^2$ and passes it through with probability $|a|^2$.

A $|\uparrow\rangle$ filter followed by a $|\rightarrow\rangle$ filter absorbs all photos. This is similar to our $|1\rangle$ blocker followed by the $|0\rangle$ blocker.

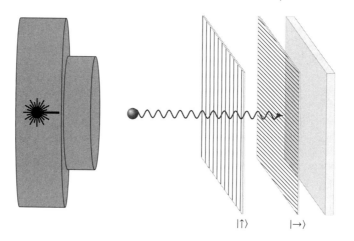

If we now add a middle filter for $|\nearrow\rangle$ in the $\{|\nearrow\rangle, |\searrow\rangle\}$ basis, the photon will reach the wall with probability $\frac{|a|^2}{4}$ ($\{|\nearrow\rangle, |\searrow\rangle\}$ play the role of $\{|+\rangle, |-\rangle\}$ from above.)

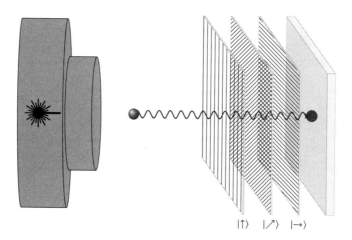

This experiment was first proposed in a 1930 textbook by Paul Dirac. [8] It demonstrates quantum states, superposition, and alternative sets of basis kets.

Question 11.3.3

Instead of having the middle filter at 45° to the others, imagine you can rotate it from the horizontal to the vertical. How does the probability of a photon reaching the wall change? Can you express this mathematically?

To learn more

Given the polarization of light as a two-state quantum system, you might think that you might be able to use photons in the implementations of qubits. You would be correct, and this is the basis for significant academic research and several startup companies.

To see how the techniques satisfy DiVincenzo's Criteria from section 11.2, see the articles by Knill, O'Brien, and others. [17] [21] [22]

11.4 Decoherence

There are three important measurements that quantum computing researchers use to measure coherence time: T_1, T_2, and its cousin T_2^*. Let's begin with T_1. They are single qubit measurements and so we can use the Bloch sphere to discuss them. Their use goes back to Felix Bloch's work on nuclear magnetic resonance (NMR) in the 1940s. [1]

11.4.1 T_1

T_1 goes under several names, all of them connected to the physics of various underlying quantum processes:

- relaxation time,
- thermal relaxation,
- longitudinal relaxation,

- spontaneous emission time,
- amplitude damping, and
- longitudinal coherence time.

It is related to the loss of energy as the quantum state decays from the higher energy $|1\rangle$ state to the $|0\rangle$ ground state. This energy is transmitted to, or leaked into, the environment and lost from the qubit. T_1 is measured in seconds or some fraction thereof such as microseconds. A microsecond is one-millionth of a second, 10^{-6} seconds.

For the computation of T_1, the qubit is moved via an **X** gate from $|0\rangle$ to $|1\rangle$. The decay toward the lower energy state $|0\rangle$ is exponential and follows the rule

$$P\left(|1\rangle = e^{-t/T_1}\right)$$

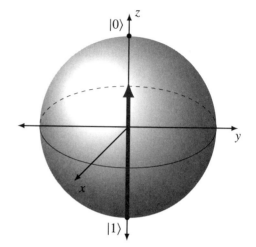

for some constant T_1. Informally, the larger the value of T_1, the longer the qubit is staying closer to $|1\rangle$.

An outline of a scheme to compute T_1 looks like this:

1. Initialize a counter c in $\mathbb{Z}$ to 0. Initialize time t to 0. Initialize the number of runs per time increment to some integer n. For example, $n = 1024$ is reasonable. Choose some very small ϵ in $\mathbb{R}$.
2. Set the time increment between measurements to some small value Δt. (The Greek letter Δ, "Delta," is commonly used for the incremental difference between some value and the next.)
3. Add Δt to t.
4. Initialize the qubit to $|0\rangle$, apply **X**, and wait t seconds.
5. Measure. If we see $|1\rangle$, add 1 to c.
6. Go back to step 4 and repeat $n - 1$ times. If we have done this already, go to the next step.
7. Compute $p_t = \frac{c}{n}$ as the percentage of times we have seen $|1\rangle$.
8. Save and plot the data (t, p_t).
9. Reset c to 0, go back to step 3 and repeat until $p_t < \epsilon$.

The decay of the quantum states for two qubits q_1 and q_2 is shown in this graph:

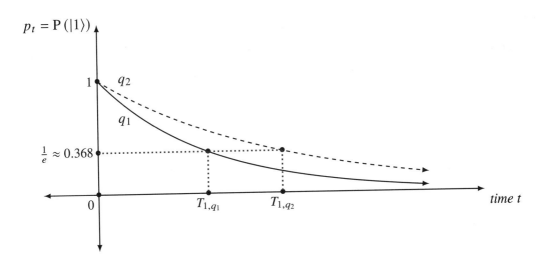

T_1 is the value of t when $p_t = 1/e \approx 0.368$. In this example, q_2 has a larger T_1 and *so has a larger longitudinal coherence time.*

It's interesting to ask at this point when my qubit is in state $|0\rangle$ due to the decay. That is, if we wait long enough after T_1 will the qubit have gone completely to the lower energy state? Theoretically, it is just getting asymptotically closer and the probability of seeing $|0\rangle$ is increasing to 0.9999+.

In practice, if you want the qubit back in an initial $|0\rangle$ state, you wait a time which is a few integer multiple of T_1. You now either assume that statistically you are at $|0\rangle$ or you do a measurement. If you get $|0\rangle$, you are done. If you get $|1\rangle$, apply an **X**.

This last step depends on your ability to do such a conditional action in your hardware and control software. If you can do it, you don't have to wait past T_1, you can just measure and conditionally move to $|0\rangle$ when you wish. See, for example, the $|0\rangle$ reset operation in subsection 7.6.13.

Given

$$P(|1\rangle) = e^{-t/T_1},$$

this means that

$$P(|0\rangle) = 1 - e^{-t/T_1}.$$

If you wait $4T_1$ seconds, the probability of your getting $|0\rangle$ is $1 - e^{-4T_1/T_1} = e^{-4} \approx 0.98168$. If you wait $10T_1$, the probability is $e^{-10} \approx 0.99995$.

11.4.2 T_2 and T_2^*

If T_1 concerned itself with going from the south pole to the north, T_2 adds in the extra element of what is happening at the equator. As a reminder, when I say "the equator," I mean the intersection of the xy-plane with the Bloch sphere.

Like T_1, T_2 and its related metric T_2^* go by several names:

- dephasing time,
- elastic scattering time,
- phase coherence time,

- phase damping, and
- transverse coherence time.

In a perfect world, the circuit

$$q_0: |0\rangle - \boxed{H} - \boxed{ID} - \boxed{H} - \boxed{\nearrow} - |m_0\rangle \ = |0\rangle$$

would return $|0\rangle$ with a probability of 1.0 every time. Though it does nothing in the logical circuit, think of the **ID** as a point in the circuit where we might wait a short period of time before doing the final **H**.

The circuit should bring us from $|0\rangle$ to $|+\rangle$ and back to $|0\rangle$. We move to the equator and then go back to $|0\rangle$. Note that neither $|0\rangle$ nor $|1\rangle$ has a phase component, so we need to experiment elsewhere and the equator is the obvious choice.

This doesn't happen with a physical qubit. Instead, once we move to the equator we begin drifting a little around the xy-plane.

In this example on the Bloch sphere, once we get to $|+\rangle$ we move counterclockwise by some small angle φ in some small time increment. That is, the qubit state is not stable.

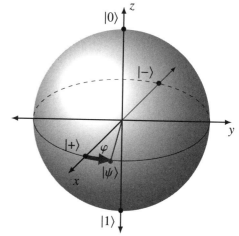

We also expect the state to be moving toward lower energy but I am focusing on what is happening with the phase. Similarly, when we look at T_1, there can be phase drift but we are looking the longitudinal decoherence.

This is an idealized and simplified version of the circuit where we see only the phase shift:

$$q_0: |0\rangle \boxed{H} \boxed{R_\varphi^z} \boxed{H} \boxed{\nearrow} |m_0\rangle$$

Consider the matrix version:

$$\mathbf{H}\mathbf{R}_\varphi^z\mathbf{H}\,|0\rangle = \begin{bmatrix} \frac{\sqrt{2}}{2} & \frac{\sqrt{2}}{2} \\ \frac{\sqrt{2}}{2} & -\frac{\sqrt{2}}{2} \end{bmatrix} \begin{bmatrix} 1 & 0 \\ 0 & e^{\varphi i} \end{bmatrix} \begin{bmatrix} \frac{\sqrt{2}}{2} & \frac{\sqrt{2}}{2} \\ \frac{\sqrt{2}}{2} & -\frac{\sqrt{2}}{2} \end{bmatrix} \begin{bmatrix} 1 \\ 0 \end{bmatrix}$$

$$= \begin{bmatrix} \frac{1}{2}+\frac{1}{2}e^{\varphi i} & \frac{1}{2}-\frac{1}{2}e^{\varphi i} \\ \frac{1}{2}-\frac{1}{2}e^{\varphi i} & \frac{1}{2}+\frac{1}{2}e^{\varphi i} \end{bmatrix} \begin{bmatrix} 1 \\ 0 \end{bmatrix} = \begin{bmatrix} \frac{1}{2}+\frac{1}{2}e^{i\varphi} \\ \frac{1}{2}-\frac{1}{2}e^{i\varphi} \end{bmatrix}$$

$$= \left(\frac{1}{2}+\frac{1}{2}e^{i\varphi}\right)|0\rangle + \left(\frac{1}{2}-\frac{1}{2}e^{i\varphi}\right)|1\rangle$$

$$= \frac{1}{2}\Big((1+e^{i\varphi})\,|0\rangle + (1-e^{i\varphi})\,|1\rangle \Big)$$

When $\varphi = 0$ we get $|0\rangle$, as expected. When $\varphi = \pi$, $\mathbf{R}_\varphi^z = \mathbf{R}_\pi^z = \mathbf{Z}$, and the result is $|1\rangle$. When $0 < \varphi < \pi$, the two probability amplitudes are non-zero.

In particular, the amplitude of $|0\rangle$ is

$$\frac{1}{2}\left(1+e^{i\varphi}\right) = \frac{1}{2}\Big(1+\cos(\varphi)+i\sin(\varphi)\Big).$$

So the probability of getting $|0\rangle$ is the square of the absolute value of this.

$$\frac{1}{4}|1+\cos(\varphi)+i\sin(\varphi)|^2 = \frac{1}{4}\Big(\sin{(\varphi)}^2 + \cos{(\varphi)}^2 + 2\cos{(\varphi)} + 1\Big)$$

$$= \frac{1+\cos(\varphi)}{2}$$

Note that this value is between 0 and 1, inclusive, as it should be.

If the phase change continued at a constant rate over time, you might expect the graph to look like this:

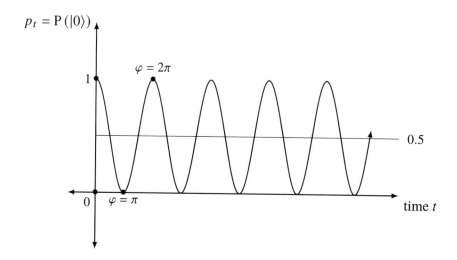

Instead, there is decay toward a probability of 0.5 but with the same short-term periodic behavior.

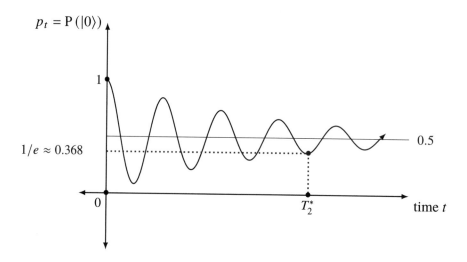

It is only short term because eventually the quantum state will fully decay to $|0\rangle$ and the final **H** will place the state on the equator. At that point the probability of getting $|0\rangle$ is exactly 0.5.

The measurement here is called T_2^* and the circuit is a *Ramsey experiment*. We are considering the wait time between the **H** gates. During this time some rotation φ around the z-axis occurs.

I want to emphasize again that what I have shown here assumed only one kind of noise, the phase drift, and even that was constant. In a real qubit, other noise and irregularities will affect the coherence, including longitudinal decoherence.

Moreover, by using $\mathbf{R}^z_\varphi$ to illustrate the phase drift, I've given the impression that the noise looks like a nice unitary transformation. It does not, as we see in the next section.

The high-level scheme to measure T^*_2 is:

1. Initialize a counter c in $\mathbb{Z}$ to 0. Initialize time t to 0. Initialize the number of runs per time increment to some integer n. For example, $n = 1024$ is reasonable. Choose some very small ϵ in $\mathbb{R}$.
2. Set the time increment between measurements to some small value Δt.
3. Add Δt to t.
4. Initialize the qubit to $|0\rangle$, apply $\mathbf{H}$, and wait t seconds. Apply $\mathbf{H}$ again.
5. Measure. If we see $|1\rangle$, add 1 to c.
6. Go back to step 4 and repeat $n - 1$ times. If we have done this already, go to the next step.
7. Compute $p_t = \frac{c}{n}$ as the percentage of times we have seen $|1\rangle$.
8. Save and plot the data (t, p_t).
9. Reset c to 0, go back to step 3 and repeat until $0.5 - \epsilon < p_t < 0.5 + \epsilon$.

For small enough Δt, T^*_2 is the largest time t where $p_t \leq 1/e$.

A related metric called T_2 is gotten via a *Hahn Echo* with a similar circuit.

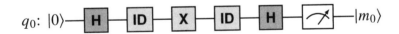

The difference from T^*_2 is that instead of waiting the full time t before the final $\mathbf{H}$, we wait half that long, do an $\mathbf{X}$, wait the remaining time, and then conclude with the $\mathbf{H}$ and measurement. By doing this we are canceling some of the phase drift but keeping the effects of other noise. This technique is called "refocusing." [15]

Generally, $T^*_2 \leq T_2 \leq 2T_1$. [29]

11.4.3   Pure versus mixed states

Each of the quantum states we have considered so far is a *pure state*. These are single linear combinations of basis kets where the probability amplitudes are complex numbers. The sum of the squares of their absolute values add up to 1. Each such square of an absolute value is the probability that we will see the corresponding basis ket when we measure.

There are times when we need to represent a collection, an *ensemble*, of pure states in which our qubit register may reside. This is different from a superposition where we are looking at a quantum phenomenon related to the complex coefficients of the basis kets.

If $\{|\psi\rangle_1, |\psi\rangle_2, \ldots, |\psi\rangle_k\}$ is a collection of pure quantum register states, we define a *mixed state* as

$$\sum_{j=1}^{k} p_j |\psi\rangle_j \text{ for } 0 \leq p_j \leq 1$$

with p_j in $\mathbb{R}$ and

$$\sum_{j=1}^{k} p_j = 1.$$

In a sense, we have a double layer of probabilities with the probability amplitudes on the pure quantum states and the classical probabilities creating the mixed state.

A pure state is a special trivial case of a mixed state where there is only one quantum state in the ensemble and so one p_j.

The density matrix of a mixed state is the sum of the density matrix of the pure states weighted by the p_j. If $\rho_j = |\psi_j\rangle\langle\psi_j|$ is the density matrix of a pure state in the ensemble, then

$$\rho = \sum_{j=1}^{k} p_j \rho_j$$

is the density matrix of the mixed state.

A density matrix ρ corresponds to a pure state if and only if $\text{tr}(\rho^2) = 1$. Otherwise, $\text{tr}(\rho^2) < 1$ and we have a non-trivial mixed state.

Consider the ensemble

$$\left\{ |\psi\rangle_1 = |+\rangle = \frac{\sqrt{2}}{2}(|0\rangle + |1\rangle), \ |\psi\rangle_2 = |-\rangle = \frac{\sqrt{2}}{2}(|0\rangle - |1\rangle) \right\}.$$

The sum $p|+\rangle + (1-p)|-\rangle$ is a non-trivial mixed state if $p \neq 0$ and $p \neq 1$.

We first compute the density matrices

$$\rho_1 = |+\rangle\langle+| = \begin{bmatrix} \frac{1}{2} & \frac{1}{2} \\ \frac{1}{2} & \frac{1}{2} \end{bmatrix} \quad \text{and} \quad \rho_2 = |-\rangle\langle-| = \begin{bmatrix} \frac{1}{2} & -\frac{1}{2} \\ -\frac{1}{2} & \frac{1}{2} \end{bmatrix}$$

and so

$$\rho = p \begin{bmatrix} \frac{1}{2} & \frac{1}{2} \\ \frac{1}{2} & \frac{1}{2} \end{bmatrix} + (1-p) \begin{bmatrix} \frac{1}{2} & -\frac{1}{2} \\ -\frac{1}{2} & \frac{1}{2} \end{bmatrix} = \begin{bmatrix} \frac{1}{2} & p - \frac{1}{2} \\ p - \frac{1}{2} & \frac{1}{2} \end{bmatrix}.$$

The square of ρ is

$$\begin{bmatrix} p^2 - p + \frac{1}{2} & p - \frac{1}{2} \\ p - \frac{1}{2} & p^2 - p + \frac{1}{2} \end{bmatrix}$$

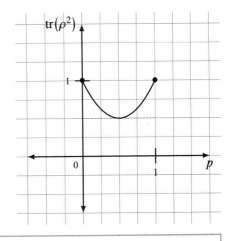

with

$$\text{tr}\left(\rho^2\right) = 2p^2 - 2p + 1.$$

If $p = 0$ or $p = 1$, then this is a trivial mixed state. Plotting the graph for $0 \le p \le 1$, we see that it is always less than 1 otherwise.

Question 11.4.1

Prove algebraically that $2p^2 - 2p + 1 < 1$ when $0 < p < 1$.

11.5 Error correction

In section 2.1 and section 6.4 we looked at some of the basic ideas around classical repetition codes: if you want to send information, create multiple copies of it and hope that enough of them get through unscathed so that you can determine exactly what was sent.

For the quantum situation, the No-Cloning Theorem (subsection 9.3.3) says that we can't copy the state of a qubit, and so traditional repetition is not available.

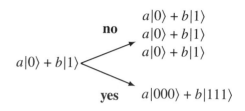

What we can do is entanglement, and it turns out that this is powerful enough when combined with aspects of traditional error correction to give us quantum error correction, or *QEC*.

How can we go from $|\psi\rangle = a|0\rangle + b|1\rangle$ to $a|000\rangle + b|111\rangle$? As you start thinking about such questions, there are two good starting points: "would applying an **H** change the situation into something I know how to handle," and "how might a **CNOT** and entanglement affect things?"

Since I already let on that entanglement is part of the solution, note that a simple

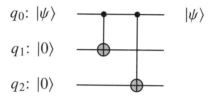

takes $|\psi\rangle|0\rangle|0\rangle$ to $a|000\rangle + b|111\rangle$. Each **CNOT** changes the $|0\rangle$ in q_1 and q_2 to $|1\rangle$ if the amplitude b of $|1\rangle$ in $|\psi\rangle$ is non-zero. Similarly, the **CNOT** does nothing if the amplitude a of $|0\rangle$ in $|\psi\rangle$ is non-zero.

Given this, how can we fix bit flips?

11.5.1 Correcting bit flips

In the classical case of a bit, only one thing can go wrong: the value changes from 0 to 1 or vice versa. Of course, noise may cause multiple bits to change, but for one bit there is only one kind of error.

In the quantum case, a bit flip interchanges $|0\rangle$ and $|1\rangle$ so that a general state $a|0\rangle + b|1\rangle$ becomes $a|1\rangle + b|0\rangle$ instead. From subsection 7.6.1, this is what an **X** does, but when we are thinking about noise and errors, we are saying that a bit flip *might* happen, not that it definitely does.

If we absolutely knew that an **X** was applied, we could just do another one and fix the problem. Therefore we need to get more clever. Consider the circuit

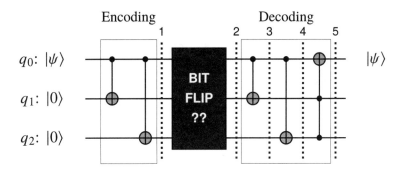

We start the analysis by looking at the possible quantum states at each numbered vertical line. We begin with $|\psi\rangle = |0\rangle$.

	Step 1	Step 2	Step 3	Step 4	Step 5
No errors	$|000\rangle$	$|000\rangle$	$|000\rangle$	$|000\rangle$	$|000\rangle$
One error	$|000\rangle$	$|001\rangle$	$|001\rangle$	$|001\rangle$	$|001\rangle$
	$|000\rangle$	$|010\rangle$	$|010\rangle$	$|010\rangle$	$|010\rangle$
	$|000\rangle$	$|100\rangle$	$|110\rangle$	$|111\rangle$	$|011\rangle$
Two errors	$|000\rangle$	$|011\rangle$	$|011\rangle$	$|011\rangle$	$|111\rangle$
	$|000\rangle$	$|101\rangle$	$|111\rangle$	$|110\rangle$	$|110\rangle$
	$|000\rangle$	$|110\rangle$	$|100\rangle$	$|101\rangle$	$|101\rangle$
Three errors	$|000\rangle$	$|111\rangle$	$|101\rangle$	$|100\rangle$	$|100\rangle$

Question 11.5.1

Create the table for $|\psi\rangle = |1\rangle$.

Question 11.5.2

What is the role of the Toffoli gate between vertical lines 4 and 5?

This circuit corrects up to one bit flip error in q_0. When the error is not corrected, the result of the circuit is always a bit flip in q_0.

If the probability of getting a bit flip error is p, the probability of no error is $1 - p$. We previously worked through the full probability of our being able to fix at most a single error in section 6.4.

11.5.2 Correcting sign flips

A sign flip is a π phase error that switches $a|0\rangle + b|1\rangle$ and $a|0\rangle - b|1\rangle$. As we saw in subsection 7.6.2, a **Z** gate does this.

Changing between the usual computation basis of $|0\rangle$ and $|1\rangle$ and the Hadamard basis of $|+\rangle$ and $|-\rangle$ has the extremely useful property of interchanging bit flips and sign flips. To fix possible sign flips, we only have to insert some **H** gates into the above circuit. This is a consequence of **H X H = Z** and the equivalent **H Z H = X**.

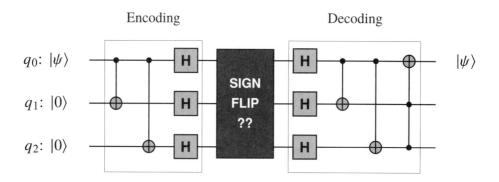

Can we combine these ideas to fix at most one bit, a sign flip, or both? This would be equivalent to fixing an errant **X**, **Z**, or **Y** gate.

11.5.3 The 9-qubit Shor code

To correct either one sign flip or one bit flip, we need eight additional qubits beyond the qubit we are trying to maintain. The circuit in Figure 11.1 is based on work published in 1995 by Peter Shor, the same year he published his breakthrough factoring paper. [27]

As we saw earlier, any 2 by 2 unitary matrix can be written as a complex unit times a linear combination of I_2 and the three Pauli matrices. I_2, σ_x, σ_y, and σ_z are the matrices for the **ID**, **X**, **Y**, and **Z**, respectively. Since we can repair a single error for these gates (where **ID** is really no error), we can repair any single error that is a linear combination of them.

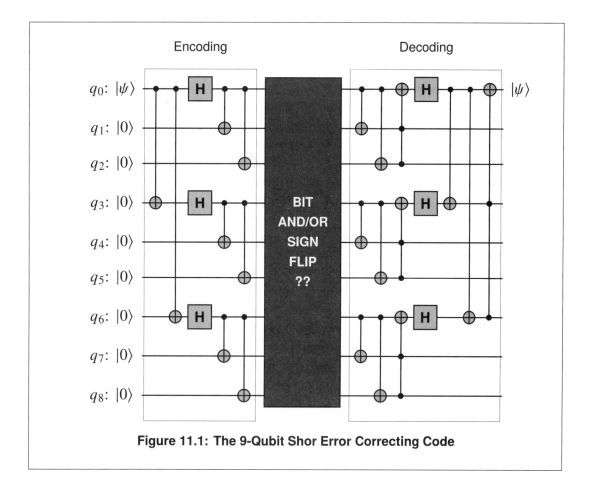

Figure 11.1: The 9-Qubit Shor Error Correcting Code

The Shor 9-bit code can correct any single qubit error corresponding to a unitary gate.

If we have n qubits then we can use $9n$ qubits to duplicate the above circuit n times. The best error correcting code that can correct like the Shor code uses only 5 qubits, and this is as good as is possible. [18]

What happens if an error occurs in one of the **CNOT** or Toffoli gates in the circuit?

11.5.4 Considerations for general fault tolerance

It appears we have a problem: to correct errors we need more qubits, and then we need more qubits to correct the error correcting qubits, and so on. Does it ever stop?

Yes, but we need to use different methods such as *surface codes* based on group theory and a branch of mathematics called topology. In essence, if we can get the error rates of our qubits and gates close to a certain threshold, then we can use entanglement and a lot of qubits to give us fault tolerance. We especially need our two-qubit gate fidelities to improve since those are usually much worse than for single gates.

How many? At the 2019 error rates of the best quantum computing systems, we need approximately 1000 physical qubits to create one logical qubit. Also at this time, the largest working quantum computers have 53 or 54 qubits. More pessimistic estimates put the number of physical qubits we need close to 10, 000.

One logical qubit does not do anyone much good. We'll need hundreds and thousands of these for the most advanced applications people foresee for quantum computers. For example, for Shor's algorithm to factor a 2048 bit number, somewhere around 100 million = 10^8 physical qubits may be needed. This translates to 10^5 = 100, 000 logical qubits.

Other than being used in computation, fault-tolerant qubit technology might also be used for quantum memory and storage. Note, however, there is a twist to this. Unlike in classical computing where you can copy data from memory into a working register for use, you cannot copy a qubit per the No Cloning Theorem from subsection 9.3.3.

But you can use teleportation (subsection 9.3.4)! It's very strange indeed to imagine a computing system with teleportation but no data replication.

I believe that error correction will be introduced partially and incrementally. As we become better able to build quantum devices with lower qubit and gate error rates, we'll be able to perform some degree of limited error correction, so as to at least extend coherence times for some qubits. This will allow us to intelligently use these and the remaining raw physical qubits to implement algorithms that may not look quite so clean and elegant as what we saw in chapter 9 and chapter 10.

By "us" I really mean "our optimizing compilers." As the architecture of quantum devices gets more sophisticated, we need compilers to smartly map our quantum application code in an optimal way to the number and kinds of qubits and available circuit depths.

The formal description of single-qubit errors is in terms of density matrices, mixed states, and probabilities and gives rise to the same for two qubits and the gates that operate upon them.

This is then part of the machinery that has given rise to the theory of large-scale quantum error correction and, we hope, its eventual implementation. [20, Chapter 8]

To learn more

For an overview of error correction techniques and to see early qubit estimates for implementing Shor's algorithm, see Fowler et al. [12] Over time, calculations for such estimates tighten up and improve.

Now that you have almost finished this book, you can read a "beginners guide" to error correction. [7] There is significant current research on error correction and potential hardware implementations. [3] and [14]

11.6   Quantum Volume

How powerful is a gate- and circuit-based quantum computer? How much progress is being made with one qubit technology versus another? When we say we can find a solution on a "powerful enough" quantum computer, what does that mean? When will we know we have arrived?

While it is certainly useful to know how well a given qubit is doing in terms of decoherence and error rates, it tells you nothing about the system in total and how well the components are working together. You may have one or two spectacular, connected, and low error rate qubits, but other aspects of your system may make it unusable for executing useful algorithms.

Implementing hundreds of very bad qubits does not give you an advantage in the circuit model over having far fewer but excellent qubits with good control and measurement. Therefore, we need a whole-system, or "holistic," metric that can tell us at a glance the relative of our quantum computer.

Such an architecture-independent metric is called Quantum Volume and was devised by scientists at IBM Research in 2017. [6]

Quantum Volume is reported as 2^v where the best performance of your system is seen on a test circuit area v qubits wide by v gates deep. In 2019, IBM announced that it had reached a Quantum Volume of $16 = 2^4$ for its superconducting transmon quantum computing systems and expected to be able to at least double it year over year for the next decade. [13]

Though IBM came up with this metric, there is nothing brand-specific about it. Indeed, other researchers have proposed generalizations of the metric to decouple the width and depth of the circuits being tested. [2]

The range of factors taken into account by Quantum Volume include

- **calibration errors**

  How well the electronic controls for programming and measuring qubits are calibrated to ensure accurate operation and error mitigation.
- **circuit optimization**

  How well an optimizing compiler improves the layout and performance of a circuit across its depth and qubit register width.
- **coherence**

  How long the qubit stays in a usable state as discussed in section 11.4.
- **connectivity and coupling map**

  How qubits are connected to other qubits and in what patterns.
- **crosstalk**

  How the state of one qubit affects nearby qubits, either in gate operations or through more passive entanglement.
- **gate fidelity**

  At what error rates do gate operations move qubits from their current to new states.
- **gate parallelism**

  How many gate operations can be run on qubits in parallel, that is, at the same time. This is different from the concept of quantum parallelism.
- **initialization fidelity**

  How accurately we can set the initial qubit state to something known, usually $|0\rangle$.
- **measurement fidelity**

  How accurately we can collapse a qubit state and read it out as $|0\rangle$ or $|1\rangle$.
- **number of qubits**

  More can be better, but not always. In any case, you need a sufficient number.
- **spectator errors**

  How much is a qubit that is supposed to be idle affected during a 1- or 2-qubit gate operation on a physically connected qubit.

As a metric, gate fidelity varies between 0.0 and 1.0, inclusively. A value of 1.0 says the gate perfectly implements the intended logical unitary transformation. Two-qubit gates typically have gate error rates much higher than single qubit gates, on the order of ten times worse.

Having qubits connected to more qubits can be a good thing, as long as that coupling does not introduce additional errors across the connected qubits.

This is one of the reasons why Quantum Volume will not increase solely by adding more and increasingly coupled qubits.

Simple Grid

All-to-All

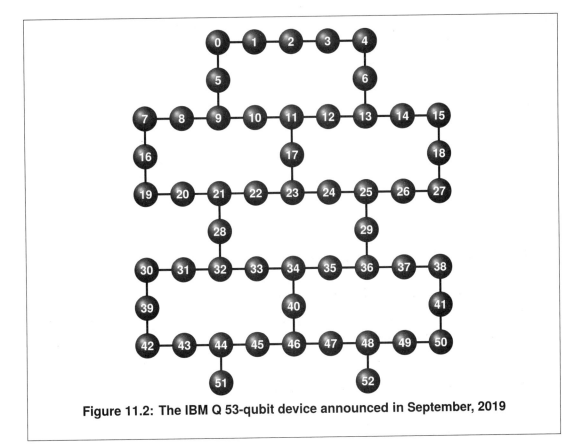

Figure 11.2: The IBM Q 53-qubit device announced in September, 2019

Where a quantum device is laid out in a physical medium like silicon, it can use non-square and non-rectangular patterns. Regular polygonal tiling patterns can be used, with the qubits not necessarily at the corners. Moreover, patterns can vary in different regions of the device and along the edges.

Question 11.6.1

In Figure 11.2, what's the maximum number of qubits to which one qubit is connected? Minimum? What is the average number of qubits to which each qubit is connected? [19]

Having a well-performing optimizing compiler can increase Quantum Volume by reducing the number of gates and rearranging how and when they are applied to qubits. By avoiding qubits with shorter coherence times and greater one- and two-qubit error rates, and using native gates on well connected qubits, quantum volume may be increased significantly by the compiler.

The theory and implementation of optimizing compilers for classical computers is well known and has been studied over decades. Quantum "transpiling" (transformational compiling) is in its infancy but fast progress is already being seen in Qiskit through the open source community of computer scientists, quantum researchers, and software developers from academia and industry.

> The number of qubits is a very bad and inaccurate metric of the quality and performance of a quantum computer. Rather, use Quantum Volume to get a whole-system measure of what your hardware, software, system, and tools for software development can do.

11.7 The software stack and access

One way of accessing a quantum computing system looks like this:

- You have downloaded and installed software development tools like the Qiskit open source quantum computing framework to your laptop or workstation. [24]
- You develop your quantum code in a programmer's editor or a Jupyter^{TM} notebook. [23]

- When run, part of your application connects to a quantum simulator on your computer or remotely to real quantum hardware or a simulator.
- The remote connection is via the Internet/cloud.
- Your application invokes one or more processes that runs on the quantum hardware or a simulator.
- Ultimately, your application makes use of the results of the quantum computation and does something valuable within your use case.

There are at least two other similar scenarios:

1. Instead of developing locally, you do so via a web browser-based environment where your Jupyter notebooks are edited, tested, stored, and executed on the cloud.
2. You have runnable code that you place in a container in the cloud and it accesses quantum computers either close to or remote from the classical cloud servers.

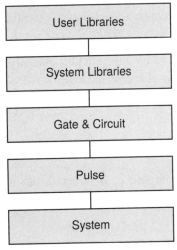

For developers, the software stack looks like the diagram on the left. On the bottom is the System level and this can include APIs for hooking in higher-level services and benchmarking.

The Pulse level, for quantum computing systems that use microwave control systems, allows direct definition and use of the microwave pulses that configure quantum gates.

Above that is the support for using builtin gates and creating new circuits. This allows implementing them within quantum registers and building out the algorithms we discussed in chapter 7 through chapter 10.

At the System Libraries level, the coder can re-use previously defined and optimized circuits within their own circuits or application code. When using these libraries, the developer mindset is still in terms of how quantum algorithms work but is well above anything that could be described as "assembly language."

Finally, at the User Libraries level we have convenience libraries to accelerate general software development. Here a developer might not even be aware that a quantum computing system is used at all.

This stack representation gives you a rough idea of where functionality is provided from deep down and close to the quantum computer all the way to the most abstract access points. Any particular framework may structure its developer software stack in more or fewer levels.

Before we leave this topic I want to give my opinion on whether we need new programming languages just for quantum computing or if the functionality should be embedded within existing languages. For various scientific applications in the past, I've done both.

For many of us, it's fun to create new languages. That might not be *your* idea of fun, but so be it! It's very tempting to start building a quantum programming environment and then find you are spending too much time doing classical computer science engineering, most of which has likely been done before – and better.

Modern programming languages like Python, Swift, Go, Java, and C++ are superb for software development. For me, it makes much more sense to build the quantum gate, circuit, and execution support within the type systems of these kinds of languages. In this way, you can quickly use the entire tool chain and support for the established languages and concentrate on quickly and comprehensively adding quantum software development.

Moreover, if your development framework is open source, you have a much larger potential set of contributors who know the existing languages than if you need to teach them a new one.

As an example, Qiskit uses Python as the main developer language but provides extensive libraries at all levels of the software stack. In some cases, lower-level access or optimized code was written and then provided with Python library interfaces.

11.8   Simulation

Is it possible to simulate a quantum computer on a classical computer? If we could do it, "quantum computing" would be only another technique for coding software on our current machines.

In this section we look at what you have to take into account if you want to write a simulator for manipulating logical qubits.

If you have a simulator handy, such as one that is provided by Qiskit or the IBM Q Experience, you can use it for small problems. Here we look at how you might build a simulator, in general terms. I offer no complete code in any particular programming language but more of a list of what you need to take into account. You can skip this section if you are not interested in such concerns.

11.8.1   Qubits

Your first decision when thinking about building a quantum computing simulator is how you represent qubits. With this and your other choices, you can build a general model or you can specialize it. We're going for the general case. Once you complete this section you can go back and think about how each of the interconnected parts can be optimized individually and together.

We want to work with more than one qubit and so we do not use a Bloch sphere. The state of a qubit is held in two complex numbers and you can store them in an ordered list, array, or similar structure. If your language or environment has a mathematical vector type, use it.

While we might here use an exact value like

$$\frac{\sqrt{2}}{2}|0\rangle + \frac{\sqrt{2}}{2}|1\rangle,$$

your language or environment probably uses floating point numbers for the real and imaginary parts of a complex number. So the above might be

$$0.7071067811865476\,|0\rangle + 0.7071067811865476\,|1\rangle$$

and the sum of the squares of the probability amplitudes would be 1.0000000000000002. You need to keep track of and possibly control this error if you use many qubits or gates.

For a very small simulator you should consider using a symbolic mathematics library such as SymPy. [28] The time and memory overhead might be too much for your use. Also, symbolic expressions can get complicated and messy quickly, so your system's ability to simplify these expressions is important.

If you have an n-bit quantum register, you need to represent a vector of 2^n complex numbers. If $n = 10$ then a list of 1024 complex numbers uses 9024 bytes. For 20 qubits it is $8,697,464$ bytes or approximately 8.3MB. Just adding two more qubits brings this to $35,746,776$, or 34MB.

Think about this: for a single state of a quantum register with 22 qubits you need 34MB to represent it. It gets exponentially bigger and worse than this as we add more qubits. We are over a gigabyte per state at 27 qubits. It's not only storage that is increasing, the time required to manipulate all those values is getting big fast. Your choice of algorithms at every level is critical.

It gets even worse: the size of the matrix for a gate is the *square* of the number of entries in the qubit state ket.

With optimizations we can get these numbers lower but the exponential nature of growth will eventually catch up with you. You might be able to do a few more qubits but simulation will eventually run out of steam. One way of doing this is to use single precision floating point numbers instead of double. It saves memory, briefly.

My guess is general-purpose quantum computation simulation will get too big and time-impractical even for supercomputers somewhere in the mid-40 number of qubits. If you have a very specific problem you are trying to simulate, then you may be able to simplify the mathematical formulas that represent the circuit. Just as $\sin^2(x) + \cos^2(x)$ reduces to the much simpler 1, the math for your circuit may get smaller. Even then I think it is likely that special purpose simulation will top out around 70 to 80 qubits.

Simulation is good for experimentation, education, and debugging part of a quantum circuit. Once we have powerful and useful real quantum computers of more than 50 qubits or so, the need for simulators will decline, probably along with the commercial market for them, too.

Consider using a sparse representation for qubits and ket vectors. Do you really need $2^{50} = 1,125,899,906,842,624$ numbers to represent the state $|0\rangle^{\otimes 50}$? After all, there are only two pieces of significant information there, 0 and 50. Add a little overhead for representing the sparse ket and you can fit it all into a few bytes.

11.8.2 Gates

If qubits are vectors, then gates are matrices. The most straightforward way of implementing multi-wire circuits is to construct the tensor product matrix for two gates. These matrices get quite large: if you have n qubits then your matrices will be 2^n by 2^n in size.

For the subcircuit

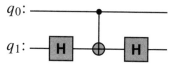

we have the product of the matrices corresponding to $\mathbf{ID} \otimes \mathbf{H}$

$$
\begin{bmatrix} 1 & 0 \\ 0 & 1 \end{bmatrix} \otimes \begin{bmatrix} \frac{\sqrt{2}}{2} & \frac{\sqrt{2}}{2} \\ \frac{\sqrt{2}}{2} & -\frac{\sqrt{2}}{2} \end{bmatrix} = \begin{bmatrix} \frac{\sqrt{2}}{2} & \frac{\sqrt{2}}{2} & 0 & 0 \\ \frac{\sqrt{2}}{2} & -\frac{\sqrt{2}}{2} & 0 & 0 \\ 0 & 0 & \frac{\sqrt{2}}{2} & \frac{\sqrt{2}}{2} \\ 0 & 0 & \frac{\sqrt{2}}{2} & -\frac{\sqrt{2}}{2} \end{bmatrix},
$$

the **CNOT**

$$
\begin{bmatrix} 1 & 0 & 0 & 0 \\ 0 & 1 & 0 & 0 \\ 0 & 0 & 0 & 1 \\ 0 & 0 & 1 & 0 \end{bmatrix}
$$

and then another $\mathbf{ID} \otimes \mathbf{H}$. This is

$$
\begin{bmatrix} 1 & 0 & 0 & 0 \\ 0 & 1 & 0 & 0 \\ 0 & 0 & 1 & 0 \\ 0 & 0 & 0 & -1 \end{bmatrix}.
$$

As we know, it flips the sign of $|11\rangle$. Had we identified this as a standard pattern, we could have done less matrix manipulation.

11.8.3 Measurement

Once you come up with a final quantum register state like

$$
\frac{\sqrt{2}}{2}|0\rangle - \frac{\sqrt{2}}{2}i\,|1\rangle
$$

or

$$0.3872983346207417\,i\,|00\rangle - 0.6082762530298219|01\rangle -$$
$$0.5099019513592785|10\rangle + 0.469041575982343\,i\,|11\rangle,$$

how do you simulate measurement? We use the simulated sampling method using random numbers from section 6.5.

If there is one ket with a non-zero amplitude then that must be the result of the measurement. We now assume there are two or more non-zero amplitudes.

Compute the probabilities corresponding to each standard basis ket: if

$$|\psi\rangle = \sum_{j=0}^{2^n-1} a_j|j\rangle_n$$

then let $p_j = |a_j|^2 = a_j\overline{a_j}$. Subject to a little round off error,

$$1.0 = \sum_{j=0}^{2^n-1} p_j.$$

These are the `example_probabilities` we use in Listing 6.1. In the output

```
Results for 1000000 simulated samples

Event        Actual Probability       Simulated Probability
  0               0.15                     0.1507
  1               0.37                     0.3699
  2               0.26                     0.2592
  3               0.22                     0.2202
```

$E_0 = |0\rangle$, $E_1 = |1\rangle$, and so on.

As another example, we can look at simulated measurements for a balanced superposition of four qubits. In this case, each of the amplitudes is 0.25 and its probability is 0.0625. Here is a sample run of 1000000 iterations:

```
example_probabilities = [1.0/16 for _ in range(16)]

Results for 1000000 simulated samples

Event        Actual Probability       Simulated Probability
  0               0.0625                   0.0627
```

1	0.0625	0.0628
2	0.0625	0.0627
3	0.0625	0.0619
4	0.0625	0.0622
5	0.0625	0.0624
6	0.0625	0.0624
7	0.0625	0.0626
8	0.0625	0.0622
9	0.0625	0.0624
10	0.0625	0.0624
11	0.0625	0.0626
12	0.0625	0.0631
13	0.0625	0.0623
14	0.0625	0.0625
15	0.0625	0.0628

Again recall, for example, that getting E_7 means we are getting a result of $|7\rangle$ on measurement.

11.8.4 Circuits

To simulate a circuit, you need to represent it in some way. Think about the wire model and then the horizontal steps from left to right where you place and execute gates.

Multi-qubit gates span wires, so you need to specify wire inputs and outputs. You need to do error checking along the way to ensure two gates in the same step do not involve the same input and output wires.

I recommend that you start with an API, an application programming interface to a collection of software routines that sits on top of your internal circuit representation. If you start by developing a new language in which to write circuits, you'll likely spend more early coding cycles on the language rather than the simulator itself.

11.8.5 Coding a simulator

Should you decide to code a quantum simulator, here is some advice:

- Unless you want to do it as an educational project or you have a brilliant new idea, don't bother. There are plenty of simulators out there, many of them open source such as those in Qiskit.
- Don't start by optimizing circuits. It can be difficult enough to debug code that is supposed to do the sequence of operations you wish.

- When you do start optimizing, look for the easy stuff like getting rid of consecutive gates that do nothing. Three examples of this are **H H**, **X X**, and **Z Z**.
- Don't start tensoring matrices together until you have a wire-spanning operation like **CNOT**.
- Build efficient subroutines that simulate standard gate combinations. Do not, for example, build a **CNOT** from a Toffoli gate but do provide the Toffoli gate in your collection.
- Go much deeper into how quantum gates are designed from more primitive gates. Learn about Clifford gates and how to simulate them, for example. Note, this will require deeper knowledge of quantum computing and computer science. [4] [5]

To learn more

To re-emphasize my point about not necessarily coding your own, a web search of "list of quantum simulators" will turn up dozens of simulators in multiple programming languages, many of them open source.

11.9 The cat

In this section we look at a well known discussion from the 1930s. We'll use it to show an example of simulating quantum physics with quantum computing.

In 1935, physicist Erwin Schrödinger proposed a thought experiment that would spawn close to a century of deep scientific and philosophical thought, as well as many bad jokes. Thought experiments are common among mathematicians and scientists.

The basic premise is that the idea is not something you would really *do*, but something you want to think through to understand the implications and consequences.

This was his attempt to show how the *Copenhagen interpretation* promoted by Niels Bohr and Werner Heisenberg in the late 1920s could lead to a ridiculous conclusion for large objects. This is one of the popular theories for how and why quantum mechanics works, though there are others.

Niels Bohr in 1922. Photo is in the public domain. ℗

By "large" here I mean "large as a cat."

Question 11.9.1

What does Copenhagen have to do with quantum mechanics?

The setup

In a large steel box that contains more than enough air for a cat to breathe for several hours, we place a small amount of radioactive material that has a 0.5 probability of having a single atom decay and emitting a particle per hour.

We also add a device that is a Geiger counter that can detect that single emission, plus a connected hammer that can smash open a closed vial of cyanide poison. If the Geiger counter detects anything, the hammer swings and the cyanide is released into the air.

We now place the otherwise charming but confused-looking cat into the box and seal the top.

Feel free to replace the cat with something else that would not survive in the presence of cyanide.

The wait

While we let time go by, we wonder about the state of the cat. Is he doing well or has he cast off his mortal coil? Has the radioactive emission occurred and triggered the hammer?

Until we look, we don't know. As far as our knowledge goes, the cat is in a superposition of being dead and alive. By our opening the top, by our observing what has happened in the box, we have caused the superposition to collapse to $|\mathbf{dead}\rangle = |0\rangle$ or $|\mathbf{alive}\rangle = |1\rangle$. This is according to the Copenhagen interpretation.

In the *Many Worlds* interpretation, two realities were created when the opportunity for a choice was created. In one world the cat is dead, in the other it is not.

Erwin Schrödinger in 1933. Photo is in the public domain. ℗

Now let's express this situation in the language of a quantum circuit.

A circuit

Consider this simple circuit with two **CNOT** gates:

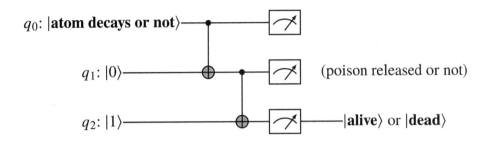

For q_0, an input state of $|0\rangle$ means no atom decays at the time the circuit is run, while $|1\rangle$ means that a particle is emitted.

q_1 is set to an initial state of $|0\rangle$ but is flipped to $|1\rangle$ if an atom decays. This causes the hammer to break the vial and the cyanide to enter the air.

For q_2, the cat starts out in $|\textbf{alive}\rangle = |1\rangle$. The final state of the cat only switches to $|\textbf{dead}\rangle = |0\rangle$ if the poison is released.

Question 11.9.2

Experiment with Bell states to introduce entanglement. Do you learn anything new about the components of this experiment?

11.10 Summary

In this chapter we connected the "logical" idea of qubits and circuits with the "physical" idea of how you might build a quantum computer. We looked at the polarization of light to show a physical quantum system, and used our ket notation to understand the unusual effect we observe when using three filters. Through this we could see that the theory of quantum mechanics does seem to provide a good model for what we experimentally observe, at least in this case.

Real qubits don't survive forever and decoherence explains the several ways in which quantum states wander over time. While we do not yet have large enough systems to implement fault tolerance, we examined what error correction might be able to do in the quantum realm.

We need to be able to measure the progress we are making and answer questions like "how powerful is your quantum computer?". The definition of quantum volume looks at many factors that can affect performance.

We mentioned several technologies that researchers are exploring to implement qubits. Superconducting transmon qubits are used by IBM, Google, and others, but ion traps and photonic techniques also hold promise, according to their proponent scientists and engineers.

Our cat-in-box was translated into quantum terms and got its own circuit.

References

[1]   F. Bloch. "Nuclear Induction". In: *Physical Review* 70 (7-8 Oct. 1946), pp. 460–474.

[2]   Robin Blume-Kohout and Kevin C. Young. *A volumetric framework for quantum computer benchmarks*. URL: https://arxiv.org/abs/1904.05546.

[3]   H. Bombin and M. A. Martin-Delgado. "Optimal resources for topological two-dimensional stabilizer codes: Comparative study". In: *Physical Review A* 76 (1 July 2007), p. 012305.

[4]   S. Bravyi and D. Gosset. "Improved Classical Simulation of Quantum Circuits Dominated by Clifford Gates". In: *Physical Review Letters* 116.25, 250501 (June 2016), p. 250501.

[5]   S. Bravyi et al. "Simulation of quantum circuits by low-rank stabilizer decompositions". In: (July 2018).

[6]   Andrew Cross et al. *Validating quantum computers using randomized model circuits*. Nov. 2018. URL: https://arxiv.org/abs/1811.12926.

[7]   Simon J. Devitt, William J. Munro, and Kae Nemoto. "Quantum error correction for beginners". In: *Reports on Progress in Physics* 76.7, 076001 (July 2013), p. 076001.

[8]   P. A. M. Dirac. *The Principles of Quantum Mechanics*. Clarendon Press, 1930.

[9]   David P DiVincenzo. *Looking back at the DiVincenzo criteria*. 2018. URL: https://blog.qutech.nl/index.php/2018/02/22/looking-back-at-the-divincenzo-criteria/.

[10]   David P DiVincenzo. "The physical implementation of quantum computation". In: *Fortschritte der Physik* 48.9-11 (2000), pp. 771–783.

[11]   Richard P. Feynman. *QED: The Strange Theory of Light and Matter*. Princeton Science Library 33. Princeton University Press, 2014.

[12]   Austin G. Fowler et al. "Surface codes: Towards practical large-scale quantum computation". In: *Phys. Rev. A* 86 (3 Sept. 2012), p. 032324.

[13]   Jay Gambetta and Sarah Sheldon. *Cramming More Power into a Quantum Device*. 2019. URL: https://www.ibm.com/blogs/research/2019/03/power-quantum-device/.

[14]   Jay M. Gambetta, Jerry M. Chow, and Matthias Steffen. "Building logical qubits in a supercon-
       ducting quantum computing system". In: *npj Quantum Information* 3.1 (2017).

[15]   Gambetta, Jay M. (question answered by). *What's the Difference between T_2 and T_2^*?* URL: https:
       //quantumcomputing.stackexchange.com/questions/2432/whats-the-difference-between-t2-and-t2.

[16]   IBM. *20 & 50 Qubit Arrays*. 2017. URL: https://www-03.ibm.com/press/us/en/photo/53377.wss.

[17]   E. Knill, R. Laflamme, and G. J. Milburn. "A scheme for efficient quantum computation with
       linear optics". In: *Nature* 409.6816 (2001), pp. 46–52.

[18]   Raymond Laflamme et al. "Perfect Quantum Error Correcting Code". In: *Physical Review Letters*
       77 (1 July 1996), pp. 198–201.

[19]   Doug McClure. *Quantum computation center opens*. 2019. URL: https://www.ibm.com/blogs/research/
       2019/09/quantum-computation-center/.

[20]   Michael A. Nielsen and Isaac L. Chuang. *Quantum Computation and Quantum Information*.
       10th ed. Cambridge University Press, 2011.

[21]   Jeremy L. O'Brien. "Optical Quantum Computing". In: *Science* 318.5856 (2007), pp. 1567–1570.

[22]   Jeremy L. O'Brien, Akira Furusawa, and Jelena Vučković. "Photonic quantum technologies". In:
       Nature Photonics 12.3 (2009), pp. 687–695.

[23]   Project Jupyter. *Project Jupyter*. URL: https://jupyter.org/.

[24]   Qiskit.org. *Qiskit: An Open-source Framework for Quantum Computing*. URL: https://qiskit.org/
       documentation/.

[25]   A.I.M. Rae. *Quantum physics: Illusion or reality?* 2nd ed. Canto Classics. Cambridge University
       Press, Mar. 2012.

[26]   Erwin Schrödinger. *What is Life? The Physical Aspect of the Living Cell*. Cambridge University
       Press, 1944.

[27]   Peter W. Shor. "Scheme for reducing decoherence in quantum computer memory". In: *Physical
       Review A* 52 (4 Oct. 1995), R2493–R2496.

[28]   SymPy Development Team. *SymPy symbolic mathematics library*. 2018. URL: https://www.sympy.
       org/en/index.html.

[29]   X. R. Wang, Y. S. Zheng, and Sun Yin. "Spin relaxation and decoherence of two-level systems".
       In: *Physical Review B* 72 (12 Sept. 2005), p. 121303.

12

Questions about the Future

We can only see a short distance ahead,
but we can see plenty there that needs to be done.

Alan Turing [2]

How will quantum computing evolve over the coming years and decades? It's critical not to say that quantum computing **will** do this or that, but rather **may**. Until someone does this or that, it's speculation, hype, or a work in progress.

Via a series of motivating questions, I give you a framework to check in on the progress of the full software, hardware, and systems stack. These questions also deal with how, where, and when you might start using, teaching, or learning about quantum computing.

The state of the art will be changing rapidly. Returning to these questions and their answers every few months will help you gauge what has been done and why it is significant. They will allow you to understand whether quantum is ready for you, and if you are ready for quantum.

Topics covered in this chapter

12.1   Ecosystem and community

"Ecosystem" is an overused and often vague word regarding groups of people who have some connection to an activity. I now try to be more precise in describing what I consider the breadth of the quantum computing ecosystem. I touched upon some parts of the ecosystem above, especially in the education section.

The goal of the ecosystem is to reach *Quantum Advantage*, the point where quantum computing can do significantly better than classical computing on important problems for business, science, and government.

For the sections that follow, think about what role or roles you have or want to have in the quantum computing ecosystem and then answer those questions relevant to you.

1. What role or roles do you play in the quantum computing ecosystem are you?

a) Algorithm developer
b) Business development / sales
c) Business or technology executive
d) Business partner
e) Cloud access provider
f) Communications
g) Community leader or participant
h) Consultant
i) Development tools provider
j) Educator
k) Full stack provider
l) Hardware provider
m) Industry analyst
n) Industry use case expert

o) Journalist
p) Marketing
q) Quantum application software provider
r) Quantum hardware engineer
s) Quantum software engineer
t) Quantum software platform provider
u) Scientist or researcher
v) Student
w) Support
x) Systems integrator
y) Venture capitalist or other investor
z) Other

To clarify some of these role definitions:

- Hardware provider – supplies the quantum computing hardware
- Software platform provider – supplies the development tools and runtime software
- Application software provider – supplies the top-level applications that implement industry use cases
- Cloud access provider – supplies the cloud services that allow you to use remote quantum computers
- Full stack provider – supplies all of the above

2. How do you interact with other members of the quantum computing ecosystem doing similar work?
3. How do you interact with other parts of the ecosystem who do complementary work?
4. How could these interactions get started or be made more productive?
5. In what ways *must* these interactions become broader and richer as quantum computing develops?
6. Are you part of an open source community developing quantum computing software?
7. What are you doing personally to improve the quality and reach of the quantum computing community?
8. If you are part of a startup, how can vendors better support you?
9. If you are an analyst or a consultant, how can you best get the information you need to advise your clients?
10. How should we all work together to achieve Quantum Advantage faster?

12.2 Applications and strategy

What do we mean by a "quantum application"? It's not software where the only computer used is a quantum one. That is not possible today, nor will it be necessary or possible for many decades, or even centuries. Rather, a quantum application is a hybrid classical-quantum solution that use both kinds of hardware and software.

Industry use cases, as I touched upon briefly in chapter 1, will drive the creation of these applications. Over the years, the definition of the use cases will change as we better understand how quantum computing systems can and can't help us. Benchmarks will be important, but only in measuring progress.

Together with those in the other sections, these questions will help you think about use cases for quantum computing and your plan for matching them to quantum solutions:

1. Where is your classical computing taking too long?
2. Where is your classical computing too inaccurate?
3. Do you currently use HPC, High Performance Computing?

4. If so, what are the bottlenecks in your solutions?
5. Can you pinpoint areas where there is an exponential growth of necessary memory or computation time?
6. Is your application data intensive or computation intensive?
7. In what ways do you want to scale the computation in your system?
8. Do you want to do what you are doing now faster, or do you want greater computing capacity to examine more possibilities and scenarios?
9. Are there use cases in industries similar to your own that are looking at pathways to using quantum computing?
10. Are there proposed NISQ quantum solutions for your use cases or will you need fault tolerance? A "NISQ quantum solution" is one that can operate with noisy qubits and short depth circuits (section 11.1).
11. How will quantum computing fit into your existing workflow?
12. Do you understand your quantum strategy well enough so you know which applications will be possible in the short, medium, and long terms?
13. Do you work with vendors, industry analysts, management consultations, and system integrators to hone your quantum strategy?
14. What is their expertise in understanding the current state of quantum hardware and software?
15. Have you established your quantum education, experimentation, and implementation roadmap?

12.3 Access

"Access" refers to how you connect to a quantum computing system. Connecting via the cloud can give you all the benefits of cloud computing regarding security, elastic resources, and software and hardware upgrades.

1. Can you get the quantum computing capacity you need through the cloud?
2. What are your security requirements for such remote access?
3. Do you require a special kind of hosted quantum cloud data center for legislative, national, or military reasons?
4. Can you use a remote quantum computer in another country?
5. From which countries *can you* and *can you not* access quantum computers?
6. What are your quality of service requirements for quantum computing, including up time, prioritization, and scheduling?

7. Will you need access to more than one quantum computer at a time?

8. In terms of Quantum Volume, how powerful a machine, or machines, do you need?

9. Does your quantum computing provider have a roadmap to provide you access to their newest and most powerful systems?

10. Can you get free cloud access to smaller quantum computers and then move on to commercial quality systems?

11. Can you access quantum computers on the same vendor cloud on which you run classical applications?

12. Do you anticipate a need to own a quantum computer versus accessing one via the cloud?

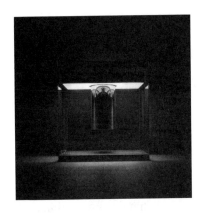

The IBM Q System One, introduced in 2019. Photo subject to use via the Creative Commons Attribution-NoDerivs 2.0 Generic license ⓒ①⊜. [1]

12.4 Software

If a quantum computer is to be programmable, it must have software. More than that, your chosen system must have a full stack of development tools and runtime facilities.

1. Do you prefer working with new, semi-proprietary languages for quantum computing or would you rather reuse existing skills in languages like Python?

2. Does your staff already have software engineering talent in Python?

3. Have you surveyed the current development platforms for quantum computing?

4. Have you assessed the breadth of functionality and algorithms implemented in the User Libraries level of the stack?

5. Do you need ready-made optimized circuits in the System Libraries level?

6. Will you be implementing new circuits?

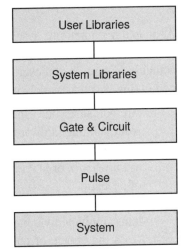

7. Does the development environment include a visual circuit designer for implementing algorithms such as those shown in this book?

8. What is the level of abstraction with which you can create circuits?

9. Does the programming environment offer direct access to the operating characteristics of the hardware?

10. Can you design new gates and directly control the qubit hardware in the Pulse level or its equivalent?

11. Does the development environment and programming language only target quantum simulators?

12. Can you write code and have it run on both quantum hardware and simulators?

13. Is your quantum software development platform open source? What license is used?

14. If so, is the software supported by a large, vibrant, and multidisciplinary group of developers?

15. What documentation and media support is available to help you start developing?

16. For how long has the software platform been under development?

17. Is there a public roadmap for planned future development?

18. What is the software development tool chain in your quantum platform?

19. Does your platform contain an optimizing compiler to get the best performance of your application on the hardware?

20. What facilities are there for debugging your code?

21. How easily and graphically can you see and understand the results of your circuits?

12.5 Hardware

A quantum computing system needs real quantum computing hardware. While simulators may be useful for learning, experimenting, and debugging small problems, the sooner you use actual hardware, the quicker you'll take advantage of its potential. You are not doing quantum computing if you are solely using classical hardware.

1. Are you sure your vendor is providing you access to real quantum hardware, or is it only a simulator?

2. Is the quantum hardware general purpose (also known as *universal*), or is it designed to solve only one kind of problem?

3. Can your quantum hardware be used to solve problems significantly better than classical technologies?

4. Does the choice of qubit technology make a difference to you?

5. What qubit technology is producing systems with the best Quantum Volumes?

6. How is qubit technology scaling?

An early IBM Q four qubit chip. Photo subject to use via the Creative Commons Attribution-NoDerivs 2.0 Generic license ⊜①⊜. [1]

7. Can you use the same quantum software development environment to target different quantum hardware?
8. Does your quantum computing system offer physical qubits, logical qubits, or both?
9. Can your quantum hardware, with its associated software, be used to entangle many qubits?
10. Are you confident that the provider of your quantum computing hardware will provide you with long-term support and continue to enhance the technology?

12.6 Education

I split the questions about education for quantum computing and coding between teaching or training, and learning.

Teaching

1. Do you currently teach or plan to teach a class involving quantum computing?
2. Do you teach a single class on the topic or do you teach parts of the subject in different classes?
3. Where would you augment the following kinds of classes with quantum computing?

   a) Artificial Intelligence
   b) Chemistry
   c) Computer Science
   d) Engineering
   e) Materials Science

   f) Philosophy
   g) Physics
   h) Pure and Applied Mathematics
   i) Quantitative Economics
      and Finance

4. Do you supplement the material with hands-on homework or labs using a software development environment?
5. Do your students download the development environment or use it on the cloud through a browser?
6. Do you use a textbook, your own material, or both?
7. Do you supplement your material with content, videos, and exercises found online?
8. Are you part of a community developing teaching materials?
9. How do you augment student activities in quantum computing outside the classroom?
10. Do you yourself take online classes on quantum computing to learn how to better teach the subject?

Learning

1. Have you taken a class on quantum computing in school, through your organization, or online?
2. What classes have you taken that could have been extended with quantum computing content?
3. Have you asked your teachers or professors to include that content?
4. Are you willing to learn about quantum computing outside the classroom?
5. Have you participated in a quantum computing hackathon at your school or in your organization?
6. Are you willing to host such a hackathon, with help from a quantum computing vendor?
7. Have you visited an exhibit about quantum computing at a science museum?
8. How important is it to you to get badges for your resumé that show your proficiency in quantum computing?
9. Do you know how to code, specifically Python? Python is one of the most commonly used programming language for scientific and AI applications.
10. Are you interested in learning how to code quantum computing applications?
11. Do you plan to do quantum computing research or software engineering, or is this just an interest of yours?

The author speaking at the Boston Museum of Science in April, 2019. Photo by Carol Lynn Alpert. Used with permission.

12.7   Resources

1. Do you have staff in your organization to guide you in your quantum business and technical strategy?
2. Is "quantum computing" part of the skills profile for your employees?
3. What is your recruiting plan to build quantum expertise in your organization?
4. Have you targeted colleges and universities which include quantum computing in their curricula for recruiting?
5. Do you attend quantum computing conferences to recruit?
6. Do you look for badges showing quantum computing proficiency for your employees and job applicants?
7. Will you reimburse employees who take quantum computing classes or training programs?

8. Are quantum computing conferences on the list of approved expenses for employees, including management?
9. What new kinds of jobs will be created in your organization when quantum computing goes mainstream?
10. Do you have staff with graduate degrees in physics who can code?
11. Do you need help building out your skills and recruiting strategy?
12. If you have one, is your Chief Technical Officer (CTO) "quantum savvy"?

12.8 Summary

In this final chapter of the book, we examined questions that allow you to think about the development of quantum computing and how you can use it. The answers will help you gauge progress in the field. This will likely be uneven as temporary scientific and engineering roadblocks are discovered and then bypassed or overcome. You should revisit the questions and your answers at least every six months.

References

[1] Creative Commons. *Creative Commons Attribution-NoDerivs 2.0 Generic (CC BY-ND 2.0)* ⓒⒾ⊖. URL: https://creativecommons.org/licenses/by-nd/2.0/legalcode.

[2] A. M. Turing. In: *Computers & Thought*. Ed. by Edward A. Feigenbaum and Julian Feldman. MIT Press, 1995. Chap. Computing Machinery and Intelligence, pp. 11–35.

Afterword

Quantum Volume will be how we measure the computing capacity of quantum computing systems until we have full fault tolerance for qubits. It may not increase smoothly over the next few years and decades, but might instead jump like a step function as we improve this or that part of the hardware, software, or system.

Quantum computing is not a niche: I predict it will first be used to complement classical computing systems and then grow in power and importance as the 21st century progresses. It could very well become the most important computing technology of our lifetimes.

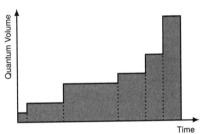

In this book I have tried to give you a solid foundation for understanding quantum computing. Mathematics is needed to see what is really going on, but I have provided what you need to get well into the subject. In various places I have pointed out how you can learn more about a topic. I encourage you to do so and you are now equipped for independent study. You have the basis to read more advanced texts on the subject as well as research papers. Through other reading, you may need to supplement what we have discussed here.

Quantum computing has something for everyone, no matter what your scientific or technical background. I hope you go forth and help shape the future of computing.

Appendices

A

Quick Reference

A.1 Common kets

One qubit

Computational basis (Z)

$$|0\rangle = \begin{bmatrix} 1 \\ 0 \end{bmatrix} \qquad |1\rangle = \begin{bmatrix} 0 \\ 1 \end{bmatrix}$$

Hadamard basis (X)

$$|+\rangle = \frac{\sqrt{2}}{2} \begin{bmatrix} 1 \\ 1 \end{bmatrix} \qquad |-\rangle = \frac{\sqrt{2}}{2} \begin{bmatrix} 1 \\ -1 \end{bmatrix}$$

Circular basis (Y)

$$|i\rangle = |\circlearrowleft\rangle = \frac{\sqrt{2}}{2} (|0\rangle + i|1\rangle) = \frac{\sqrt{2}}{2} \begin{bmatrix} 1 \\ i \end{bmatrix}$$

$$|-i\rangle = |\circlearrowright\rangle = \frac{\sqrt{2}}{2} (|0\rangle - i|1\rangle) = \frac{\sqrt{2}}{2} \begin{bmatrix} 1 \\ -i \end{bmatrix}$$

Two qubits

Computational basis

$$|00\rangle = \begin{bmatrix} 1 \\ 0 \\ 0 \\ 0 \end{bmatrix} \qquad |01\rangle = \begin{bmatrix} 0 \\ 1 \\ 0 \\ 0 \end{bmatrix} \qquad |10\rangle = \begin{bmatrix} 0 \\ 0 \\ 1 \\ 0 \end{bmatrix} \qquad |11\rangle = \begin{bmatrix} 0 \\ 0 \\ 0 \\ 1 \end{bmatrix}$$

Bell state basis

$$|\Phi^+\rangle = \tfrac{\sqrt{2}}{2}|00\rangle + \tfrac{\sqrt{2}}{2}|11\rangle \qquad |\Psi^+\rangle = \tfrac{\sqrt{2}}{2}|01\rangle + \tfrac{\sqrt{2}}{2}|10\rangle$$

$$|\Phi^-\rangle = \tfrac{\sqrt{2}}{2}|00\rangle - \tfrac{\sqrt{2}}{2}|11\rangle \qquad |\Psi^-\rangle = \tfrac{\sqrt{2}}{2}|01\rangle - \tfrac{\sqrt{2}}{2}|10\rangle$$

A.2   Quantum gates and operations

Name	Qubits	Matrix	Circuit Symbol
CNOT / CX	2	$\begin{bmatrix} 1 & 0 & 0 & 0 \\ 0 & 1 & 0 & 0 \\ 0 & 0 & 0 & 1 \\ 0 & 0 & 1 & 0 \end{bmatrix}$	
CY	2	$\begin{bmatrix} 1 & 0 & 0 & 0 \\ 0 & 1 & 0 & 0 \\ 0 & 0 & 0 & -i \\ 0 & 0 & i & 0 \end{bmatrix}$	
CZ	2	$\begin{bmatrix} 1 & 0 & 0 & 0 \\ 0 & 1 & 0 & 0 \\ 0 & 0 & 1 & 0 \\ 0 & 0 & 0 & -1 \end{bmatrix}$	
Fredkin / **CSWAP**	3	$\begin{bmatrix} 1 & 0 & 0 & 0 & 0 & 0 & 0 & 0 \\ 0 & 1 & 0 & 0 & 0 & 0 & 0 & 0 \\ 0 & 0 & 1 & 0 & 0 & 0 & 0 & 0 \\ 0 & 0 & 0 & 1 & 0 & 0 & 0 & 0 \\ 0 & 0 & 0 & 0 & 1 & 0 & 0 & 0 \\ 0 & 0 & 0 & 0 & 0 & 0 & 1 & 0 \\ 0 & 0 & 0 & 0 & 0 & 1 & 0 & 0 \\ 0 & 0 & 0 & 0 & 0 & 0 & 0 & 1 \end{bmatrix}$	

Hadamard **H** or **H**$^{\otimes 1}$	1	$\frac{\sqrt{2}}{2}\begin{bmatrix} 1 & 1 \\ 1 & -1 \end{bmatrix}$	
Hadamard **H**$^{\otimes 2}$	2	$\frac{1}{2}\begin{bmatrix} 1 & 1 & 1 & 1 \\ 1 & -1 & 1 & -1 \\ 1 & 1 & -1 & -1 \\ 1 & -1 & -1 & 1 \end{bmatrix}$	
ID	1	$\begin{bmatrix} 1 & 0 \\ 0 & 1 \end{bmatrix}$	
Measurement	1		
Pauli **X**	1	$\sigma_x = \begin{bmatrix} 0 & 1 \\ 1 & 0 \end{bmatrix}$	
Pauli **Y**	1	$\sigma_y = \begin{bmatrix} 0 & -i \\ i & 0 \end{bmatrix}$	
Pauli **Z**	1	$\sigma_z = \begin{bmatrix} 1 & 0 \\ 0 & -1 \end{bmatrix}$	
R$^x_\varphi$	1	$\begin{bmatrix} \cos\left(\frac{\varphi}{2}\right) & -i\sin\left(\frac{\varphi}{2}\right) \\ -i\sin\left(\frac{\varphi}{2}\right) & \cos\left(\frac{\varphi}{2}\right) \end{bmatrix}$	
R$^y_\varphi$	1	$\begin{bmatrix} \cos\left(\frac{\varphi}{2}\right) & -\sin\left(\frac{\varphi}{2}\right) \\ \sin\left(\frac{\varphi}{2}\right) & \cos\left(\frac{\varphi}{2}\right) \end{bmatrix}$	
R$^z_\varphi$	1	$\begin{bmatrix} 1 & 0 \\ 0 & e^{\varphi i} \end{bmatrix}$	

$\mathbf{S} = \mathbf{R}^{\mathbf{z}}_{\frac{\pi}{2}}$ 　　　　1 　　　　$\begin{bmatrix} 1 & 0 \\ 0 & i \end{bmatrix}$ 　　　　

$\mathbf{S}^{\dagger} = \mathbf{R}^{\mathbf{z}}_{\frac{3\pi}{2}} = \mathbf{R}^{\mathbf{z}}_{-\frac{\pi}{2}}$ 　　　　1 　　　　$\begin{bmatrix} 1 & 0 \\ 0 & -i \end{bmatrix}$ 　　　　

$\sqrt{\mathbf{NOT}}$ 　　　　1 　　　　$\frac{1}{2}\begin{bmatrix} 1+i & 1-i \\ 1-i & 1+i \end{bmatrix}$ 　　　　

$\mathbf{SWAP}$ 　　　　2 　　　　$\begin{bmatrix} 1 & 0 & 0 & 0 \\ 0 & 0 & 1 & 0 \\ 0 & 1 & 0 & 0 \\ 0 & 0 & 0 & 1 \end{bmatrix}$ 　　　　

$\mathbf{T} = \mathbf{R}^{\mathbf{z}}_{\frac{\pi}{4}}$ 　　　　1 　　　　$\begin{bmatrix} 1 & 0 \\ 0 & \frac{\sqrt{2}}{2} + \frac{\sqrt{2}}{2}i \end{bmatrix}$ 　　　　

$\mathbf{T}^{\dagger} = \mathbf{R}^{\mathbf{z}}_{\frac{7\pi}{4}} = \mathbf{R}^{\mathbf{z}}_{-\frac{\pi}{4}}$ 　　　　1 　　　　$\begin{bmatrix} 1 & 0 \\ 0 & \frac{\sqrt{2}}{2} - \frac{\sqrt{2}}{2}i \end{bmatrix}$ 　　　　

Toffoli / $\mathbf{CCNOT}$ 　　　　3 　　　　$\left[\begin{array}{cccccc|cc} 1 & 0 & 0 & 0 & 0 & 0 & 0 & 0 \\ 0 & 1 & 0 & 0 & 0 & 0 & 0 & 0 \\ 0 & 0 & 1 & 0 & 0 & 0 & 0 & 0 \\ 0 & 0 & 0 & 1 & 0 & 0 & 0 & 0 \\ 0 & 0 & 0 & 0 & 1 & 0 & 0 & 0 \\ 0 & 0 & 0 & 0 & 0 & 1 & 0 & 0 \\ \hline 0 & 0 & 0 & 0 & 0 & 0 & 0 & 1 \\ 0 & 0 & 0 & 0 & 0 & 0 & 1 & 0 \end{array}\right]$

B

Symbols

B.1 Greek letters

Name	Lowercase	Uppercase	Name	Lowercase	Uppercase
alpha	α	A	nu	ν	N
beta	β	B	xi	ξ	Ξ
gamma	γ	Γ	o	o	O
delta	δ	Δ	pi	π	Π
epsilon	ϵ	E	rho	ρ	P
zeta	ζ	Z	sigma	σ	Σ
eta	η	H	tau	τ	T
theta	θ	Θ	upsilon	υ	Υ
iota	ι	I	phi	φ	Φ
kappa	κ	K	chi	χ	X
lambda	λ	Λ	psi	ψ	Ψ
mu	μ	M	omega	ω	Ω

B.2 Mathematical notation and operations

Short name	Notation	Description
addition mod 2	$\oplus$	Addition of integers or bits modulo 2.
adjoint	$\mathbf{v}^\dagger$	Complex transpose of vector $\mathbf{v}$.
adjoint	$A^\dagger$	Complex transpose of matrix A.
bra	$\langle v|$	Row vector in Dirac notation.
Cartesian product	$V \times W$	Cartesian product of vector spaces V and W.
ceiling	$\lceil x \rceil$	Smallest integer greater than or equal to x.
complex numbers	$\mathbb{C}$	Complex numbers.
conjugate	$\bar{z}$	Complex conjugate of z.
direct sum	$V \oplus W$	Direct sum of vector spaces V and W.
dot product	$\mathbf{v} \cdot \mathbf{w}$	Dot product of vectors $\mathbf{v}$ and $\mathbf{w}$.
e	e	Base of the natural logarithms.
floor	$\lfloor x \rfloor$	Largest integer less than or equal to x.
i	$i = \sqrt{-1}$	Square root of -1.
inner product	$\langle v|w \rangle$	Inner product of a bra and a ket in Dirac notation.
integers	$\mathbb{Z}$	Integers.
ket	$|v\rangle$	Column vector in Dirac notation.
logarithm	$\log_{10}$	Base 10 logarithm.
logarithm	$\log_2$	Base 2 logarithm.
logarithm	$\log$	Natural logarithm.
natural numbers	$\mathbb{N}$	Natural numbers.
outer product	$|v\rangle\langle w|$	Outer product of a ket and a bra in Dirac notation.
product	$\displaystyle\prod_{j=k}^{n} f(j)$	Product $f(k) \times f(k+1) \times \cdots \times f(n)$.
rationals	$\mathbb{Q}$	Rational numbers.

reals	$\mathbb{R}$	Real numbers.
summation	$\displaystyle\sum_{j=k}^{n} f(j)$	Sum $f(k) + f(k+1) + \cdots + f(n)$.
tensor product	$\lvert v\rangle \otimes \lvert w\rangle$	Tensor product of kets $\lvert v\rangle$ and $\lvert w\rangle$.
tensor product	$\mathbf{v} \otimes \mathbf{w}$	Tensor product of vectors $\mathbf{v}$ and $\mathbf{w}$.
tensor product	$A \otimes B$	Tensor product of matrices A and B.
tensor product	$V \otimes W$	Tensor product of vector spaces V and W.
transpose	$\mathbf{v}^{\mathsf{T}}$	Transpose of vector $\mathbf{v}$.
transpose	A^{T}	Transpose of matrix A.
vector	$\mathbf{v}$	Vector $\mathbf{v}$.
whole numbers	$\mathbb{W}$	Whole numbers.

C

Notices

C.1 Creative Commons Attribution 3.0 Unported (CC BY 3.0)

You are free to:

> **Share** – copy and redistribute the material in any medium or format
> **Adapt** – remix, transform, and build upon the material for any purpose, even commercially.

The licensor cannot revoke these freedoms as long as you follow the license terms.

Under the following terms:

> **Attribution** – You must give appropriate credit, provide a link to the license, and indicate if changes were made. You may do so in any reasonable manner, but not in any way that suggests the licensor endorses you or your use.
> **No additional restrictions** – You may not apply legal terms or technological measures that legally restrict others from doing anything the license permits.

C.2 Creative Commons Attribution-NoDerivs 2.0 Generic (CC BY-ND 2.0)

You are free to:

> **Share** – copy and redistribute the material in any medium or format for any purpose, even commercially.

The licensor cannot revoke these freedoms as long as you follow the license terms.

Under the following terms:

> **Attribution** – You must give appropriate credit, provide a link to the license, and indicate if changes were made. You may do so in any reasonable manner, but not in any way that suggests the licensor endorses you or your use.
> **NoDerivatives** —- If you remix, transform, or build upon the material, you may not distribute the modified material.
> **No additional restrictions** – You may not apply legal terms or technological measures that legally restrict others from doing anything the license permits.

https://creativecommons.org/licenses/by-nd/2.0/legalcode

C.3 Creative Commons Attribution-ShareAlike 3.0 Unported (CC BY-SA 3.0)

You are free to:

> **Share** – copy and redistribute the material in any medium or format.
> **Adapt** – remix, transform, and build upon the material for any purpose, even commercially.

The licensor cannot revoke these freedoms as long as you follow the license terms.

Under the following terms:

> **Attribution** – You must give appropriate credit, provide a link to the license, and indicate if changes were made. You may do so in any reasonable manner, but not in any way that suggests the licensor endorses you or your use.

ShareAlike – If you remix, transform, or build upon the material, you must distribute your contributions under the same license as the original.

No additional restrictions – You may not apply legal terms or technological measures that legally restrict others from doing anything the license permits.

C.4 Los Alamos National Laboratory

"Unless otherwise indicated, this information has been authored by an employee or employees of the Los Alamos National Security, LLC (LANS), operator of the Los Alamos National Laboratory under Contract No. DE-AC52-06NA25396 with the U.S. Department of Energy. The U.S. Government has rights to use, reproduce, and distribute this information. The public may copy and use this information without charge, provided that this Notice and any statement of authorship are reproduced on all copies. Neither the Government nor LANS makes any warranty, express or implied, or assumes any liability or responsibility for the use of this information."

C.5 Trademarks

- IBM, IBM Q, IBM Q Experience, and IBM Q Network are registered trademarks of the International Business Machines Corporation. IBM Q System One is a trademark of the International Business Machines Corporation.
- MATLAB is a registered trademark of The MathWorks, Inc.
- Mathematica is a registered trademark of Wolfram Research, Inc.
- Polaroid is a registered trademark of Polaroid Corporation.
- Python is a registered trademark of the Python Software Foundation.
- Wikipedia is a registered trademark of the Wikimedia Foundation, Inc.

D

Production Notes

The source content for this book was written in LaTeX markup. I used many packages including amsmath, amssymb, biblatex, bookmark, ccicons, enumitem, framed, geometry, graphicx, hyperref, listings, minitoc, multicol, tcolorbox, and xifthen. Information about these packages is available at CTAN, the Comprehensive TeX Archive Network.

The diagrams and graphs were created with pgf/tikz, its libraries, and associated packages such as circuitikz. I am especially indebted to Alastair Kay for his brilliant quantikz package. All quantum circuits were created using this package.

I prepared the text in the open source Visual Studio Code editor from Microsoft and others. James Yu's LaTeX Workshop extension made creating this book much easier. tex4ht and make4ht were used with custom Python and sed scripts to produce the eBook.

Files were stored in Dropbox folders and version control was handled by git and github.

If you cite this book with BibTeX, please use

```
@BOOK{Sutor:2019:DwQ,
    AUTHOR = {Sutor, Robert S.},
    PUBLISHER = {Packt Publishing},
    DATE = {2019},
    EDITION = {1},
    ISBN = {978-1-83882-736-6},
    TITLE = {Dancing with Qubits},
    SUBTITLE = {How quantum computing works and how it can change the world}
}
```

Other Books You May Enjoy

If you enjoyed this book, you may be interested in these other books by Packt:

Mastering Quantum Computing with IBM QX

Dr. Christine Corbett Moran

ISBN: 978-1-78913-643-2

- Study the core concepts and principles of quantum computing
- Uncover the areas in which quantum principles can be applied
- Design programs with quantum logic
- Understand how a quantum computer performs computations
- Work with key quantum computational algorithms including Shor's algorithm and Grover's algorithm
- Develop the ability to analyze the potential of quantum computing in your industry

Python Machine Learning - Third Edition

Sebastian Raschka, Vahid Mirjalili

ISBN: 978-1-78995-575-0

- Master the frameworks, models, and techniques that enable machines to 'learn' from data
- Use scikit-learn for machine learning and TensorFlow for deep learning
- Apply machine learning to image classification, sentiment analysis, intelligent web applications, and more
- Build and train neural networks, GANs, and other models
- Add machine intelligence to web applications
- Clean and prepare data for machine learning
- Classify images using deep convolutional neural networks
- Best practices for evaluating and tuning models
- Predict continuous target outcomes using regression analysis
- Uncover hidden patterns and structures in data with clustering
- Dig deeper into textual and social media data using sentiment analysis

Leave a review – let other readers know what you think

Please share your thoughts on this book with others by leaving a review on the site that you bought it from. If you purchased the book from Amazon, please leave us an honest review on this book's Amazon page. This is vital so that other potential readers can see and use your unbiased opinion to make purchasing decisions, we can understand what our customers think about our products, and our authors can see your feedback on the title that they have worked with Packt to create. It will only take a few minutes of your time, but is valuable to other potential customers, our authors, and Packt. Thank you!

Index

Symbols

$\sqrt{-1}$, 97

Δ, 407, 416, 463

Ω, 348, 463

$\Omega(\ )$, 348

Φ, 279, 310, 463

Π, 463

Ψ, 279, 463

Σ, 463

χ, 386, 463

δ, 238, 463

$\delta_{j,k}$, 361

ϵ, 210, 416, 463

γ, 408, 463

κ, 188, 463

λ, 193, 406, 463

μ, 215

ν, 406, 463

ω, 359, 463

φ, 123, 279, 463

π, 100, 463

ψ, 231, 279, 463

ρ, 129, 238, 463

σ, 215, 463

σ^2, 215

σ_x, 185, 254, 426

σ_y, 185, 257, 426

σ_z, 185, 256, 426

θ, 123, 463

$\prod$ (product), 287, 464

$\sum$ (summation), 286, 464

$\approx$, 180

$\lceil\ \rceil$, 77, 464

$\cdot$, 177, 340, 350

$\lfloor\ \rfloor$, 77, 464

$\Rightarrow$, 102

$\mapsto$, 108

$\oplus$, 37, 306, 347

$\rightarrow$, 109

$|v\rangle\langle w|$, 230

$|0\rangle$, 227, 231, 265, 417

$|00\rangle$, 277

$|01\rangle$, 277

$|1\rangle$, 227, 231

$|10\rangle$, 277

$|11\rangle$, 277

$|+\rangle$, 232

$|-\rangle$, 232

$|-i\rangle$, 231

$|\Phi^-\rangle$, 279, 310

$|\Phi^+\rangle$, 279, 310

$|\Psi^-\rangle$, 279, 310

$|\Psi^+\rangle$, 279, 310

$|i\rangle$, 231

$|\psi\rangle$, 231

$|w\rangle\langle v|$, 231

W

X

Z